Springer Series on Atomic, Optical, and Plasma Physics

Volume 122

The Springer Series on Atomic, Optical, and Plasma Physics covers in a comprehensive manner theory and experiment in the entire field of atoms and molecules and their interaction with electromagnetic radiation. Books in the series provide a rich source of new ideas and techniques with wide applications in fields such as chemistry, materials science, astrophysics, surface science, plasma technology, advanced optics, aeronomy, and engineering. Laser physics is a particular connecting theme that has provided much of the continuing impetus for new developments in the field, such as quantum computation and Bose-Einstein condensation. The purpose of the series is to cover the gap between standard undergraduate textbooks and the research literature with emphasis on the fundamental ideas, methods, techniques, and results in the field.

Rashid A. Ganeev

High-Order Harmonics Generation in Plasmas

Resonance Processes, Quasi-Phase-Matching, and Nanostructures

Springer

Rashid A. Ganeev
University of Latvia
Riga, Latvia

ISSN 1615-5653 ISSN 2197-6791 (electronic)
Springer Series on Atomic, Optical, and Plasma Physics
ISBN 978-3-031-09039-4 ISBN 978-3-031-09040-0 (eBook)
https://doi.org/10.1007/978-3-031-09040-0

This Springer imprint is published by the registered company Springer Nature Switzerland AG
The registered company address is: Gewerbestrasse 11, 6330 Cham, Switzerland

To my parents, wife, son and daughter

Preface

The attention to the formation of the sources of coherent radiation in the Extreme Ultraviolet (XUV) range is attributed to different potential applications of short-wavelength radiation in various fields of physics, chemistry, biology, etc. The High-Order Harmonics Generation (HHG) in gases seems the most suitable mechanism allowing the creation of such sources. Previously, HHG in gases and during specular reflection from surfaces of extremely intense laser pulses were the main approaches in this direction. The application of plasma formed during ablation of targets as the medium for frequency conversion did not show the perspectives during the experiments carried out in the 90s. However, the new trends in this direction allowed finding a niche in those studies, which led to the development of this method of harmonics generation. Resonance enhancement of single harmonic, quasi-phase-matching in multi-jet plasmas, applications of quantum dots and large nanoparticles, and other peculiarities distinguished HHG in plasmas from the harmonics generation in gases and during specular reflection from surfaces. The significance and timeliness of these topics are related to the spectroscopy of materials using their ablation and analysis of the harmonics generation ability. More details are presented in the Introduction chapter of this book.

There is a limited list of monographs on this topic providing a detailed overview of this process. A previous book describing different topics of HHG (R. A. Ganeev, *High-Order Harmonic Generation in Laser Plasma Plumes*, Imperial College Press, London, 2012) served as an introduction in the field of the frequency conversion of laser sources in the Laser-Induced Plasmas (LIP) while limiting in its description to some specific processes developed at that moment. The present book (*High-Order Harmonics Generation in Plasmas: Resonance Processes, Quasi-Phase-Matching, and Nanostructures*) comprises a detailed consideration of the novel approaches developed in this area during the last few years. This book, which can be considered as a continuation of the above monograph, comprises the main topics and the most recent advancements in the HHG in LIP. The major difference in the case of the present book with respect to the previous one and some other books comprising separate areas of the developments in this field is related to the broader consideration of different nonlinear optical aspects of light-plasma interactions. We start from the lowest harmonic generation and move toward rather complicated processes of frequency conversion in plasmas. Particularly, this book contains the collection of the most recent studies of nanoparticles and quantum dots as the perspective media for HHG. We also discuss the role of resonances in the enhancement of single harmonic while emphasizing the particular properties of ablated material in each particular case. The consideration of the quasi-phase-matching concept in the growth of the groups of harmonics in different ranges of XUV is presented. The necessity in this collection of various recent publications prompted us in the compilation of various HHG studies disseminated in different journals in a single book. The collection of these studies helps the reader to learn about the most recent achievements in this field developed by different research groups. Thus, this book allows providing insight into the complex processes of HHG in LIP reported during the recent few years. It would certainly help the audience be able to quickly become involved in this field and could also help as a reference monograph for certain topics.

This book is aimed to support researchers and other readers in their development in the area of high-order nonlinear spectroscopy. Particularly, the advanced studies of nanoparticles and quantum dots for the formation of new sources of radiation in different short-wavelength spectral ranges show the ways to further implement the specific features of the small-sized species in a relatively new field of studies—laser ablation-induced high-order harmonics generation spectroscopy. The researchers involved in the development of new methods of frequency conversion will be benefited by finding the most recent novelties in this field. The undergraduate students can reveal interesting information about the recent findings in plasma harmonic research. Additionally, the usefulness of the book could be proven by the potential applications of the new knowledge developed during recent years for ultrafast pulses generation using the proposed schemes of plasma-light interaction. Thus, the audience may also include those researchers involved in the state-of-the-art developments in attophysics.

The primary groups of readers comprise researchers in the fields of nanomaterial science, plasma physics, frequency conversion, and nonlinear optics. The book would be interesting for both academics and professionals in the applied fields, as well as for the high education students and postdocs, who are aimed at further development of their career in the fields of laser-matter interaction. Additionally, any professionals interested in the application of the advanced techniques for material science could be also benefited from updating their knowledge of new methods of material studies using high-order nonlinear spectroscopy. The potential collective users could be the universities and scientific institutes where the nonlinear optical studies have a connection with plasma physics, ultrashort pulses generation, nano-optics, coherent processes analysis, and laser-matter interactions.

Results presented in this book comprise the studies reported during the last few years. All these amendments of the HHG studies in plasmas could not be realized without the fruitful collaboration of different groups involved in similar research activities. I would like to thank H. Kuroda, T. Ozaki, P. D. Gupta, P. A. Naik, J. P. Marangos, J. W. G. Tisch, M. Castillejo, C. Guo, H. Zacharias, O. V. Ovchinnikov, A. S. Alnaser, V. V. Strelkov, G. S. Boltaev, V. V. Kim, P. V. Redkin, K. Zhang,

M. Venkatesh, Y. Fu, S. K. Maurya, W. Yu, W. Li, S. R. Konda, S. Y. Stremoukhov, A. V. Andreev, K. L. Ishikawa, and M. Iqbal for the long-lasting collaboration.

The inspirers of this activity are my wife Lydia, son Timur and daughter Dina, who help me overcome various obstacles of life.

Riga, Latvia
November 2022

Rashid A. Ganeev

Introduction

Maximizing the Impact of Laser Ablation-Induced High-Order Harmonics Spectroscopy on Materials Science

The introduction of new advanced methods in plasma spectroscopy using the High-Order Harmonics Generation (HHG) approach (for example, using filaments, radiation sources in the mid-infrared region, external two-color pumping, solid-state nanostructures, clusters, quantum dots possessing high values of lower-order optical nonlinearities, etc.), along with previously developed methods will allow a deeper understanding of the physical and optical properties of materials through the interaction of the laser with ablated samples. This knowledge is also extremely useful for further improving the conversion efficiency of HHG and the formation of effective sources of coherent Extreme Ultraviolet Radiation (XUV) for practical needs.

From the point of view of the variety of topics implemented during the development of this approach, HHG in plasma can be considered as a multi-panel problem, including the interaction of laser with the matter, atomic spectroscopy, the synthesis and application of nanomaterials, laser physics, low-density plasma studies, quantum, and nonlinear optics. As follows from the objectives of this approach, it is aimed at studying materials due to the synergy of the above areas. The subjects of these studies are almost all non-radioactive solid materials of the periodic table. The solid elements treated in previous HHG studies are marked by circles in Fig. 1a [1]. Note that the commonly used HHG in gases is limited to a few, usually noble, gases (squares in Fig. 1a). Additionally, HHG conversion efficiency in Laser-Induced Plasmas (LIP) can be in the case of some ablating targets larger than the one in gases (Fig. 1b [1]). The potential of HHG spectroscopy in ablated materials suggests its application to

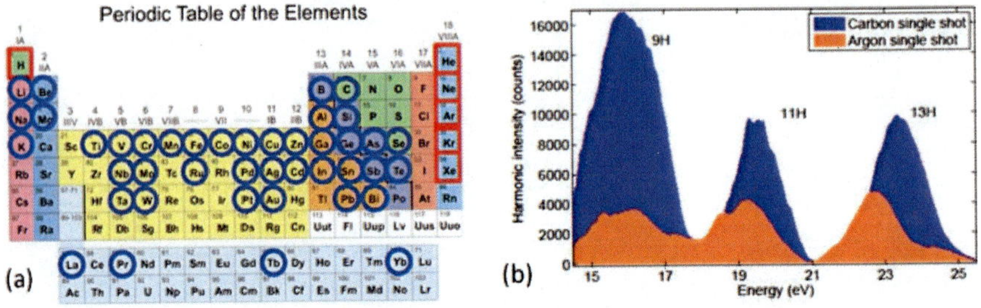

Fig. 1 **a** Map of the elements of the periodic table used in previous studies of HHG in plasmas (blue circles) and gases (red squares). **b** Harmonic spectra from the carbon plasma produced by the 10 ns pulse and argon gas at the optimal conditions of harmonic generation for both media. Reproduced from [1] with permission from American Physical Society

thousands of solids (metals, metalloids, semiconductors, dielectrics, complex structures like large organic molecules, alloys, nanostructured species, such as clusters, quantum dots, nanoparticles, thin films, etc.). Thus, this HHG method significantly expands the fields of material characterization using new approaches.

The use of the plasma formed during laser ablation of solids as a medium for HHG is a very attractive and promising approach since it leads to the realization of advantages that allow the analysis of plasma components in various ways [2]. Plasma media provide various possibilities for creating effective coherent sources in the XUV region using HHG when the femtosecond pulses pass through them. The efficiency of converting laser energy into harmonics, in this case, can be higher than in gaseous media. The use of extended LIPs, resonances in atomic or ionic systems, clusters, quantum dots, and nanoparticles in plasma formations, two- and three-color pumping of LIPs, creation of optimal conditions for energy transfer in spatially modulated plasma structures, and other technical and fundamental advantages of this HHG approach allow us to expect high conversion efficiency to XUV ($\sim 10^{-4}$ [3]), which can lead to the practical applications of such coherent radiation sources. Similar to HHG in gases, harmonics from plasma plumes may possess the unique temporal characteristics being in the attosecond time domain, which makes it possible to simultaneously generate ultrashort pulses and diagnose various processes related to the harmonic emitters.

Further steps in the development of unique advanced research in the field of nonlinear optical spectroscopy of various ablated materials involve a combination of a number of fundamentally new and improved methods that have already been developed, which will increase the efficiency of HHG in plasma media when analyzing the spectroscopic properties of ablated materials. The main efforts will be aimed at finding ways to efficiently convert laser radiation propagated through the LIPs into the XUV range, taking into account collective macro-processes in plasma (meeting the conditions for phase matching during HHG in atomic and multi-particle spatially modulated structures for the formation of heterogeneous media) in combination with

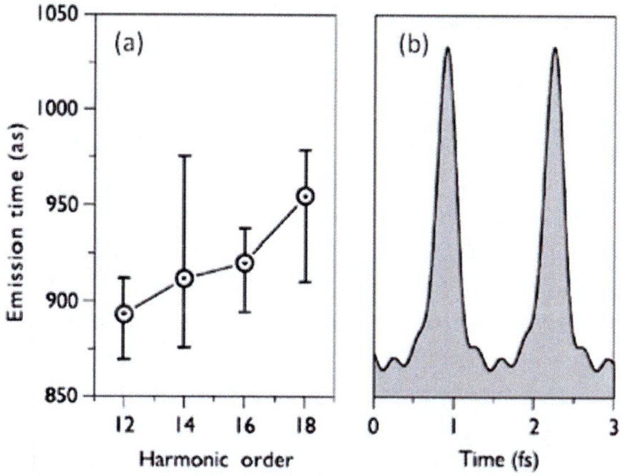

Fig. 2 **a** Emission times and **b** temporal profile corresponding to the 11th to the 19th harmonic orders. The reconstructed temporal profile of the harmonic emission forms an attosecond pulse train, with each pulse of 285 as duration. Reproduced from [4] with permission from the Optical Society of America

micro-processes (resonant amplification of individual harmonics, the use of poly-atomic structures, etc.). In addition to the potential applications of short-wavelength sources, the possibility of generating attosecond pulses during HHG using specially formed LIP structures is of the greatest interest (Fig. 2a and b [4]).

At the same time, conducting research in various combinations of nonlinear media (gas, plasma, and solid state), previously used separately for HHG, will allow us to find the most effective approaches to studying the advantages of each of them. One of the most important strategies in this direction is the implementation of new unexplored principles of this process, such as the use of plasma filaments, pumping sources in the mid-infrared region, gas+plasma schemes [5], external multicolored pumping, solid-state nanostructured films, clusters, quantum dots with high values of low-order optical nonlinearities, etc., for effective HHG in the LIP.

The method is original because it uses various spectroscopic approaches of HHG in laser plasma, as well as from the point of view of systematic application of unique experimental methods, including those used for the first time, for the effective implementation of this process in a specially selected plasma-based nonlinear optical medium. This approach can be considered as the spectroscopy of materials using HHG or, more precisely, the Laser Ablation-Induced High-Order Harmonics Generation Spectroscopy (LAIHOHGS).

A Brief Overview of HHG in LIP

Coherent short-wave radiation is becoming increasingly important for a wide range of fundamental and applied research in various fields of science and technology. Monochromatic and coherent light pulses in the XUV spectral range with the femtosecond and attosecond duration will increase the resolution of optical lithography from several micrometers to several tens of nanometers, which will facilitate the optical visualization of nanoscale and subcellular systems, as well as allow monitoring of the movement of electrons in atoms, molecules, and nanostructured systems in real time. In addition, such radiation sources will contribute to the creation of a new analytical approach—resonance spectroscopy in XUV.

The most effective approach for generating coherent radiation in the XUV region is HHG using ultrashort laser pulses. The most commonly used method since the late 80s is HHG in gases [6–9]. Another method, reflection from surfaces, also allows obtaining high-order harmonics in the range of XUV [10, 11]; however, this approach requires exceptional characteristics of laser radiation ("relativistic" intensities (10^{18-19} W/cm^2 and higher), as well as the extremely high contrast of laser pulses). The use of these methods is limited by the relatively low efficiency of converting the main wave into higher harmonics (i.e., into harmonics exceeding the thirtieth orders of commonly used near-infrared lasers, respectively, less than 10^{-5} and less than 10^{-6} for HHG in gases and from surfaces), which, despite considerable efforts and financial costs, apparently have not significantly improved over the past 30 years. Relatively recently, the HHG method has aroused great interest in the passage of laser pulses through solid-state structures [12]. However, the method still allows this process to be carried out only using long-wavelength lasers emitting in the range of several micrometers, which does not allow reaching the XUV region (i.e., wavelengths below 100 nm) due to significant absorption of generating harmonics in solids. The fourth method is the use of laser plasma for HHG, which began its development in the 90s of the last century [13–18]. However, the maximum observed harmonics generated in those experiments were below the twentieth order due to some impeding factors in the highly excited LIPs.

HHG in laser-induced plasma or, in short, studies of "plasma harmonics" has been more widely developed since 2005, when a new approach to the use of low-energy and low-ionized LIPs was proposed and implemented [19]. This approach allowed significantly amending HHG and increasing the generating harmonics above the hundredth order. Since then, along with numerous theoretical groups that have contributed to these studies, an increasing number of experimenters from different countries (Japan [20], Canada [21], the United Arab Emirates [22], the United States of America [23], Israel [24], Uzbekistan [25], France [4], China [26], Spain [27], Russia [28], Germany [29], India [30], Great Britain [31], and others) have implemented practical aspects of this topic. The use of LAIHOHGS has made it possible to develop new approaches to the study of large molecules, clusters, and nanoparticles subjected to ablation, as well as significantly expanded the range of research subjects compared to the generation of harmonics in gases.

Numerous metals, metal alloys, and semiconductors were analyzed by the method of ablation and generation of plasma harmonics. As already mentioned, most of the laser-ablated solid elements showed the ability to HHG. In addition, various molecular structures were used to study plasma harmonics. Some metal ions have demonstrated resonant amplification of a single high-order harmonic of 800 nm class lasers. HHG in the LIPs containing nanoparticles allowed increasing the harmonic output compared to a plasma containing monomers of the same elemental consistency (Fig. 3a [32]). Quasi-phase-matching is another mechanism that allows amplifying a group of harmonics in different ranges of XUV (Fig. 3b [33]). Plasma harmonics allow analyzing such processes as quantum path interference when the short and long trajectories of accelerating electrons coherently contribute to the spectrum of harmonics (Fig. 4a [34]). Finally, some elements of the periodic table, particularly carbon and silver, have shown exceptional properties as effective harmonic emitters.

Usually, one laser can be used in these studies. Harmonic radiation is generated by focusing femtosecond pulses in the plasma formed on the surface of the target. The laser beam after the regenerative amplifier is divided into two parts, and the larger part is compressed and used for harmonics generation, while the lower one is used to form LIP. In the case of a Two-Color Pumping (TCP) scheme, a nonlinear crystal

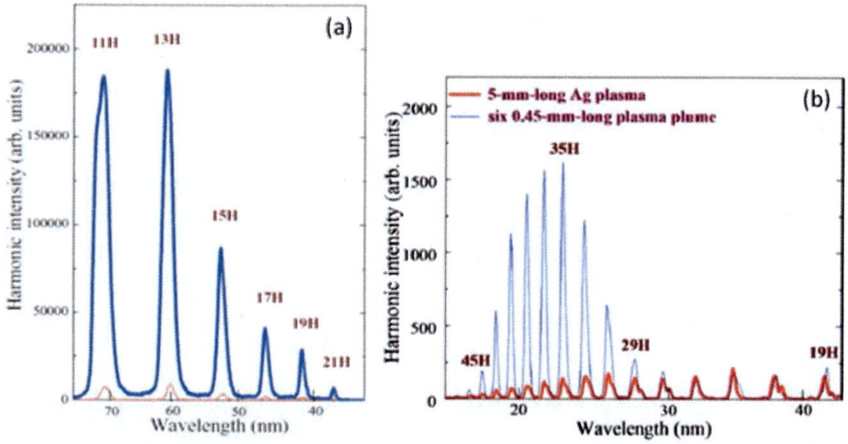

Fig. 3 **a** Relative intensities of the harmonics generating in aluminum nanoparticles (thick blue curve) and aluminum atoms (thin red curve). Adapted from [32] with permission from the American Physical Society. **b** Spectral distributions of harmonics from silver plasma in the case of 5-mm-long plasma (thick curve) and six 0.38-mm-long plasma plumes (thin curve). The enhancement factor of maximally enhanced 35th harmonic compared with the case of using the line 5-mm-long plasma was 12×. Adapted from [33] with permission from the American Institute of Physics

is inserted into the vacuum chamber along the axis of propagation of femtosecond pulses. The generating short-wavelength harmonic spectrum is analyzed using an XUV spectrometer. The details of this scheme will be discussed in more detail in the next section.

As we have already mentioned, LAIHOHGS is not only an alternative approach for generating coherent XUV radiation but can also serve as a tool for various spectroscopic and analytical applications, some of which have begun to appear in recent years. The analysis of coherent XUV radiation from various ablated species can be used in physical, chemical, and biological studies. In addition to the above-mentioned fields of application, they include femtosecond coherent diffraction imaging with time resolution, photoinduced processes on surfaces and nanoparticles, lithography, plasma diagnostics, and a number of others. In general, the development of nonlinear spectroscopy based on the generation of harmonics in plasma is perfectly suited for material science.

The harmonic spectrum in the XUV range may contain certain information about the electronic structure of the ionized atomic or molecular medium. The application of few-cycle driving pulses allows generating unusual spectral distributions of harmonics like the one shown in Fig. 4b [35]. One can see the broadband 33rd harmonic generating in manganese plasma in the vase of 4 fs driving pulses. Other harmonics in the vicinity of this enhanced harmonic are suppressed. The application of multi-cycle pulses (35 fs) drastically changed the harmonic distribution (Fig. 4c [35]). Accordingly, the study of plasma harmonics will contribute to the understanding of the fundamental laws in atomic and molecular systems. Plasma harmonics allow us to analyze molecules, from simple diatomic to more complex samples (aromatic species, peptides, DNA components, etc.) to study their ionization, dissociation, and fragmentation under the action of a short-wavelength field.

The advantages of generating higher harmonics in LIP can be largely realized using weakly ionized plasma since limiting processes, such as phase mismatch and absorption, which affect the dynamics of converting the laser radiation wavelength to the XUV range, will play an insignificant role in such an environment [36]. This

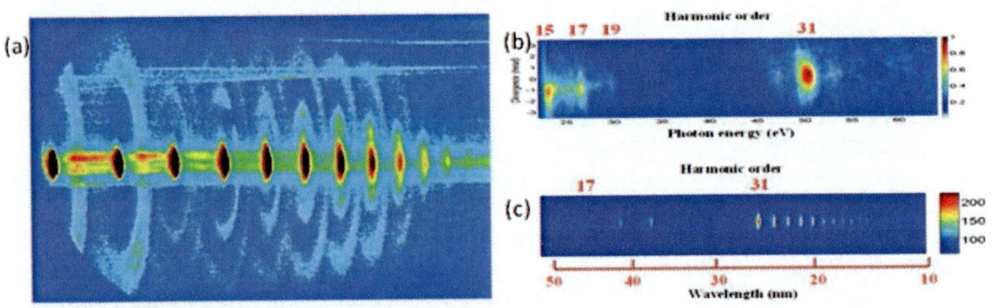

Fig. 4 **a** Spectrally and spatially resolved harmonic spectrum from aluminum plasma in the case of focusing after the plasma area. Lowest order harmonic is the 17th one (on the left side). Adapted from [34] with permission from the American Physical Society. **b, c** Harmonic spectra from Mn plasma in the case of 4 fs **b** and 35 fs **c** driving pulses. Adapted from [35] with permission from the American Physical Society

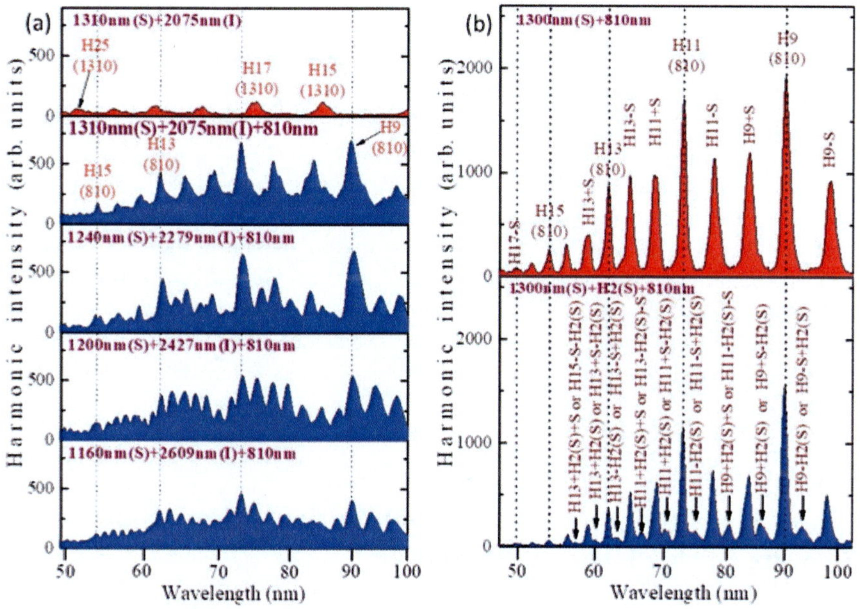

Fig. 5 **a** Harmonic spectra from silver plasma in the case of (upper panel) driving signal (S, 1310 nm) and idler (I, 2075 nm) pulses; (second panel) 1310, 2075, and 810 nm pulses at the conditions of the overlap of the 1310 and 810 nm waves; and (three bottom panels) 810 nm and spectrally tuned signal and idler pulses. Dotted lines in this and other figures correspond to the odd harmonics (H9–H15) of the 810 nm wave. The sum and difference components between H9 (810 nm) and H11 (810 nm) are marked in the fourth panel. Similar labeling can be applied for other frequencies between each pair of neighboring odd harmonics of the 810 nm wave. **b** Harmonic spectra using two-color (1300 nm + 810 nm; upper panel) and three-color (1300 nm + 650 nm + 810 nm; bottom panel) pumps. The crystal was adjusted for the phase matching of H2 of the MIR pulses. Arrows in bottom panel show additional frequency components appeared after insertion of BBO on the path of 1300 and 810 nm pumps. Particularly, arrows at both sides of H9 show two options of sum and difference frequency generation. Reproduced from [36] with permission from the Optical Society of America

assumption has been confirmed in a number of innovative studies of HHG under various conditions of plasma formation. A significant increase in the orders of generating harmonics [37], the appearance of an extended plateau in the energy distribution of harmonics [38], high conversion efficiency toward XUV harmonics in a number of plasma structures [39–41], the implementation of resonant amplification of individual harmonics [42–48], a significant amplification of generating harmonics in the LIPs containing nanoparticles of various materials [6, 49], the analysis of quantum dots as effective harmonic emitters [50], the implementation of quasi-phase-matching of interacting waves to increase the output of harmonics in various ranges of XUV [51], the use of extended plasma structures [52], two- and three-color pumps of LIPs (Fig. 5 [36]), and other properties revealed in a number of subsequent experimental and theoretical studies performed in many laboratories around the world over the past decade [53–76] have demonstrated the advantages of LAIHOHGS in generating low- and high-order harmonics for studying a whole range of ablated solids compared with the generation of harmonics in several gases.

The next steps of high-order nonlinear spectroscopy for creating effective sources of ultrafast pulses in the XUV range require further systematic studies of the influence of various plasma parameters on the HHG process, as well as the use of new approaches in this direction. The search for new plasma media that will contribute to the amplification of higher harmonics will further increase the efficiency of HHG. Thus, the combination of fundamental research in this field and new approaches to HHG in laser-induced plasma will lead to even greater success in creating practical coherent sources in the field of XUV, which will be useful for nonlinear spectroscopy of materials. In the following sections, we address the most interesting achievements of LAIHOHGS.

Role of Ionic Resonances in the Modification of Harmonics Intensity During HHG in Plasmas

The harmonics generated in the LIPs strongly depend on the oscillator strength of nearby ion transitions. The proof of this assumption was demonstrated in the case of an indium plasma (Fig. 6a [3]), which allows an exceptional increase (with a coefficient of 80×) of a single (13th) harmonic of a laser with a wavelength of 800 nm due to the high oscillator strength of the In II transition in the spectral range of 62 nm [2,3]. This discovery opened the door for numerous studies, both theoretical [44, 46, 53, 56, 57, 59, 61, 77–80] and experimental [4, 27, 33, 37, 38, 42, 45, 47, 48, 65, 67, 71, 72, 76, 81–84], on the participation of ion transitions in changing the harmonic distribution in the XUV range.

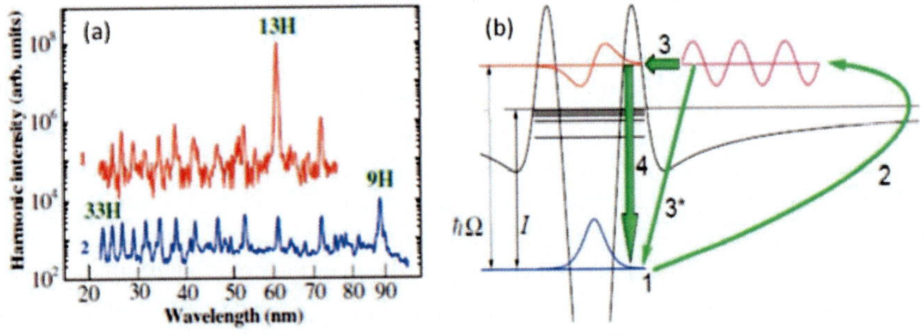

Fig. 6 **a** High-order harmonic spectra from (1) indium and (2) silver LIPs. Reproduced from [3] with permission from the Optical Society of America. **b** The three-step model includes ionization (step 1), free electronic motion in the field (2), and recombination (3*) in the suggested model [59] the latter stage is replaced by the radiationless transition to the autoionizing state (3), and relaxation with XUV emission (4). Reproduced from [59] with permission from American Physical Society

Notice that none of the harmonics in the HHG scheme, which includes a specular reflection from a "moving mirror", can have such properties. Moreover, even in the case of harmonic generation in gases, no resonance amplification associated with ion transitions was reported. This scheme supports the conditions when only neutral particles participate in HHG, in order to exclude the influence of free electrons on the phase matching between the exciting and harmonic fields. Atomic transitions are noticeably less strong compared to ionic ones, so even if a certain harmonic order coincides with the energy levels of noble gas atoms, these "resonant" transitions do not change the distribution of harmonics in the plateau range. On the contrary, LIP ions play a crucial role in changing the envelope of the distribution of plasma harmonics in the XUV region.

Various methods have been proposed to describe the phenomenon of resonant amplification of a single harmonic in the LIP. The HHG approach (the so-called four-stage model, Fig. 6b [59]), describing the harmonic amplification coinciding with the transition between the ground and autoionizing states (AIS) of ions, was developed in [59]. The third (recombination) stage of the three-stage HHG scenario [11] was divided into two stages: the capture of an electron accelerated by a laser into the AIS of the initial ion, followed by radiation relaxation of this state to the ground state with the emission of a harmonic photon. Probably, this scenario of resonant HHG can also be applied to ionized large particles.

Other models are based on coupled transitions [85–87], as well as on the coupling of multi-electron excited states with increased harmonic output [53, 57, 88]. The theory of high-order harmonic generation, which generalizes the strong-field approximation for a resonant case when the frequency of harmonics is close to the frequency of the transition from the ground state to the AIS of the generating system, is presented in [89].

The existing theories of the resonance-induced amplification of a single harmonic in an atomic/ion plasma are mainly based on (i) a four-stage model; (ii) harmonic generation in the presence of a shape resonance using time-frequency analysis of intensity and phase, which emphasized the resonance amplification regardless of the pulse duration and supported a four-stage model; and (iii) an approach generalized for the case when the capture in the AIS is replaced by the excitation of the ground state into this state caused by a field. The usual three-stage scenario, in which the factorization formula is applied without additional assumptions [44], also explains the described processes that enhance resonance in different LIPs. Each of these models describes to some extent the majority of experiments with HHG related to resonance in plasma.

The basic scheme commonly used for HHG in LIP is as follows (Fig. 7a). The bulk target is ablated by laser pulses to form a LIP, followed by the generation of harmonics during the propagation of femtosecond laser pulses through the formed plasma plume. Part of the uncompressed radiation of a femtosecond laser is usually used as a Heating Pulse (HP) for ablation of the target.

Focused compressed femtosecond Excitation Pulses (DP) propagate through the LIP at a distance of several hundred micrometers above the target surface. In addition to Single-Color Pumping (SCP) from a laser, TCP (for example, in the case of Ti:

sapphire laser, 800 nm and its second harmonic (400 nm)) is used to generate odd and even harmonics. Harmonic radiation is analyzed using an XUV spectrometer, which usually contains a cylindrical mirror with a gold coating and a flat-field grating with variable line spacing. The spectrum is recorded using various XUV radiation recorders, in particular, a microchannel plate detector with a phosphor screen, the image of which is captured by a CCD camera.

To increase the stability of HHG in LIP, a rotating target configuration is often used (Fig. 7b [32]). This approach makes it possible to stabilize the plasma characteristics and optimize the harmonic output. The use of rotating targets is especially necessary in the case of brittle targets, such as nanoparticle powders.

Modification of Harmonic Intensity Distribution Along the XUV Region Using the Quasi-Phase-Matching Approach in LIP

The Phase Mismatch (PMM) process suppresses the high-order harmonics of ultrashort laser pulses in both gas and plasma media. PMM is associated with a dephasing between the propagated harmonic field and the laser-induced polarization ($\Delta k = qk_1 - k_q$), caused mainly by the dispersion of the medium, which is increasing in the presence of free electrons. The difference in the velocities of these waves leads to conditions when, at some distance from the beginning of the medium, the phase shift between the waves becomes close to π. After the propagation of this distance (coherence length, $L_{coh} = \pi/\Delta k$), the constructive accumulation of harmonic photons reverts to the destructive one, when the newly generated XUV photons in the reverse phase compensate for the previously generated photons, thereby reducing the harmonic output.

Quasi-Phase-Matching (QPM) allows reducing this limitation of the harmonic conversion efficiency in various ways. In particular, QPM was demonstrated using multiple gas puffs [90–92] and multi-jet plasma [22, 51, 70, 81]. However, in the

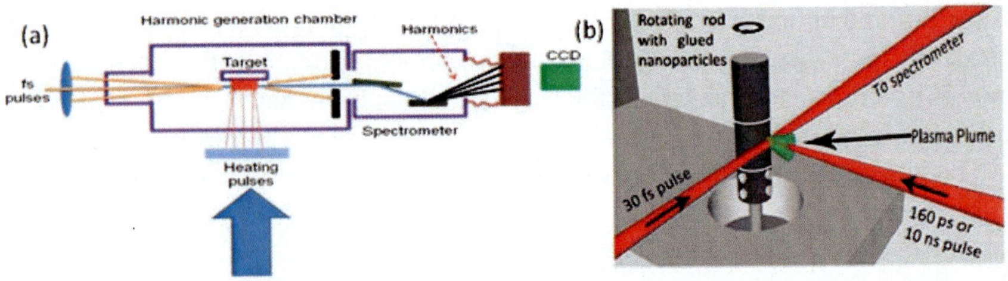

Fig. 7 **a** The typical experimental setup of harmonic generation in laser-induced plasma. **b** Experimental setup for ablation of a rotating rod coated by a layer of metallic nanoparticles, with further harmonic generation in the plasma plume. Reproduced from [32] with permission from American Physical Society

latter case, the separate role of two groups of free electrons in the QPM mechanism remained unclear: those that are generated during ablation of the target and those that come as a result of tunneling ionization in the driving field. The PMM induced by the plasma dispersion for the qth harmonic is determined by the ratio [93]

$$\Delta k_{disp} = qN_e e^2\lambda/4\pi m_e \varepsilon_0 c^2, \tag{1}$$

where λ is the wavelength of the driving radiation; N_e, m_e, and e are the density, the mass, and the charge of the electron; c is the light velocity; and ε_0 is the vacuum permittivity. The coherence length of this harmonic is equal to

$$L_{coh} = \pi/\Delta k \approx \pi\,\Delta k_{disp} \approx 4\pi^2 m_e \varepsilon_0 c^2/qN_e e^2\lambda. \tag{2}$$

From this expression, the coherence length (in mm) at the conditions of using the 800 nm driving laser could be presented as

$$L_{coh} \approx 1.4 \times 10^{18}/N_e q_{\mathrm{qpm}}. \tag{3}$$

Here q_{qpm} is the harmonic order in the QPM region showing the highest enhancement and N_e is the electron density in the plasma jets measured in cm^{-3}. This simple formula allows defining the electron density by knowing the coherence length (which is in fact the size of a single plasma jet at the conditions of the QPM in the multi-jet structure) and the maximally enhanced harmonic order. The decrease of heating pulse fluence on the surface of the ablating target leads to a decrease of electron density (due to less ablation- and tunnel-induced electrons) in the plasma plume followed by the shift of q_{qpm} toward the shorter-wavelength region. Similarly, one can anticipate that, for the plasma jets of different sizes, the maximally enhanced harmonics will also be tuned along the XUV spectrum.

In the studies reported in Ref. [51], silver was used as the target for ablation. The size of the target where the ablation using a cylindrical focusing lens occurred was 5 mm (Fig. 8a [51]). To create multi-jet plasma, a Multi-Slit Mask (MSM) was used. The size of the slits was 0.3 mm with the distance between them also 0.3 mm. The MSM was installed between the focusing cylindrical lens and the target in such a way as to divide the imperforated 5-mm-long plasma into eight 0.3-mm-long plasma jets with a separation of 0.3 mm.

The spectrum of harmonics of silver plasma with a length of 5 mm at a heating radiation fluence of 0.9 J cm^{-2} showed a featureless plateau-like shape with a gradual decrease in the harmonic intensity in the region of shorter wavelengths (Fig. 8b, upper panel [51]). The other panels in Fig. 8b show the harmonic spectra obtained using the insertion of MSM in the path of the heating beam. The MSM was tilted at different angles (0°, 15°, 30°, and 45°) relative to the heating beam. The slope of the MSM led to a decrease in the size of single jets and an increase in the number of jets. In all these cases, a characteristic QPM-induced amplification of harmonic groups in different spectral ranges was observed. The maximum increase of harmonic yield was achieved for the H37, H41, H43, and H51, respectively. The maximally increased harmonic orders approximately corresponded to the rule when the product $L_{coh}q_{qpm}$ (where L_{coh} corresponded to the sizes of single jets) was kept constant at a fixed electron density in accordance with Eq. 3 ($L_{coh}q_{qpm} \approx 1.4\times10^{18}/N_e$). A significant increase in harmonic groups compared to the 5-mm-long expanded plasma was due to the participation of multi-jet plasma in the generation of harmonics, which made it possible to achieve QPM.

The increase in the cross section of tunnel ionization, which leads to the appearance of additional free electrons in a multi-jet medium, did not worsen the QPM conditions. Figure 8c (empty circles [51]) shows a close similarity of q_{qpm} in a wide range of changes in the energy of the driving pulse. Significantly different behavior of QPM was observed when the energy of the heating pulse changed, which led to a change in the electron density of the plasma (Fig. 8c, filled squares [51]).

Comparative studies of the harmonic spectra generated by an eight-jet silver plasma at different heating pulse fluencies on the target surface were performed by changing the heating radiation energy using calibrated filters. The energy of the heating pulse was varied to analyze the effect of this parameter on the dynamics and shape of the envelope of the harmonic distribution using the plasma with the same amount of separated jets. The group of harmonics gradually adjusts toward lower orders (i.e., longer wavelength region) with an increase in the energy of the heating

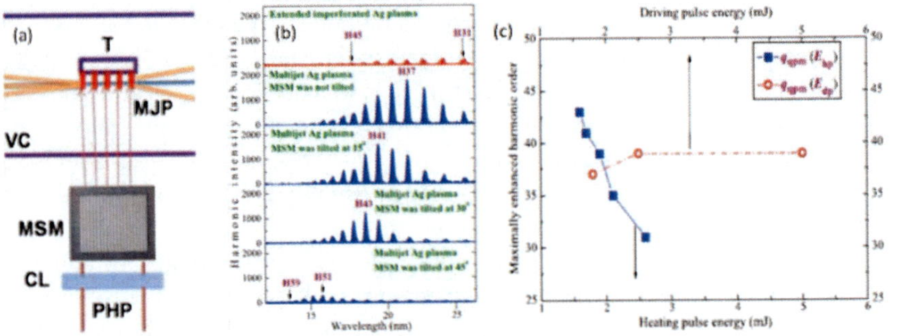

Fig. 8 **a** Experimental scheme for the harmonic generation in the multi-jet plasmas. PHP, picosecond heating pulse; CL, cylindrical lens; MSM, multi-slit mask; VC, vacuum chamber; T, target; MJP, multi-jet plasma. **b** Harmonic spectra generated in different multi-jet silver plasmas produced by tilting the MSM. Upper panel shows the harmonic spectrum generated in the imperforated 5-mm-long Ag plasma. **c** Variations of maximally enhanced harmonic order at different energies of heating (filled squares) and driving (empty circles) pulses. One can see that the threefold growth of driving pulse energy (from 1.85 to 5 mJ) did not significantly change the q_{qpm}. In the meantime, the q_{qpm} was strongly dependent on the variation of the heating pulse energy. Adapted from [51] with permission from American Physical Society

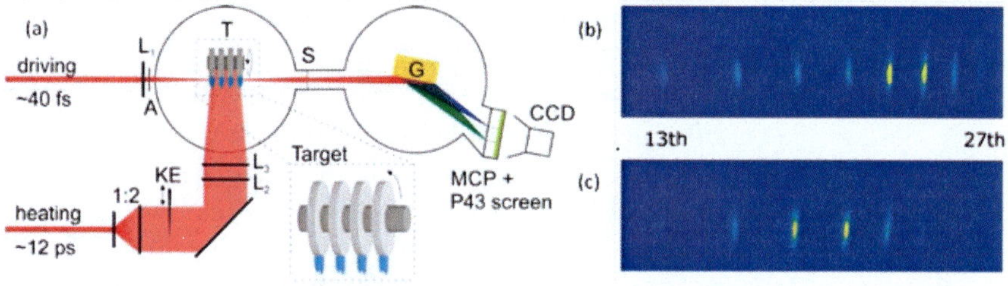

Fig. 9 **a** Experimental setup for HHG from the plasmas produced on the perforated targets. The driving pulse is focused by lens (L1) into the plasma. An aperture (A) is used for signal optimization. The heating pulse is focused by a cylindrical lens (L2) onto the target (T). A second lens (L3) is used to confine the beam and a knife edge (KE) allows for variation of the plasma width. The harmonic light passes a slit (S) before it is dispersed by a grating (G). The spectra are recorded with a charge-coupled device (CCD). **b, c** Raw spectra ranging from the 13th to 27th harmonic. The target configuration is 0.4 mm disk width and 0.7mm spacer width in (b), and 0.5 mm disk width and 0.9 mm spacer width in (c). Adapted from [70] with permission from the Optical Society of America

pulse. A decrease in N_e with a weaker excitation of the target leads to an optimization of QPM for a higher q_{qpm} in order to keep the $q_{\mathrm{qpm}}N_e$ product unchanged at fixed spatial characteristics of plasma jets, which was demonstrated in Fig. 10c. Thus, additional electrons appearing during stronger ablation of targets significantly affect the QPM conditions. However, whether these additional electrons were attributed to those formed during ablation or tunneling ionization of additional silver particles remains unclear.

Another scheme for QPM was implemented in Ref. [70]. To create expanded plasma, a cylindrical lens (L2) with a focal length of 300 mm focused the plasma beam along the target surface (Fig. 9a [70]). The targets for implementing QPM consist of up to four aluminum disks. Disks of various thicknesses (light gray) with adjustable spacers separating them (dark gray) are placed on a rotating axis. When illuminated by a plasma-generating beam, separated plasma jets are formed in front of the target. The axis of rotation is parallel to the direction of the driving beam. Moving targets are required at a heating pulse repetition rate of 1 kHz to avoid unfavorable plasma formation when the same spot is repeatedly illuminated on the target [94]. For comparison, unstructured plasma can also be created using an elongated cylindrical target. The converted radiation was analyzed by an XUV spectrometer.

To implement quasi-phase-matching, certain plasma jets with a certain separation are created using a structured target. Within individual jets, an increase in the output of harmonics is allowed, while in the free space between them, the phase coupling develops without additional generation of harmonics. If the dimensions are chosen correctly, the intensity of harmonics of a certain harmonic order will increase quadratically with the effective width of all plasma jets. Near the focal spot of the high-beam beam, the Gouy phase increases almost constantly with the propagation distance.

The application of a structured target consisting of disks with a width of 0.4 mm and a distance of 0.7 mm with four disks in total, harmonics at around the 23rd order, showed a clear sign of QPM (Fig. 9b [70]). However, the 25th harmonic was already close to the cut-off region. QPM is most effective when the coupling of harmonic fields generated in successive plasma jets becomes constructive again at the beginning of the subsequent jet. Figure 9c shows a raw image of the harmonic spectrum obtained using enlarged target dimensions with a disk width of 0.5 mm and a distance of 0.9 mm. Thus, the ability to change the spatial properties of the plasma structure provides control over the QPM process. As shown in Fig. 9b and c [70], plasma with smaller structures amplifies higher harmonic orders, while lower orders are enhanced in the case of larger jet structures. A similar conclusion can be retrieved from Fig. 8b.

Applications of Clusters, Quantum Dots, and Nanoparticles for Enhancing the Harmonic Yield

The high-order nonlinear optical properties of various inorganic nanoparticles (Ag, Au, Pd, Pt, Ru, In, $BaTiO_3$, and $SrTiO_3$) appearing in LIP were analyzed in [26, 31, 41, 66, 71, 95]. It was shown that nanoparticles significantly increase the harmonic output compared to atoms and ions of the same elemental consistency (Fig. 10 [32]). Similarly, several studies have reported the use of carbon-containing nanostructures as effective harmonic emitters [23, 96–99]. The quantum dots of metal sulfides also demonstrated an increase in the efficiency of HHG conversion and corresponding stability of LIP during laser ablation [28, 100, 101]. A correlation was observed between exceptionally large values of the third-order nonlinear susceptibility and the output of high-order harmonics of quantum dots. Clusters consisting of several (from 2 to 30) atoms also demonstrated greater harmonic conversion efficiency compared to atomic particles of the same elemental consistency [102, 103].

Thus, the extension of HHG research to multi-particle systems, i.e., clusters, quantum dots, and nanoparticles, allowed to demonstrate new ways of amplifying harmonics. In addition, the analysis and the role of nanoparticles appearing *in situ* during ablation of various solid targets on the harmonic output [104] showed the formation of conditions when, even with strong excitation of the ablating material, the growth of harmonic output induced by nanoparticles becomes the predominant mechanism that overcomes the phase mismatch of interacting waves caused by free electrons. The earlier reported increase in the conversion efficiency of HHG in gaseous media and LIPs containing nanoparticles can be explained by a change in the cross section of these particles compared to smaller particles (atoms and ions). The recombination cross section of the accelerated electron with the parent particle in the case of multi-particle systems is much larger than that of atoms. Although it is still being discussed, this harmonic amplification is often observed when using gaseous clustered targets [105, 106]. It can be expected that large nanoparticles will

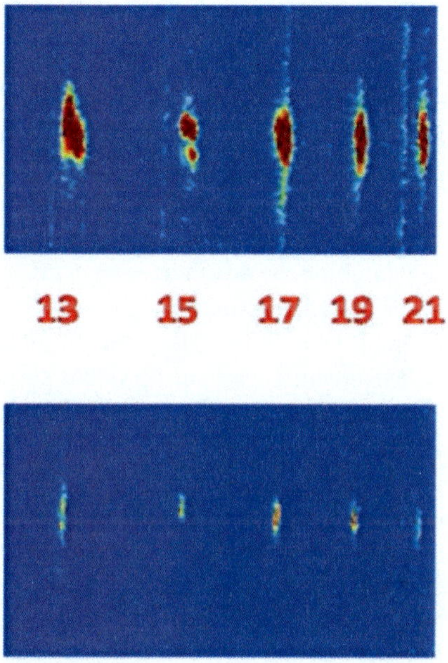

Fig. 10 Raw images of harmonic spectra and plasma emission in the case of LIPs produced during ablation of the target containing 25 nm Ag nanoparticles (upper panel) and bulk silver target (bottom panel) at optimal conditions of ablation. Adapted from [32] with permission from American Physical Society

not effectively participate in the HHG process, but rather will become a source of the formation of small clusters during laser ablation. The partial decay of large nanoparticles during heating, melting, and evaporation causes the formation of quite small clusters containing a small number of atoms that can become the main emitters of harmonics. Previous results did not provide direct evidence of the role of small clusters in achieving HHG amplification in the plasmas comprising of nanoparticles. To verify this, it is necessary to conduct a systematic study of the role of the sizes of polyatomic samples on the yield of HHG under the same experimental conditions. The crucial role of small clusters in harmonic amplification can be confirmed by identifying the types of clusters using time-of-flight mass spectrometry.

Other interesting targets for HHG are the quantum dots (1–4 nm). The use of thin films containing quantum dots that support the growth of the nonlinear optical response caused by quantum confinement due to the local field amplification can solve the problem of the optimal sizes of multi-particle components of plasma. It can be assumed that larger nanoparticles (5–30 nm) could also be suitable for these purposes, but there are some limitations in the growth of their nonlinearity with increasing particle size. Probably, it is possible to find some optimal sizes of nanoparticles in order to increase the harmonic output, thereby providing additional information for spectroscopy of these materials in the XUV range.

This approach is ideally combined with other proposals to use nanophotonic structures for various purposes. In addition to the above suggestion in using the quantum dot structures, thin perovskite films can be used for HHG. Structures like an urchin, a triangle, a rod, and cuboids, synthesized during laser ablation of several metals under aqueous ambient conditions [107–111] and deposited on a glass substrate, can also be the subject of high-order spectroscopic studies of 2D nanophotonic compounds.

Numerous studies have shown that clusters and nanoparticles can be attractive media for HHG since they combine the advantages of using atomic media and bulk solid media. Studies of LIPs containing metal nanopowders assumed the dominant role of harmonics, which mainly come from the emitters of nanoparticles, and not from isolated atoms/ions. Meanwhile, a recent study has shown that both nanoparticles and isolated monomers can dominate the harmonic spectrum, depending on which part of the plasma plume interacts with the driving laser [112]. While analyzing the plume of indium nanoparticles, it was found that the harmonic spectra from the region dominated by monomers differ markedly from the region dominated by nanoparticles. These results demonstrated that accurate determination of the contribution of nanoparticles to HHG requires an accurate choice of the interaction region of the laser plasma, a factor that was not carefully considered in previous studies [113, 114].

Structure of the Book

All the above-discussed studies allowed the demonstration of new approaches in generation of the short-wavelength radiation in laser-induced plasmas. The advanced approach in applications of HHG in LIPs provides the tools for the optical analysis of recently developed nanostructures. The developments in this field of nonlinear optics are rapidly improving to follow the needs and requirements of this area of research. One can expect further achievements in this direction in the nearest future.

This book is organized as follows. In Chap. 1, we analyze the lowest order harmonics generation of laser-induced plasmas. We discuss the third harmonic generation and optical nonlinearities of ablated nickel nanoparticles, analyze the results of studies of various particles by third harmonic generation based on laser pulse induced plasma, as well as discuss the third and fifth harmonics generation in air and nanoparticle-containing plasmas using 150 kHz fiber laser. Chapter 2 is dedicated to different aspects of high-order harmonics generation in plasmas, such as incoherent and coherent extreme ultraviolet emission from boron plasma, calculations of high-order harmonic generation in laser-produced lithium plasma, and spectral modification of converting radiation and high-order harmonics through filamentation in argon and propagation in laser-produced plasmas. In Chap. 3, we analyze the high-order harmonic generation in plasmas using high-pulse repetition rate lasers. We discuss the time-dependent optimization of laser-produced molecular plasmas through high-order harmonic generation, high-order harmonics generation in the plasmas produced on different rotating targets during ablation using 1 kHz and 100 kHz lasers, and application of 150 kHz laser for high-order harmonic generation in different plasmas. Chapter 4 describes the applications of clusters and quantum dots for HHG in LIPs. We analyze the role of carbon clusters in low- and high-order harmonics generation in graphite plasmas. Low- and high-order nonlinear optical properties of Ag_2S quantum dot thin films are discussed. The methods of the effective high-order harmonic generation from metal sulfide quantum dots are described. We also discuss the high-order harmonic generation using quasi-phase-matching and two-color pump in the plasmas containing molecular and alloyed metal sulfide quantum dots. In Chap. 5, recent studies of the HHG using carbon-contained nanoparticles are presented. We analyze the role of carbon clusters in the high-order harmonic generation of 800 nm femtosecond radiation in graphite plasmas. As an example, the efficient high-order harmonics generation using 1030 nm laser in carbon nanostructures is discussed.

HHG in metal nanoparticles is analyzed in Chap. 6. We demonstrate and discuss the application of oxide nanoparticles for HHG. Another example, which was analyzed in this chapter, is the high-order harmonic generation in Au nanoparticle-contained plasmas. Chapter 7 presents the analysis of the studies of the two-color pumps of laser-induced plasmas. We discuss the comparative studies of high-order harmonic generation in argon gas and different laser-produced plasmas using single-color and two-color pumps. High-order harmonic generation during different overlaps of two-colored pulses in laser-produced plasmas and gases is presented and analyzed. The resonance-induced enhancement of a single harmonic is discussed in Chap. 8. We describe how to control a single harmonic enhancement in laser-produced plasmas. We also present the studies of the resonance-enhanced harmonics in mixed laser-produced plasmas. In Chap. 9, we discuss the application of the quasi-phase-matching concept in laser-induced plasmas. The theory and experiment describe the current status of this research. Particularly, we discuss the analytical treatment of quasi-phase-matching of high-order harmonics in multi-jet laser plasmas: influence of free electrons between jets, intrinsic phase, and Gouy phase. The experimental examples of the application of the quasi-phase-matching concept for enhancement of high-order harmonics of ultrashort laser pulses in plasmas are presented and analyzed. Another example is the high-order harmonics generation under quasi-phase-matched conditions in silver, boron, and silver sulfide plasmas of different configurations. Finally, we summarize the main advances of current studies presented in this book and discuss the perspectives of this method of HHG.

References

1. R.A. Ganeev, T. Witting, C. Hutchison, F. Frank, P.V. Redkin, W.A. Okell, D.Y. Lei, T. Roschuk, S.A. Maier, J.P. Marangos, J.W.G. Tisch, Phys. Rev. A **85**, 015807 (2012)
2. R.A. Ganeev, *High-Order Harmonic Generation in Laser Plasma Plumes* (Imperial College Press, London, 2012)
3. R.A. Ganeev, M. Suzuki, M. Baba, H. Kuroda, T. Ozaki, Opt. Lett. **31**, 1699 (2006)
4. L. B. Elouga Bom, S. Haessler, O. Gobert, M. Perdrix, F. Lepetit, J.-F. Hergott, B. Carré, T.Ozaki, and P. Salières, Opt. Express 19, 3677 (2011)

5. R.A. Ganeev, G.S. Boltaev, V.V. Kim, M. Venkatesh, C. Guo, OSA Continuum **2**, 2381(2019)
6. M. Ferray, A. L'Huillier, X.F. Li, L.A. Lompré, G. Mainfray, G. Manus, J. Phys. B **21**, L31(1988)
7. A. McPherson, G. Ginson, H. Jara, N. Johann, I.A. McIntyre, K. Boyer, C.K. Rhodes,J. Opt. Soc. Am. B **4**, 595 (1987)
8. J.L. Krause, K.J. Schafter, K.C. Kulander, Phys. Rev. Lett. **68**, 3535 (1992)
9. P.B. Corkum, Phys. Rev. Lett. **71**, 1994 (1993)
10. P.A. Norreys, M. Zepf, S. Moustaizis, A.P. Fews, J. Zhang, P. Lee, M. Bakarezos, C.N.Danson, A. Dyson, P. Gibbon, P. Loukakos, D. Neely, F.N. Walsh, J.S. Wark, A.E. Dangor,Phys. Rev. Lett. **76**, 1832 (1996)
11. B. Dromey, M. Zepf, A. Gopal, K. Lancaster, M.S. Wei, K. Krushelnick, M. Tatarakis, N.Vakakis, S. Moustaizis, R. Kodama, M. Tampo, C. Stoeckl, R. Clarke, H. Habara, D. Neely,S. Karsch, P. Norreys, Nature Phys. **2**, 456 (2006)
12. S. Ghimire, A.D. DiChiara, E. Sistrunk, P. Agostini, L.F. DiMauro, D.A. Reis, Nature Phys.**7**, 138 (2011)
13. Y. Akiyama, K. Midorikawa, Y. Matsunawa, Y. Nagata, M. Obara, H. Tashiro, K. Toyoda,Phys. Rev. Lett. **69**, 2176 (1992)
14. S. Kubodera, Y. Nagata, Y. Akiyama, K. Midorikawa, M. Obara, Phys. Rev. A **48**, 4576(1993)
15. W. Theobald, C. Wülker, F.R. Schäfer, B.N. Chichkov, Opt. Commun. **120**, 177 (1995)
16. C.-G. Wahlström, S. Borgström, J. Larsson, S.-G. Pettersson, Phys. Rev. A **51**, 585 (1995)
17. R.A. Ganeev, V.I. Redkorechev, T. Usmanov, Opt. Commun. **135**, 251 (1997)
18. A. Krushelnick, W. Tighe, S. Suckewer, J. Opt. Soc. Am. A **14**, 1687 (1997)
19. R. Ganeev, M. Suzuki, M. Baba, H. Kuroda, T. Ozaki, Opt. Lett. **30**, 768 (2005)
20. R.A. Ganeev, M. Suzuki, H. Kuroda, Phys. Rev. A **89**, 033821 (2014)
21. R. A. Ganeev, L. B. Elouga Bom, J. Abdul-Hadi, M. C. H. Wong, J. P. Brichta, V.R. Bhardwaj, and T. Ozaki, Phys. Rev. Lett. 102, 013903 (2009).
22. G.S. Boltaev, V.V. Kim, M. Iqbal, N.A. Abbasi, V.S. Yalishev, R.A. Ganeev, A.S. Alnaser,Photonics **7**, 66 (2020)
23. Y. Pertot, S. Chen, S. D. Khan, L. B. Elouga Bom, T. Ozaki, and Z. Chang, J. Phys. B 45,074017 (2012)
24. A.H. Sheinfux,_ Z. Henis, M. Levin, and A. Zigler, Appl. Phys. Lett. 98, 141110 (2011)
25. G.S. Boltaev, R.A. Ganeev, I.A. Kulagin, T. Usmanov, J. Opt. Soc. Am. B **31**, 436 (2014)
26. Y. Fu, R.A. Ganeev, G.S. Boltaev, S.K. Maurya, V.V. Kim, C. Zhao, A. Rout, C. Guo, Nanophotonics **8**, 849 (2019)
27. M. López-Arias, M. Oujja, M. Sanz, R. de Nalda, R.A. Ganeev, M. Castillejo, Molec. Phys.**110**, 1651 (2012)
28. A.V. Andreev, R.A. Ganeev, H. Kuroda, S.Y. Stremoukhov, O.A. Shoutova, Eur. Phys. J. D**67**, 22 (2013)
29. M. Wöstmann, P.V. Redkin, J. Zheng, H. Witte, R.A. Ganeev, H. Zacharias, Appl. Phys.B **120**(1), 17 (2015)
30. R.A. Ganeev, H. Singhal, P.A. Naik, J.A. Chakera, H.S. Vora, R.A. Khan, P.D. Gupta, Phys.Rev. A **82**, 053831 (2010)
31. R.A. Ganeev, T. Witting, C. Hutchison, V.V. Strelkov, F. Frank, M. Castillejo, I.Lopez-Quintas, Z. Abdelrahman, J.W.G. Tisch, J.P. Marangos, Phys. Rev. A **88**, 033838(2013)
32. R.A. Ganeev, C. Hutchison, M. Castillejo, I. Lopez-Quintas, F. McGrath, D.Y. Lei, J.P. Marangos, Phys. Rev. A **88**, 033803 (2013)
33. R.A. Ganeev, G.S. Boltaev, B. Sobirov, S. Reyimbaev, H. Sherniyozov, T. Usmanov, M.Suzuki, S. Yoneya, H. Kuroda, Phys. Plasmas **22**, 012302 (2015)
34. R.A. Ganeev, C. Hutchison, T. Siegel, A. Zaïr, J.P. Marangos, Phys. Rev. A **83**, 063837(2011)
35. R.A. Ganeev, T. Witting, C. Hutchison, F. Frank, M. Tudorovskaya, M. Lein, W.A. Okell,A. Zaïr, J.P. Marangos, J.W.G. Tisch, Opt. Express **20**, 25239 (2012)
36. R.A. Ganeev, J. Opt. Soc. Am. B **33**, E93 (2016)
37. R. A. Ganeev, L. B. Elouga Bom, J.-C. Kieffer, and T. Ozaki, Phys. Rev. A 76, 023831(2007)

38. R.A. Ganeev, M. Suzuki, M. Baba, H. Kuroda, Appl. Phys. Lett. **94**, 051101 (2009)
39. R. A. Ganeev, L. B. Elouga Bom, J.-C. Kieffer, and T. Ozaki, J. Opt. Soc. Am. B 24, 1319(2007)
40. R. A. Ganeev, P. A. Naik, J. A. Chakera, H. Singhal, N. C. Pramanik, P. A. Abraham, N.Rani Panicker, M. Kumar, and P. D. Gupta, J. Opt. Soc. Am. B 28, 360 (2011).
41. R.A. Ganeev, C. Hutchison, M. Castillejo, I. Lopez-Quintas, F. McGrath, D.Y. Lei, J.P. Marangos, Phys. Rev. A **88**, 033803 (2013)
42. M. Suzuki, M. Baba, R. Ganeev, H. Kuroda, T. Ozaki, Opt. Lett. **31**, 3306 (2006)
43. R.A. Ganeev, P.V. Redkin, Opt. Commun. **281**, 4126 (2008)
44. M.F. Frolov, N.L. Manakov, A.F. Starace, Phys. Rev. A **82**, 023424 (2010)
45. R.A. Ganeev, V.V. Strelkov, C. Hutchison, A. Zaïr, D. Kilbane, M.A. Khokhlova, J.P. Marangos, Phys. Rev. A **85**, 023832 (2012)
46. J. M. Ngoko Djiokap, and A. F. Starace, Phys. Rev. A 88, 053412 (2013)
47. R.A. Ganeev, M. Suzuki, S. Yoneya, H. Kuroda, J. Phys. B **48**, 165401 (2015)
48. R.A. Ganeev, Z. Wang, P. Lan, P. Lu, M. Suzuki, H. Kuroda, Phys. Rev. A **93**, 043848(2016)
49. Y. Fu, R.A. Ganeev, P.S. Krishnendu, C. Zhou, K.S. Rao, C. Guo, Opt. Mater. Express **9**,976 (2019)
50. R.A. Ganeev, Opt. Spectrosc. **127**, 293 (2019)
51. R.A. Ganeev, V. Tosa, K. Kovács, M. Suzuki, S. Yoneya, H. Kuroda, Phys. Rev. A **91**,043823 (2015)
52. R.A. Ganeev, M. Suzuki, H. Kuroda, Phys. Plasmas **21**, 053503 (2014)
53. D.B. Milošević, J. Phys. B **40**, 3367–3373 (2007)
54. L. B. Elouga Bom, F. Bouzid, F. Vidal, J.-C. Kieffer, and T. Ozaki, J. Phys. B 41, 215401(2008)
55. H. Singhal, V. Arora, B.S. Rao, P.A. Naik, U. Chakravarty, R.A. Khan, P.D. Gupta, Phys.Rev. A **79**, 023807 (2009)
56. I.A. Kulagin, T. Usmanov, Opt. Lett. **34**, 2616 (2009)
57. D.B. Milošević, Phys. Rev. A **81**, 023802 (2010)
58. M. Oujja, R. de Nalda, M. López-Arias, R. Torres, J.P. Marangos, M. Castillejo, Phys. Rev.A **81**, 043841 (2010)
59. V. Strelkov, Phys. Rev. Lett. **104**, 123901 (2010)
60. Y. Pertot, L. B. Elouga Bom, V. R. Bhardwaj, and T. Ozaki, Appl. Phys. Lett. **98**, 101104(2011)
61. M. Tudorovskaya, M. Lein, Phys. Rev. A **84**, 013430 (2011)
62. L. B. Elouga Bom, Y. Pertot, V. R. Bhardwaj, and T. Ozaki, Opt. Express 19, 3077 (2011)
63. S. Haessler, L. B. Elouga Bom, O. Gobert, J.-F. Hergott, F. Lepetit, M. Perdrix, B. Carré, T.Ozaki, and P Salières, J. Phys. B 45, 074012 (2012)
64. M. Kumar, H. Singhal, J.A. Chakera, P.A. Naik, R.A. Khan, P.D. Gupta, J. Appl. Phys. **114**,033112 (2013)
65. S. Haessler, V. Strelkov, L. B. Elouga Bom, M. Khokhlova, O. Gobert, J.-F. Hergott, F.Lepetit, M. Perdrix, T. Ozaki, and P. Salieres, New J. Phys. 15, 013051 (2013)
66. H. Singhal, P.A. Naik, M. Kumar, J.A. Chakera, P.D. Gupta, J. Appl. Phys. **115**, 033104(2014)
67. N. Rosenthal, G. Marcus, Phys. Rev. Lett. **115**, 133901 (2015)
68. M.A. Fareed, N. Thire, S. Mondal, B.E. Schmidt, F. Legare, T. Ozaki, Appl. Phys. Lett. **108**,124104 (2016)
69. M.A. Fareed, V.V. Strelkov, N. Thire, S. Mondal, B.E. Schmidt, F. Legare, T. Ozaki, NatureCommun. **8**, 16061 (2017)
70. M. Wostmann, L. Splitthoff, H. Zacharias, Opt. Express **26**, 14524 (2018)
71. Z. Abdelrahman, M.A. Khohlova, D.J. Walke, T. Witting, A. Zair, V.V. Strelkov, J.P. Marangos, J.W.G. Tisch, Opt. Express **26**, 15745 (2018)
72. M.A. Fareed, V.V. Strelkov, M. Singh, N. Thire, S. Mondal, B.E. Schmidt, F. Legare, T.Ozaki, Phys. Rev. Lett. **121**, 023201 (2018)
73. I.S. Wahyutama, T. Sato, K.L. Ishikawa, Phys. Rev. A **99**, 063420 (2019)

74. M. Kumar, H. Singhal, J.A. Chakera, J. Appl. Phys. **125**, 155902 (2019)
75. M. Singh, M.A. Fareed, A. Laramée, E. Isgandarov, T. Ozaki, Appl. Phys. Lett. **115**, 231105(2019)
76. J. Liang, Y.H. Lai, W. Fu, Y. Shan, W. Yu, C. Guo, Phys. Rev. A **102**, 053117 (2020)
77. R. Taïeb, V. Véniard, J. Wassaf, A. Maquet, Phys. Rev. A **68**, 033403 (2003)
78. D.B. Milošević, J. Opt. Soc. Am. B **23**, 308 (2006)
79. I.A. Ivanov, A.S. Kheifets, Phys. Rev. A **78**, 053406 (2008)
80. J. M. Ngoko Djiokap and A. F. Starace, Phys. Rev. A 102, 013103 (2020)
81. R.A. Ganeev, J. Appl. Phys. **119**, 113104 (2016)
82. R.A. Ganeev, H. Kuroda, OSA Continuum **4**, 1545 (2021)
83. R.A. Ganeev, H. Kuroda, Photonics **8**, 29 (2021)
84. R.A. Ganeev, H. Kuroda, Appl. Sci. **11**, 2193 (2021)
85. E. S. Toma, P. Antoine, A. d. Bohan and H. G. Muller, J. Phys. B 32, 5843 (1999)
86. B. Gaarde, K.J. Schafer, Phys. Rev. A **64**, 013820 (2001)
87. C. Figueira de Morisson Faria, R. Kopold, W. Becker and J. M. Rost, Phys. Rev. A 65,023404 (2002)
88. R.A. Ganeev, S. Odžak, D.B. Milošević, M. Suzuki, H. Kuroda, Laser Phys. **26**, 075401(2016)
89. V.V. Strelkov, M.A. Khokhlova, NYu., Shubin, Phys. Rev. A **89**, 053833 (2014)
90. J. Seres, V.S. Yakovlev, E. Seres, C.H. Streli, P. Wobrauschek, C.H. Spielmann, F. Krausz,Nat. Phys. **3**, 878 (2007)
91. A. Pirri, C. Corsi, M. Bellini, Phys. Rev. A **78**, 011801 (2008)
92. V. Tosa, V.S. Yakovlev, F. Krausz, New J. Phys. **10**, 025016 (2008)
93. L. Zheng, X. Chen, S. Tang, R. Li, Opt. Express **15**, 17985 (2007)
94. C. Hutchison, R.A. Ganeev, T. Witting, F. Frank, W.A. Okell, J.W.G. Tisch, J.P. Marangos,Opt. Lett. **37**, 2064 (2012)
95. R.A. Ganeev, *Nanostructured Nonlinear Optical Materials: Formation and Characterization* (Elsevier, Amsterdam, 2018)
96. R.A. Ganeev, P.A. Naik, H. Singhal, J.A. Chakera, M. Kumar, M.P. Joshi, A.K. Srivastava,P.D. Gupta, Phys. Rev. A **83**, 013820 (2011)
97. R.A. Ganeev, C. Hutchison, T. Witting, F. Frank, S. Weber, W.A. Okell, E. Fiordilino, D.Cricchio, F. Persico, A. Zaïr, J.W.G. Tisch, J.P. Marangos, J. Opt. Soc. Am. B **30**, 7 (2013)
98. R.A. Ganeev, M. Baba, M. Suzuki, H. Kuroda, J. Appl. Phys. **115**, 183101 (2014)
99. R.A. Ganeev, V.V. Kim, K.S. Rao, C. Guo, Appl. Sci. **11**, 2143 (2021)
100. R.A. Ganeev, G.S. Boltaev, V.V. Kim, K. Zhang, A.I. Zvyagin, M.S. Smirnov, O.V.Ovchinnikov, P.V. Redkin, M. Wöstmann, H. Zacharias, C. Guo, Opt. Express **26**, 35013(2018)
101. R.A. Ganeev, G.S. Boltaev, V.V. Kim, M. Venkatesh, A.I. Zvyagin, M.S. Smirnov, O.V.Ovchinnikov, M. Wöstmann, H. Zacharias, C. Guo, J. Appl. Phys. **126**, 193103 (2019)
102. R.A. Ganeev, C. Hutchison, T. Witting, F. Frank, W.A. Okell, A. Zaïr, S. Weber, P.V.Redkin, D.Y. Lei, T. Roschuk, S.A. Maier, I. López-Quintás, M. Martín, M. Castillejo, J.W.G. Tisch, J.P. Marangos, J. Phys. B **45**, 165402 (2012)
103. R.A. Ganeev, G.S. Boltaev, K. Zhang, S.K. Maurya, M. Venkatesh, Z. Yu, V.V. Kim, P.V.Redkin, C. Guo, OSA Continuum **2**, 1510 (2019)
104. H. Singhal, R.A. Ganeev, P.A. Naik, J.A. Chakera, U. Chakravarty, H.S. Vora, A.K.Srivastava, C. Mukherjee, C.P. Navathe, S.K. Deb, P.D. Gupta, Phys. Rev. A **82**, 043821(2010)
105. C.-H. Pai, C.C. Kuo, M.-W. Lin, J. Wang, S.-Y. Chen, J.-Y. Lin, Opt. Lett. **31**, 984 (2006)
106. I.-Y. Park, S. Kim, J. Choi, D.-H. Lee, Y.-J. Kim, M.F. Kling, M.I. Stockman, S.-W. Kim,Nature Photon. **5**, 677 (2011)
107. K. Zhang, R.A. Ganeev, G.S. Boltaev, P.V. Redkin, C. Guo, Mater. Res. Express **6**, 045027(2019)

108. K. Zhang, D.S. Ivanov, R.A. Ganeev, G.S. Boltaev, P.S. Krishnendu, S. Singh, M.E. Garcia,I.N. Zavestovskaya, C. Guo, Nanomater. **9**, 767 (2019)
109. S.K. Maurya, A. Rout, R.A. Ganeev, C. Guo, J. Nanomater. **2019**, 9686913 (2019)
110. K.S. Rao, R.A. Ganeev, K. Zhang, Y. Fu, G.S. Boltaev, S.K. Maurya, C. Guo, Opt. Mater.**92**, 366 (2019)
111. G.S. Boltaev, R.A. Ganeev, P.S. Krishnendu, K. Zhang, C. Guo, Scient. Reports **9**, 11414(2019)
112. J. Liang, Y.H. Lai, W. Fu, C. Guo, W. Li, Opt. Express **29**, 23421 (2021)
113. R.A. Ganeev, L.B. Elouga Bom, M.C.H. Wong, J.-P. Brichta, V.R. Bhardwaj,P.V. Redkin, and T. Ozaki, Phys. Rev. A 80, 043808 (2009)
114. H. Singhal, R.A. Ganeev, P.A. Naik, A.K. Srivastava, A. Singh, R. Chari, R.A. Khan, J.A.Chakera, P.D. Gupta, J. Phys. B **43**, 025603 (2010)

Contents

Chapter 1
Lowest-Order Harmonics Generation Nonlinearities of Laser-Induced Plasmas

1.1 Study of Various Particles by Third Harmonic Generation Based on Laser Pulse Induced Plasma

The generation of ejected particles during laser ablation (LA) strongly depends on the conversion of electronic and vibration excitation upon the interaction of the material with laser pulses into the kinetic energy of particles motion. The LA process can be divided into three different steps including (a) bond-breaking and plasma ignition, (b) plasma expansion and cooling, and (c) particle ejection and condensation. The plasma ignition process largely depends on the width of heating pulses. For nanosecond (ns) laser pulses, the plasma ignition process mostly involves thermal vaporization, whereas, for picosecond (ps) laser pulses, it consists of both thermal and non-thermal mechanisms [1]. Meanwhile, the Coulomb's explosion is responsible for plasma ignition in the case of femtosecond (fs) laser pulses used for LA [2], which is primarily a bond-breaking mechanism. The concentration of smaller-sized particles in the plasma is found to be larger compared to large-sized particles. The components of plasma have been widely studied using various techniques [3, 4]. Those techniques have been extensively applied for the determination of the sizes, shape, and composition of plasma.

The clusters and nanoparticles have been recognized as the media possessing large optical nonlinearities. It has been proven that these multi-particle structures can be efficiently used for the generation of lower- and higher-order harmonics of ultrashort laser pulses. The harmonic generation (HG) can also be used as a characterization tool for the analysis of the composition of laser-induced plasmas. The spatiotemporal separation of emitted particles during the formation of plasma has made HG an effective technique to characterize them. There are few reports on the usage of third-harmonic generation (THG) as a characterization scheme [5–7] for the generated particles in plasma. This has been achieved by varying the arrival time of the probe pulses with respect to the heating pulses. The time delay-based measurement of third harmonic emission has shown the presence of the larger aggregates formed

R. A. Ganeev, *High-Order Harmonics Generation in Plasmas*, Springer Series on Atomic, Optical, and Plasma Physics 122, https://doi.org/10.1007/978-3-031-09040-0_1

during LA, which can be also further analyzed during their deposition on the nearby substrates [8, 9].

The purpose of discussed research was to investigate the role of physical characteristics of the bulk target on THG in various plasmas using the variation of the delay between heating and probe pulses [10]. An alternative approach of probing the ejected particle during plasma formation, other than time delay-based measurements, is the measurement of the third harmonic (TH) yield through the variation of the distance between the surface of the target and the probe pulses. Probing such particles in plasma through the measurement of TH can determine the relationship between the mass and the velocity of the ejected particle during plasma creation similar to the delay-based measurements. THG was used as a probe mechanism to characterize the components of plasmas from various bulk targets such as graphite (C), silver (Ag), nickel (Ni), chromium (Cr), and indium (In), cadmium sulfide (CdS), and zinc sulfide (ZnS). THG in various plasmas is compared with the mass of the studied materials under vacuum conditions and in the air. Furthermore, the relative efficiency of THG in various plasmas, which were created using ns and ps heating laser pulses, is analyzed.

1.1.1 Description of Experimental Arrangements

All the bulk targets for the creation of plasma were used as purchased. C (99.9%), Ag (99.95%), Ni (99.95%), Cr (99.95%), and In (99.95%) were choosen as the ablation targets. THG from wide-bandgap semiconductor materials, such as CdS (99.95%), and ZnS (99.95%), were also carried out in order to observe the influence of the ejected particle mass on THG.

The apparatus for the THG study consisted of the regenerative amplifier (Spitfire ACE, Spectra-Physics) operated at 800 nm with the pulses repetition rate of 1 kHz and 40 fs pulse duration. Two different heating schemes were employed to perform THG in various targets. The first scheme (Fig. 1.1a) consisted of 200 ps heating pulses from the amplifier as heating pulses to create plasma plume from the target. The second scheme (Fig. 1.1b) consisted of 6 ns heating pulses at 1064 nm with a repetition rate of 10 Hz (Q-smart 850, Quantel). The pulse width and beam diameter of the probe pulses at the position of the focusing lens were fixed at 40 fs and 20 mm, respectively. The beam diameters of ns and ps heating pulses were 5 and 7 mm respectively. The arrival time of the fs probe pulse in the plasma area was fixed at 38 ns from the beginning of ablation by ps heating pulses. In the case of ns heating pulses, the arrival time of fs probe pulses with respect to heating radiation was controlled using a delay generator (DG535, Stanford Research Instruments), which was operated using a trigger pulse from the amplifier running at 1 kHz.

The heating pulses were focused on the targets using a 300 mm focal length lens, whereas the probe pulses were focused using a 500 mm focal length lens and propagated parallel to the surface of the target and perpendicular to the heating pulses. In the case of the ps heating scheme, the focus point of the fs probe beam was set

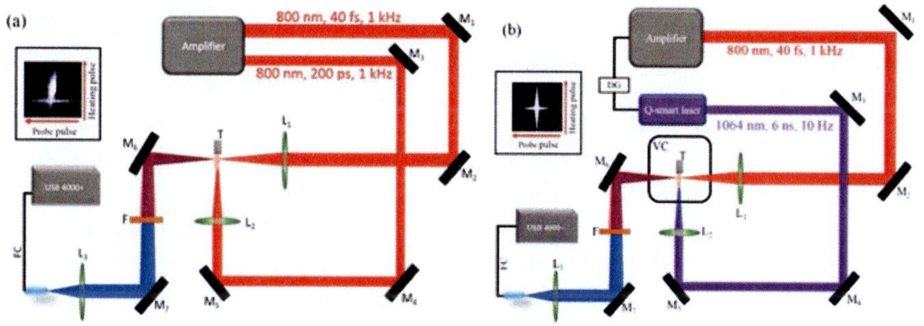

Fig. 1.1 Experimental setups for third-harmonic generation in **a** air and **b** vacuum conditions. M: mirrors, L: lenses, F: filter, T: target, VC: vacuum chamber, DG: delay generator. The insets show the spatial arrangements of the heating and probe laser pulses at the sample position in each schematic of the experimental setup. Reproduced from [10] with permission from Elsevier

to be 2 mm away from the focal point of the heating beam as shown in the inset of Fig. 1.1a. In the case of ns heating pulses, the focal points of both heating and probe pulses were set to be at the same point (see inset in Fig. 1.1b). The energy of the heating pulses was fixed at 0.56 mJ for ps pulses and 3 mJ for ns pulses. The energy of probe pulses was varied by using a pair of a polarizer and a half-wave plate with a maximum employed energy of 0.5 mJ. The pressure in the vacuum chamber was kept at 0.1 Pa throughout the experiments for all samples. The generated TH radiation from the excited plasma was separated from the co-propagating fundamental beam using a prism and then was registered by a fiber spectrometer (Ocean Optics, USB 4000 +). The target sample was moved with constant speed to get the stable TH intensity. In order to establish the contribution of plasma species to the THG, the plasma spectra were collected using a 100 mm fused silica lens, which focused the plasma emission on the tip of the optical fiber interfaced with a spectrometer.

A systematic study was performed for the THG in various plasmas, where the dependence of the TH emission on the power of probe pulses, position of the target with respect to the probe pulses, and TH yield as a function of the arrival time of the probe pulses with respect to the heating pulses were carried out. The discussed study consisted of two parts. The measurement of THG in the air was performed using the scheme shown in Fig. 1.1a, whereas the schematic in Fig. 1.1b was used to study the time evolution of TH in various plasmas by varying the arrival time of probe pulses with respect to heating pulses.

1.1.2 THG from the Metal Plasmas Produced in Air

THG in the plasmas created on C, Ag, and Ni targets in the presence of air was carried out in order to compare the role of ejected particles based on their mass. Prior to the measurement of TH in respective plasmas, THG in the air was measured as a function of the probe pulse energy up to 0.5 mJ, which showed the cubic dependence of harmonic yield on the probe pulse energy. The intensity of TH from plasma was determined by subtracting the TH yield generated in the air from TH yield from plasma in the presence of air. As shown in Fig. 1.1a, the wavelengths of the probe and heating pulses were kept at 800 nm with their respective pulse widths of 40 fs and 200 ps. The energy of the heating pulses was maintained at 560 μJ to create the plasmas. The arrival time of fs probe pulses was fixed at 38 ns after the beginning of plasma formation using ps heating pulses. The acquisition time for the TH collection was kept at 10 s. The surface of the bulk target was placed ~2 mm away from the axis of propagation of the probe pulse. The spatial arrangement of the probe and the heating pulses at the target position is shown in the inset of Fig. 1.1a, which can anticipate a significant difference in TH intensity with the variation of the distance between the probe pulses and the target.

Figure 1.2 shows the variation of TH intensity as a function of the distance between the axis of propagation of probe pulses and the target (C, Ag, Ni) position. TH intensity dependences show similar behavior with respect to the target position with maxima at ~2 mm for each of these targets. The trend in the TH intensity variations with this distance is related to the variation of heating laser fluence as the target moves away from the axis of probe pulse propagation due to the varying spot size of heating pulses on the target surface. The spot size variation changes the fluence of heating pulses on the target surface during ablation thus eventually producing the variable concentration of ejected particles during plasma movement towards the path of the probe beam. The major difference was seen for the case of carbon plasma, which allowed observation of TH up to almost 4 mm away from the target position. This difference in TH yields from different plasmas is attributed to the difference in the masses of the ejected particles, which are assumed to be lighter for C compared to the heavier masses of the ejected particles from Ag and Ni targets. In order to support the effect of the mass of the ejected particle on TH intensity, the comparison of TH intensities in C plasma with respect to other heavier metal targets was analyzed under vacuum conditions, which is discussed later.

Figure 1.3 shows the variation of TH intensity as a function of probe pulse energy

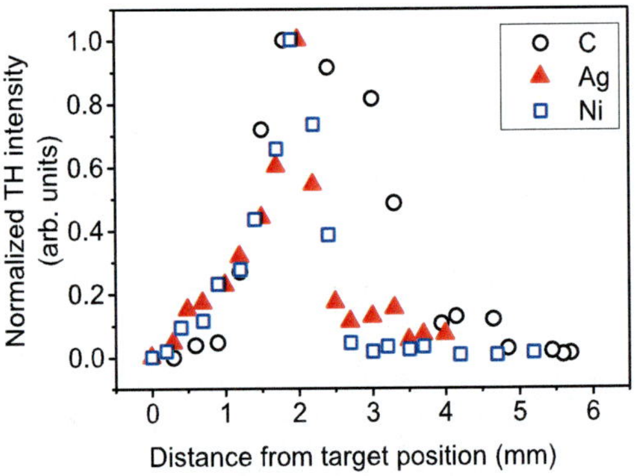

Fig. 1.2 TH intensity as a function of the distance between the axis of propagation of probe pulses and the C, Ag, and Ni surfaces at constant ps heating pulse energy (0.56 mJ) and fs probe pulse energy (0.045 mJ). Reproduced from [10] with permission from Elsevier

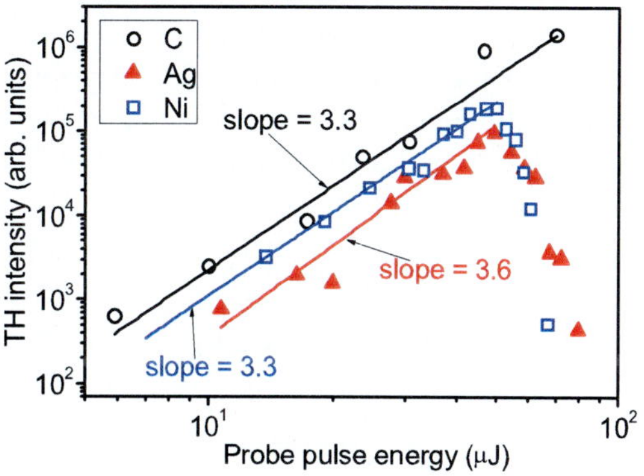

Fig. 1.3 TH intensity as a function of probe pulse energy at the constant ps heating pulse energy of 0.56 mJ on the surfaces of ablating C, Ag, and Ni targets. Reproduced from [10] with permission from Elsevier

at constant heating pulses energy of 0.56 mJ. The energy of probe pulses was varied between 5 and ~70 μJ. The surfaces of the bulk targets were placed 2 mm away from the probe pulses propagation in order to measure the influence of the energy of these pulses on TH yield. The approximately cubic dependence of the yield of TH up to the probe pulse energy of 50 μJ was observed. As one increased the energy of probe pulses, the decrease of TH yield was observed in the case of Ag and Ni targets.

This notable deviation from the cubic dependence can be attributed to the appearance of a large number of free electrons due to the interaction of intense fs probe pulses with the ejected particles during ablation, which led to the phase mismatch between the probe (800 nm) and TH (266 nm) waves in the plasma area. However, at this energy of probe pulses (70 μJ), TH in C plasma still showed the cubic dependence. This difference in the slopes of TH yields of three plasmas shows that the 70 μJ probe pulse energy employed for the generation of TH in carbon plasma was far from the threshold limit of the generation of a large number of free electrons. The absence of a decrease in TH intensity upon the increase of probe pulse energy beyond 50 μJ for C as compared to Ag and Ni, can be related to the ionization energies of used atoms, which is higher for C (11.26 eV) and relatively smaller for Ag (7.57 eV) and Ni (7.6 eV). Notice that the second ionization potentials of C, Ag and Ni are 24.38 eV, 21.48 eV, and 18.48 eV, which also points out the smaller probability of further ionization of the former species.

The employed intensity of 560 μJ, 200 ps heating pulses was ~6×10^{10} W cm^{-2}. At this intensity of heating pulses, the degree of ionization was greater than 1.5, which indicates the appearance of free electrons in the created Ag and Ni plasmas. Under these conditions, the plasma consists of singly charged ions with a small fraction of doubly charged ions. At the same time, the 50 μJ probe pulse energy corresponds to the peak intensity of ~2.7×10^{14} W cm^{-2}. The appearance of free electrons at this intensity is mostly due to further ionization of singly ionized species rather than the ionization of neutral species in plasma plume. The intensity provided by the probe pulses is close to the tunnel ionization threshold for the singly-charged Ag and Ni ions. However, the tunnel ionization threshold for singly-charged C ions is higher than the employed probe pulse intensity. Correspondingly, carbon plasma should demonstrate a lesser influence of free electrons on TH yield compared with Ag and Ni plasmas at similar conditions of experiments. As a consequence of this difference between the studied plasmas, in the latter two cases, the increase in free-electron density led to a decrease in the intensity of TH due to the phase mismatch of the probe and TH waves.

1.1.3 THG from Metal Plasma in the Vacuum

In this subsection, THG study in various bulk targets, such as C, Cr, Ag, In, CdS, and ZnS, ablated under a vacuum pressure of 0.1 Pa is reported. The schematic of experiments is shown in Fig. 1.1b, where 1064 nm, ns heating pulses were used for LA to create the plasma plumes. The application of ns pulses for LA provided precise control over the time delay between the heating and probe pulses using the delay generator. Prior to these studies, TH yield from four plasmas (Ag, C, Cr, and In) at variable probe pulse energies was measured using the fixed heating pulse energy (3 mJ, Fig. 1.4).

The measurements of TH intensity were performed by setting the target position 0.6 mm away from the axis of probe pulses propagation. The variations of TH intensity in three plasmas showed approximately cubic dependences with the variation of probe pulses energy up to the maximum used pulse energies of 500 μJ. This observation allows concluding that the threshold limit for the generation of free electrons in those plasmas under vacuum conditions was far from maximal applied pulse energies, which has earlier been considered to be the dominating restricting factor at reasonably high energies of the probe pulses [11]. The THG efficiency was found to be maximal for Ag plasma, whereas C plasma showed the lowest efficiency.

Figure 1.5 shows the variations of TH intensity as the functions of the distance between the target and the probe pulses during ablation of Ag, Cr, In, and C targets at vacuum conditions. The generation of TH during ablation of C was observed up to ~4 mm away from the position of a target, whereas, for the other three bulk targets, the TH intensity was achieved only up to the distance of 1 mm. This study reveals

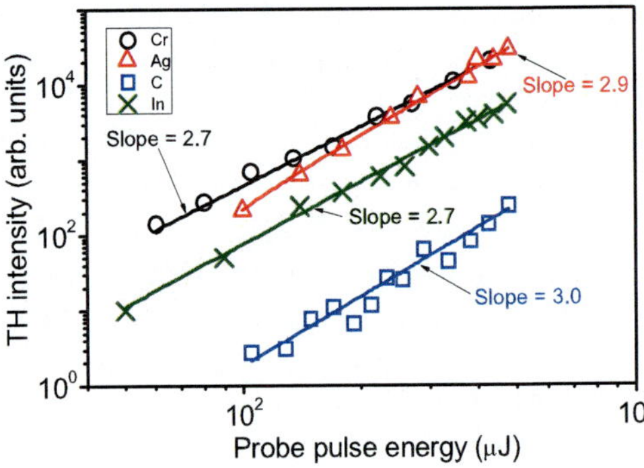

Fig. 1.4 TH intensity as a function of the energy of the probe pulses propagating through Ag, Cr, C, and In plasmas formed under vacuum conditions at the constant energy of ns heating pulse (3 mJ). Reproduced from [10] with permission from Elsevier

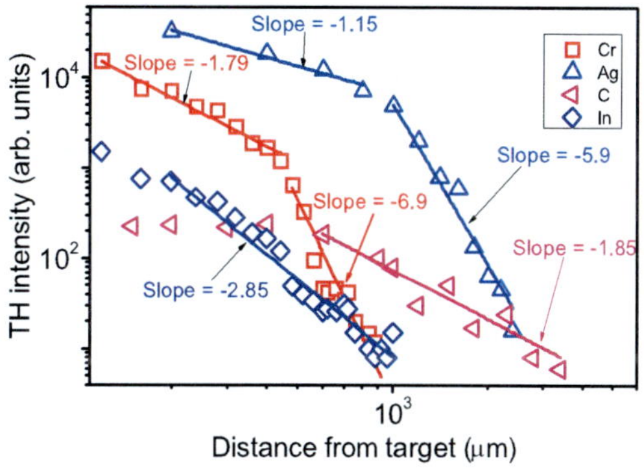

Fig. 1.5 TH intensity as a function of the distance between the surface of bulk target (Ag, Cr, In, and C) and fs probe pulses at the constant heating pulse energy (3 mJ) and probe pulse energy (0.5 mJ). Reproduced from [10] with permission from Elsevier

the characteristics of the particles, which were ejected during LA of different bulk targets and participated in THG. It was assumed that TH intensity depends on the distance between the probe pulses propagation and target (x) in accordance with the relation $I_{TH} \propto x^{-n}$, where n was considered to be dependent on the heating pulse intensity. From Fig. 1.5, TH intensity in the case of C and In was proportional to the distance from the target with the power low of $n = -1.85$ and -2.85 respectively. However, a completely different scenario was observed for the case of Cr and Ag plasmas, which showed two regions with different slopes (-1.15 and -5.9 in the case of Ag plasma, and -1.79 and -6.9 in the case of Cr plasma).

TH intensity dependences at the distances close to target surfaces (up to ~0.6 mm) show a similar trend for all targets. As the distance from targets further increases, TH intensity decreases with two different regimes of x variations. This difference can be attributed to TH intensity changes due to both the density and the length of the nonlinear medium variations, which demonstrate the inverse relationship with each other as the ejected particles propagate away from the surface of the target. In order to understand the TH intensity variations with the growth of distance from the target, the statistical distribution of each group of particles during plasma formation should be analyzed. The dependence of TH intensity on probe distance from the surface of the target has shown the presence of different species with different masses and velocities during the formation of plasma that has a direct impact on TH yield. This is largely governed by the motion of the ejected particles away from the surface of the ablated target while considering the energy conservation during LA. This leads to the conclusion that during LA of the bulk target, lighter particles, such as C and Ag, are responsible for TH at larger distances from the target.

Similar to TH intensity variation in Ag and Cr plasmas, the TH intensity in CdS and ZnS plasmas shows a similar trend with the variation of the distance between target and probe pulses (Fig. 1.6). In order to explain TH intensity variations in all plasmas shown in Figs. 1.5 and 1.6, the presence of two groups of particle clouds is assumed. Their presence can be explained by the fact that the ablation of the outermost surface of targets results in atomization, whereas the inner surface is mostly responsible for the generation of molecules, clusters, and nanoparticles due to the explosion and

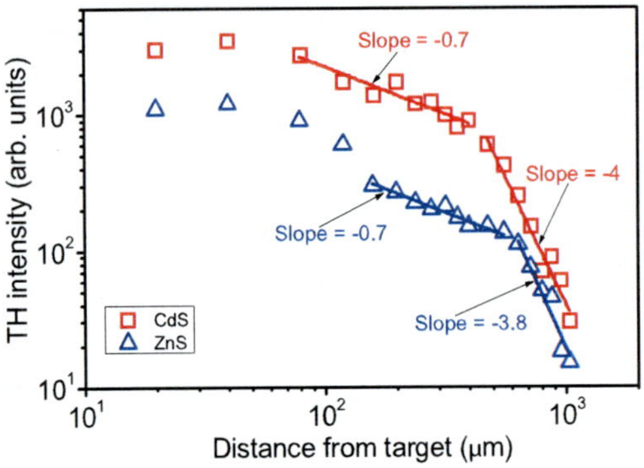

Fig. 1.6 TH intensity as a function of the distance between the surfaces of bulk CdS and ZnS and femtosecond probe pulses at the constant pulse energy of 3 mJ for ns heating pulses and 0.5 mJ for fs probe pulses. Reproduced from [10] with permission from Elsevier

mechanical fragmentation processes. These atomic and clustered clouds of particles move towards the probe beam with different velocities, which was observed during THG studies. This peculiarity can be the dominating factor for the TH intensity behavior in the case of Ag, Cr, CdS, and ZnS. From this trend of TH intensity variations, one can observe the ejected cloud of particles containing few atom-sized clusters or large nanoparticles during the creation of plasma, which move smaller distances compared to the generated atoms and ions during the creation of plasma.

TH intensity was analyzed as a function of a time delay between the heating and probe pulses under vacuum conditions. In order to measure the dependence of TH intensity on the variation of the delay between the heating and probe pulses, the target sample was placed at the focus of the heating pulses and 0.6 mm away from probe pulses. The pulse energies of the heating and the probe pulses were maintained at 3 and 0.58 mJ respectively. The comparison of TH intensity at variable delay was carried out for C, Cr, Ag, and In plasmas. Figure 1.7a shows the modulation of TH yield with the variation of the delay between heating and probe pulses. TH intensity shows a similar trend with delay variations for all samples, with TH maxima observed at 90 ns (C), 180 ns (Cr), 200 ns (Ag), and 250 ns (In). The TH intensity decreases slowly after reaching a maximum as the delay time increases. The TH intensity variations at different delays were in good agreement with the earlier reported studies [8]. The TH intensity variations show that the atomic and ionic species are the dominant components responsible for harmonic generation at discussed conditions compared with the clusters and nanoparticles in different plasmas. The optical emission spectra of these species (see Fig. 1.9a) clearly indicate the atomization of bulk materials during heating by ns pulses. It is worth mentioning that the objective of this study is to look into the impact of the type of bulk materials on TH intensity variations. One can correlate the delay times at which maximal yields of TH were achieved with their respective masses (Fig. 1.7b). The atomic mass of the ablated target has a direct effect on the arrival time of the ejected particle at the probe pulse

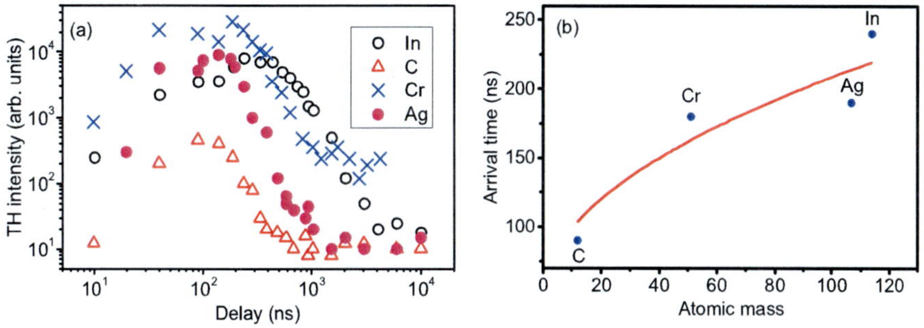

Fig. 1.7 **a** TH intensity as a function of the delay between the heating and probe pulses. **b** Arrival time of C, Cr, Ag, and In particle clouds at the area of propagation of the probe pulses when TH intensity maxima were observed. The solid curve represents the phenomenological fit of the relationship between the delay time and atomic mass ($t_d \propto M^{0.5}$). Reproduced from [10] with permission from Elsevier

axis. Thus HG in plasmas provides an assessment of the particle size distribution for different materials (Fig. 1.8).

The HG in plasma can provide information about the role of the complex composition of ablated species. The nature of HG emitters can be revealed by the evolution of harmonic yield towards the maximum at some specific delay from the beginning of laser ablation. This specific delay, at fixed heating pulse energy, can be used to characterize the propagation time required for cluster/molecules to reach the probe pulses axis. The components created during laser ablation can be characterized using their average arrival times, which can be assigned to different cluster sizes governed by conservation of average kinetic energy ($E = mv^2/2$). The relationship between employed energies and the masses of ejected particles allows determining the relation between the arrival time and the masses of ejected particles. The delay at which TH yield reaches its maximum should scale as a square root of atomic weight, e.g.,

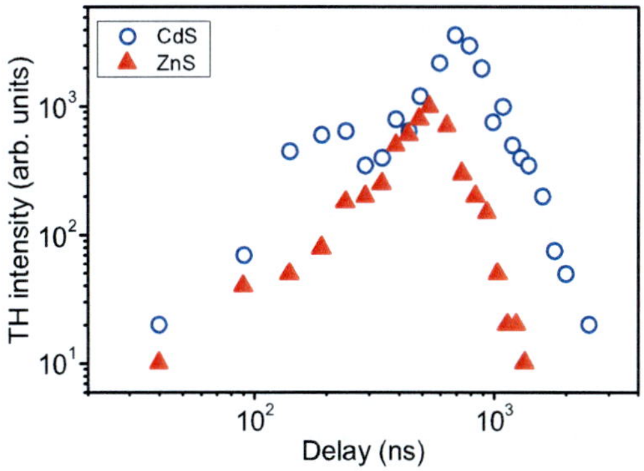

Fig. 1.8 TH intensities as the functions of the delays between ns heating and fs probe pulses in the case of CdS and ZnS plasmas. Reproduced from [10] with permission from Elsevier

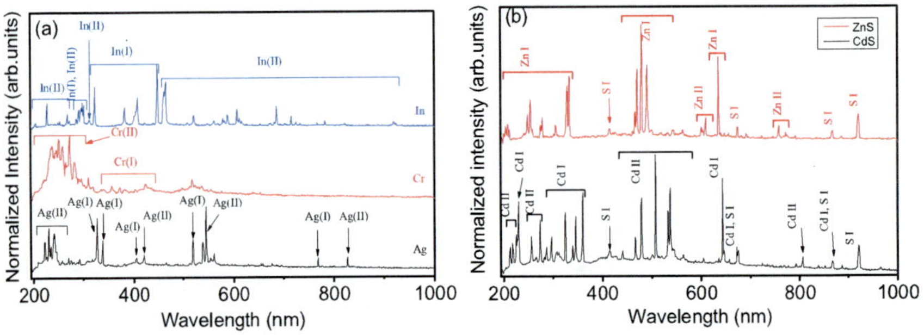

Fig. 1.9 Emission spectra of the studied plasmas excited by ns heating pulses (3 mJ). **a** Ag, Cr, and In. **b** CdS and ZnS. Reproduced from [10] with permission from Elsevier

$t_d \propto M^{0.5}$. It reflects that the lighter atoms and ions from the target surface will reach the probe pulse position before the heavier species.

Following this logic, due to the kinetic relation of particles movement, the large clusters/aggregates should appear in the laser-plasma interaction zone $M^{0.5}$ times later than single atoms and ions of respective materials. In other words, the appearance of the group of clusters in the interaction zone could be expected at delayed times compared to the lighter particles. In order to validate the relationship between the arrival time and the atomic masses of materials, the temporal evolution of the TH yield from various plasmas was analyzed, as shown in Fig. 1.7a. As mentioned above, the distance between target and femtosecond beam propagation in these experiments was maintained at ~0.6 mm. As expected, the arrival times for atomic and ionic species of Cr, In and Ag were $(M_{Cr}:M_C)^{0.5}$ ~1.27, $(M_{In}:M_C)^{0.5}$ ~3 and $(M_{Ag}:M_C)^{0.5}$ ~2.9 times longer compared with the optimal delay for C particles. This anticipated maximal yield of TH from respective plasmas was observed in the experiments as shown in Fig. 1.7b.

The evolution of TH intensity with the variation of delay between the heating and probe pulses has been studied theoretically and experimentally in [8], which has shown that the TH intensity variations at t_d <300 ns are possibly due to the presence of ions and neutral atoms. The TH intensity variations at the delays greater than 300 ns can be attributed to the presence of clusters and nanoparticles in the plasma. It can be seen from the plasma spectra of all bulk targets that, under experimental conditions, the optical emission from plasma consists of the emission attributed to the respective ions (see Fig. 1.9a), which supports the argument that the initial variation of TH intensity with delay is mostly due to the presence of atoms and ions during plasma formation.

In order to validate the relationship between the arrival times of the ejected particles and their masses, TH intensity measurements during LA of the wide-bandgap semiconductors (CdS and ZnS) at different delays between the heating and probe pulses were performed. Figure 1.8 shows that TH intensities reach the maximums at the delays of 540 ns (ZnS) and 690 ns (CdS). This observation also supports the relationship between the arrival times and atomic masses, which predicts that the expected arrival time for CdS atoms, ions, or clusters is $(M_{ZnS}:M_{CdS})^{0.5}$ ~1.22 times longer than the arrival time of ZnS species. TH in these plasmas was observed up to the delays of 2 μs.

The optical emission spectra from the CdS and ZnS plasmas revealed the existence of atomic species such as Cd, Zn, S, and their respective ions in the plasma area (Fig. 1.9b). The complex mixture of Cd(II) and Cd(I) was observed in the optical emission spectra of CdS plasma, whereas the emission spectra from ZnS plasma consisted dominantly on the emission lines attributed to Zn(I).

The TH intensity approaches its maximum at around 100–300 ns for C, Ag, Cr, and In, which is an earlier timescale compared to the timescales of 540 ns for CdS and 690 ns for ZnS. These time scales can be better explained by the fact that the created plasmas mostly contain two components. One consists of atoms and ions, which can possess the velocity at around 5×10^5 cm s^{-1}, and the other consists of clusters and nanoparticles with the respective velocities of 1×10^5 cm s^{-1}. The TH intensity variations in these semiconductor materials are probably due to the existence of large-sized clusters and nanoparticles, which mostly arrive on a longer time scale as compared to their respective atoms or ionic species. These measurements confirmed that, in semiconductor materials, THG is mostly due to the presence of their respective clusters, which arrive at a microsecond time scale as compared to their respective ions and atoms ejected during plasma formation.

1.2 Third and Fifth Harmonics Generation in Air and Nanoparticle-Containing Plasmas Using 150 kHz Fiber Laser

As it was shown in previous sections, third harmonic generation in air and laser-induced plasmas (LIP) using femtosecond pulses from Ti:sapphire class lasers is a well-studied process that allows determining the maximally available conversion efficiency of this nonlinear optical effect and diminishing the impeding mechanisms restricting this process. THG allows the formation of ~266 nm sources of femtosecond pulses at the conversion efficiency from 800-nm class lasers of up to 10^{-3}. Longer wavelength lasers allow observation of this process in the visible range, or even in the near and mid IR ranges. However, the application of these sources does not allow achieving such relatively large conversion efficiencies due to the growing restricting mechanisms, which limit the number of photons of converted radiation.

To increase the average power of such sources of coherent radiation in the shorter-wavelength range one has to apply high pulse repetition rate laser systems. Commonly used Ti:sapphire lasers do not allow operating with the rates above 1 kHz regime. The application of MHz sources of femtosecond pulses is also inefficient due to low pulse energies restricted by tens of nJ. In this connection, the millijoules-class Yb-doped fiber lasers ($\lambda \approx 1030$ nm) operating in the 100 kHz regimes can be considered as the suitable sources of driving pulses for THG ($\lambda \approx 343$ nm). This process has been studied in a few gases [12, 13]. Moreover, these lasers are suitable for the fifth harmonic generation (FHG, $\lambda \approx 206$ nm).

The use of such lasers allows increasing the average third harmonic (3H) power by about two orders of magnitude, which becomes sufficient for various applications of these UV sources in spectroscopy, photomodification of sensitive organic materials, photoluminescent studies of dyes and complex molecules, etc. These lasers can also be used for low-order nonlinear optical spectroscopy of LIPs generated on the surfaces of various materials.

Increasing the pulse energy (i.e., the number of photons in the pulse) to 10 mJ allows the generation of harmonics with pulse energy up to several tens of μJ at a laser repetition rate of 10 Hz. Alternatively, using a large number of photons per second (i.e., a high pulse repetition rate) at low pulse energy can also lead to an increase in the number of coherent photons in the UV range. In particular, 100 kHz class lasers produce ~100 W of the average power of femtosecond pulses, which in turn can be converted to a 3H with an average power of 0.1 W.

The limitation of THG conversion efficiency in the case of such fiber chirped laser pulse amplification systems is that their pulse duration is usually limited to a few hundred femtoseconds. The use of such long pulses leads to ionization at the peak of the laser pulse, which causes an increase in phase mismatch between the interacting waves. Meanwhile, for pulses shorter than 100 fs, the intensity can be significantly increased, resulting in improved THG conversion efficiency. The use of 40 fs class lasers generating pulses with a repetition rate of ~100 kHz, can open up new opportunities in THG from air and LIPs by achieving large average fluxes of coherent UV radiation.

In this section, we discuss the 3H ($\lambda = 343$ nm) and fifth harmonic (5H, $\lambda = 206$ nm) generation produced in air and different LIPs by applying the Yb-doped fiber laser generating 37 fs, 100 kHz, 1030 nm, 0.5 mJ pulses [14]. Different characteristics of these processes (power and density dependences, influence of impeding mechanisms, role of synthesized NPs in THG, variations of spectral modulation of driving and harmonic pulses, etc.) are reported. We compare the formation of NPs-containing plasmas on the surfaces of aluminum, carbon, manganese, and titanium bulk species, as well as quantum dots (QDs) of metal sulfides as the media for low-order harmonics generation at different delays between the heating and driving pulses, using infrared 250 fs (1030 nm) and 5 ns (1064 nm) pulses. The role of NPs and QDs in the enhancement of harmonic yield is discussed.

1.2.1 Scheme for THG and FHG Using High Pulse Repetition Laser

Ultrafast high-performance Yb-doped fiber laser (UFFL_300_2000_1030_300; Active Fiber System) was used in the discussed studies. The first port of this laser

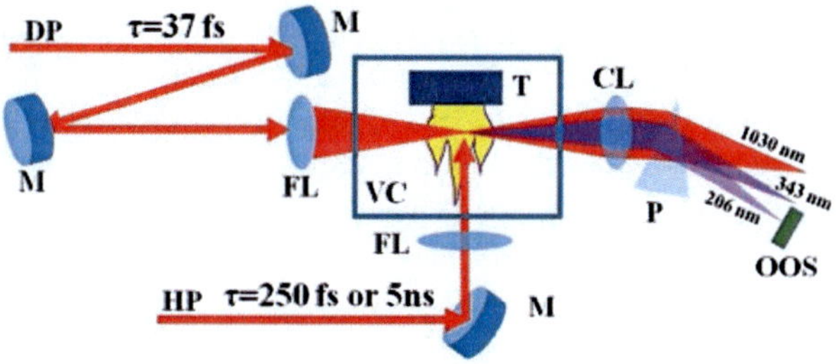

Fig. 1.10 Experimental setup for 3H and 5H generation from air and plasmas. DP, driving pulse; HP, heating pulse; M, mirrors, FL focusing lenses; VC, vacuum chamber; T, target; CL, collecting lens; P, prism, OOS, Ocean Optics spectrometer. Reproduced from [14] with permission from Springer

provided 250 fs, 1030 nm, 50–150 kHz, 0.3 mJ pulses that were used for LIP formation. The second port provided 37 fs pulses (1030 nm, 50–150 kHz, ~1 mJ, 120 nm spectral width), which were used as driving radiation for THG and FHG in air and plasma. The laser operated at 50 kHz–20 MHz pulse repetition rate; however, the discussed studies were limited to 100 kHz regime.

The 250 fs radiation from the first port was directed towards the vacuum chamber (Fig. 1.10) and focused using a 200-mm focal length spherical lens on the surface of a target into a spot of 0.3 mm size. The optimal fluence and intensity of 250-fs heating radiation on the target surface were $F = 0.5$ J cm^{-2} and $I_{hp} = 1.5 \times 10^{12}$ W cm^{-2}, respectively. At these conditions of ablation, the maximal yields of 3H and 5H from different plasmas were observed.

The nanosecond pulses from Nd: YAG laser (5 ns, 1064 nm, 10 Hz, 5 mJ) were also used for ablation of targets using approximately the same fluence to compare the harmonic yield with regard to the ablation using femtosecond pulses. The arrival time of femtosecond pulses with respect to heating radiation was controlled using a delay generator (DG535, Stanford Research Instruments), which was operated using a trigger pulse from the amplifier running at 100 kHz.

The radiation from the second port was delayed with regard to the one from the first port to propagate through the formed plasma at a maximal density of the ejected particles. This radiation was focused by a 400-mm focal length spherical lens inside the LIP. The pressure in the vacuum chamber was kept at 0.1 Pa throughout the plasma formation experiments for all samples. In the case of the THG in air, the dependence of harmonic yield on the gas pressure in the chamber was analyzed. The confocal parameter of the focused radiation was 10 mm. The intensity of the driving pulses in the plasma area was maintained below 5×10^{14} W cm^{-2}. The optical delay line between 250 and 37 fs pulses was established to maintain the optimal delay for the used LIPs and for the 0.3 mm distance between the target surface and the optical axis of the driving pulses inside the vacuum chamber. In the case of harmonic generation from the air, the vacuum chamber was filled in with air at different pressures. The generated 3H and 5H radiation from the air and plasma plume were separated from the co-propagating fundamental beam using a quartz prism and then registered by a spectrometer (USB2000, Ocean Optics).

Titanium (Ti), aluminum (Al), carbon (C), and manganese (Mn) bulk targets, as well as Ag_2S, ZnS, and CdS QDs were used for ablation and plasma formation. The sizes of Ti, Al, and Mn targets were $5 \times 5 \times 2$ mm^3. The QDs were synthesized and immersed in gelatin to dry and insert into the vacuum chamber for further ablation. The samples were placed on a computer-driven 3D coordinate movement system. The target sample was moved with constant speed to get the stable harmonic intensity.

1.2.2 THG and FHG in NP and Atomic Plasmas

Below we discuss the systematic study of the low-order harmonics generation in air and various plasmas using high pulse repetition rate femtosecond pulses. The dependences of the 3H and 5H emission yield on the air pressure and power of driving pulses at different ratios of harmonic yields were determined. The ratios of harmonic yields from metals and air were compared at different conditions of these studies. The 3H yield as a function of the arrival time of the driving pulses with respect to the heating pulses was analyzed. The THG was also analyzed by touching the solid target surface by the driving radiation. The morphology studies showed that, in most cases, the plasma consisted of the NPs produced during laser ablation of materials.

Figure 1.11a shows the spectra of 3H and 5H measured at similar conditions of laser-air interaction (at 5 W of the driving 37 fs pulses). One can see the appearance of the broadened component in the shorter wavelength side of the third harmonic. Such a blue shift of harmonic is a characteristic of the harmonic generation when different impeding mechanisms influence this nonlinear optical process. This figure demonstrates the relatively narrow low-order harmonics, without any sign of plasma emission from the filaments in the air. This condition of filaments formation can be

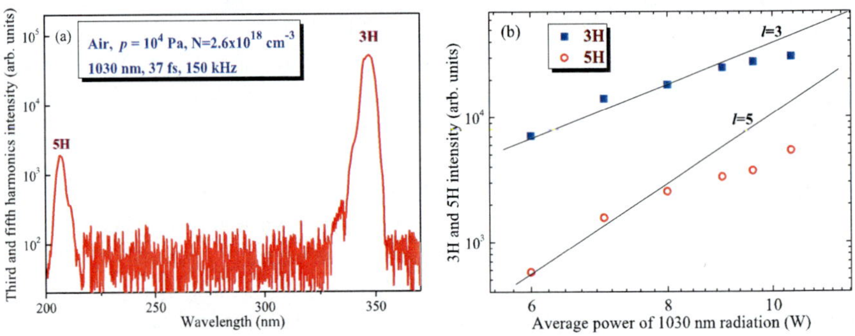

Fig. 1.11 3H and 5H generation in air. **a** Spectra of harmonics at the air pressure of 10^4 Pa. **b** Power dependences for THG and FHG. The deviation from $l = 3$ and $l = 5$ slopes was observed at the average power of driving pulse above 8 W (see text). Reproduced from [14] with permission from Springer

identified as the "optimal" from the point of view of the maximal harmonic yield and the absence of incoherent plasma emission. These experiments were carried out at 1/10th of atmospheric pressure (number of particles in cubic centimeter $N \approx 2.6 \times 10^{18}$ cm^{-3}) to diminish the role of impeding processes (pulse broadening, self-phase modulation, extended filamentation, spatial disintegration of harmonics, etc.). The ratio of 3H and 5H yields at these conditions was ~40:1. This ratio increased up to 200:1 at the normal air pressure ($N = 2.6 \times 10^{19}$ cm^{-3}) due to worsened conditions of the phase matching for 5H with regard to 3H. This deterioration of the optimal relation between the phases of driving and converted waves mostly affected the higher-order harmonic (5H), which was 3 to 6 times weaker than in the case of the optimal laser-air interaction conditions. A smaller density of air ($N = 5 \times 10^{17}$ cm^{-3} and below) allowed the generation of 5H without the influence of restricting factors. That is why the low-order harmonics from air and LIPs were compared using this range of particles concentration. Those comparative studies are presented below.

3H and 5H in the air were measured as functions of the driving pulse power up to 11 W of the average power of 100 kHz, 37 fs pulses. The energy of driving pulses was varied between 5 μJ and ~30 μJ. These dependencies showed the cubic and fifth-order dependences of harmonic yield on the driving pulse power (Fig. 1.11b) and were maintained at the pulse powers up to ~8 W after which the saturation and deviation from the cubic and fifth-order dependences were observed. This observation allows concluding that the threshold limit for the generation of free electrons in filamented air at P = 8 W, which has earlier been considered to be the dominating restricting factor of conversion efficiency [70], was far from the maximal applied pulse energies. The notable deviation from the expected dependences is attributed to the appearance of a large amount of free electrons in filaments, which led to the phase mismatch between the 1030 nm, 3H (343 nm), and 5H (206 nm) waves.

Figure 1.12 shows the comparative studies of the (a) 3H and (b) 5H in aluminum plasma and air at a similar density of particles (3×10^{17} cm^{-3}) in both media and at the intensity of driving pulses of 2×10^{14} W cm^{-2}. No special features in spectral variations of harmonics, like blue-shift, were observed at these conditions. The density of different plasmas has earlier been estimated using the code HYADES for the ablated materials during HHG at optimal conditions of plasma formation [15]. The harmonic yields from the air were normalized to be equal to 1. These measurements showed the 11- and 27-fold enhancement of the 3H and 5H yields, respectively, in this LI with regard to the air at similar conditions of experiments. The stronger harmonic yield from LIPs is attributed to the appearance of the NPs synthesized in situ prior to the arrival of the driving femtosecond pulses in the area of interaction. Their presence in LIP was confirmed by the morphology studies of the deposited debris from the used LIPs.

Similar comparative studies of harmonics were carried out using other metal and QD LIPs. Particularly, the 5H yield from Ti plasma was 7 times stronger as compared with the 5H from the air at a similar density of these two media and equal intensities of the driving pulses (Fig. 1.12c). Each LIP demonstrated stronger harmonic yields with regard to harmonic generation in air.

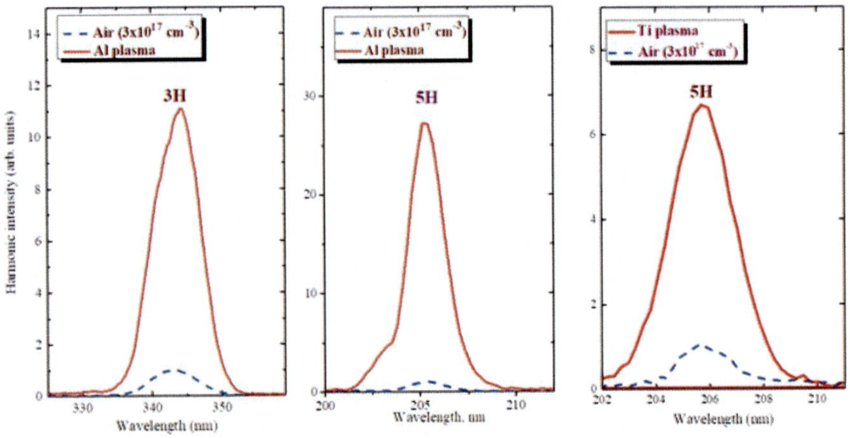

Fig. 1.12 3H (left panel) and 5H (central and right panels) spectra from Al (left and central panels) and Ti (right panel) plasmas. Dashed curves show the normalized spectra of 3H and 5H generated from air at similar density as in the case of LPPs. Reproduced from [14] with permission from Springer

The formation of optimal LIP at the moment of the driving pulse propagation through the studied ablated material is a crucial requirement for the generation of as strong as possible harmonics. The term "optimal" refers to the experimental conditions, which can be different for each ablating material, as well as the characteristics of converted and heated radiation. However, this term in each of those cases is related to the conditions when the maximal yield of harmonics can be achieved. Different parameters of such experiments play important role in the determination of the best conditions for optimal plasma formation (i.e., delay between the heating and driving pulses, fluence on the surface of ablating target, the intensity of driving wave in the plasma area, presence of the aggregates of evaporated material in LIP, etc.).

The delay between pulses allows determining the conditions when the femtosecond pulse propagates through the dense LIP spreading out from the surface at the specific velocity determined by the kinetics of plasma formation and the atomic weight of ablating material. The optimization of delay includes the determination of the time scale characterized by the distance from the target surface and the velocities of plasma components. Practically, one has to change the delay between the beginning of the plasma ignition process and the propagation of the ultrashort laser pulse at the chosen distance from the target. By estimating the velocities of atoms and ions possessing the medium atomic weight, like Mn (55 AMU), in the range of 2×10^3 m s^{-1} and by choosing the 0.2 mm distance from the target one can determine the time scale between the heating and driving pulses at which one can expect the maximal harmonic yield (~100 ns) when the central part of plasma cloud arrives in the area of driving beam propagation.

It was found that the maximal yield of 3H in the case of manganese LIP can be achieved at ~80–100 ns delay (Fig. 1.13a). Further increase of delay led to a gradual

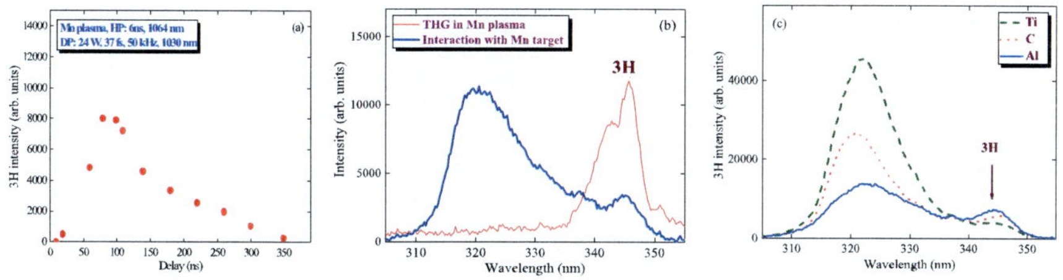

Fig. 1.13 **a**, **b** Characterization of 3H generating from Mn plasma. **a** Delay dependence between the beginning of ablation by nanosecond heating pulses (HP) and propagation of femtosecond driving pulses (DP) through the Mn LIP. **b** 3H spectrum generated during propagation through the pre-formed Mn plasma (red thin solid curve) and during interaction of the wing of driving beam with Mn solid target without the formation of plasma by the heating pulses (blue thick solid curve). **c** 3H spectra obtained during interaction of the wing of driving beam with Ti (green dashed curve), C (red dotted curve), and Al (blue solid curve) bulk targets. Reproduced from [14] with permission from Springer.

decrease of harmonic yield until it almost disappeared at >400 ns delay.

A scheme of nanosecond pulse heating and femtosecond laser pulse conversion has already been used for both low-order [16] and high-order [17] harmonic generation studies. The advantage of this approach is related to the control of the delay between the heating nanosecond pulses and the driving femtosecond pulses over a wide range. Earlier, the demonstration of large particle formation in the atmosphere of noble gases allowed the THG not only from atoms and ions of carbon materials (at small delays) but also from their aggregates (at large delays) [18].

Manganese plasma, similar to Al and Ti LIPs, allowed the generation of 3H emission, which was four times stronger with regard to THG in air. Meanwhile, the blueshift of the shorter-wavelength component of 3H in the case of Mn LIP while increasing the intensity of driving pulses from 2×10^{14} W cm^{-2} to 4×10^{14} W cm^{-2} was observed (Fig. 1.13b, thin solid red curve).

3H emission was also observed without target ablation by heating pulse while minimizing the distance between the axis of driving femtosecond beam propagation and the target surface. Once the spatial wing of the driving beam touches the Mn target along its whole length (5 mm), the 3H is generated even without the formation of plasma by the heating pulse (Fig. 1.13b, blue thick curve). The efficiency of this process was approximately 20 times lower than the THG from the preformed LPP. The maximums of 3H yields in these two cases (THG in the preformed 0.3-mm thick plasma and during touching the target) were normalized to emphasize the difference in the 3H spectra. In the latter case, a very strong, broadened, and blueshifted component at ~320 nm was dominated along the whole spectrum. The same feature was observed in the case of other targets at similar conditions of experiments (Fig. 1.13c), Ti, C, and Al) when the spectrum of harmonic radiation was also broadened and blueshifted. Notice that, at these conditions, the spectral variation of propagated laser radiation was notably smaller. One can assume that the strong 320 nm shoulder of 3H emission was enhanced due to self-phase modulation of the

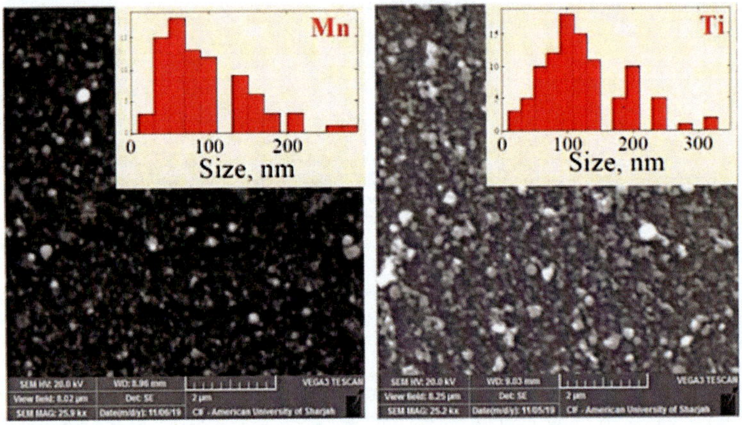

Fig. 1.14 SEM images of the deposited layers of Mn (left panel) and Ti (right panel) ablation. Histograms show the size distribution of Mn and Ti NPs. Reproduced from [14] with permission from Springer

generated UV pulse propagating through the excited extended plasma in the vicinity of the target.

The ablated and deposited debris from the carbon, titanium, and manganese bulk targets during ablation by 250 fs pulses contained the nano-sized particles. The smallest among them were the carbon NPs (~25 nm), while the mean sizes of Mn and Ti NPs were determined to be 50 and 100 nm, respectively (see SEM images and histograms of deposited Ti and Mn species, Fig. 1.14). Their formation and presence in the plasma plume can, to some extent, cause the enhancement of harmonic yield with regard to the atomic species of similar elemental consistency. There are several methods for the formation of the nanoparticles, where complete control over size and shape were achieved depending on the material characteristics. They require some procedure and time to synthesize the NPs, which can later be used for different goals. In the meantime, the laser ablation technique provides the opportunity for the formation of NPs in situ, i.e., during the experiments sensitive to the presence of the small-sized species.

1.2.3 Low-Order Harmonics Generation in QD Plasmas

The in-situ formation of nano-sized species creates the conditions for laser-NPs interaction. Though yet confirmed for low-order harmonics, stronger HHG flux from clusters, QDs, and NPs with regard to the monomers has earlier been attributed to better conditions of the formation of harmonic emitters in the case of agglomerates formation and spreading, as well as to the specific properties of NPs, such as the ionization and recombination to the same ion of the cluster, to the neighboring ions and to the whole cluster, allowing the increase of recombination cross-section of the electrons participating in HHG. Additionally, the enhancement of HHG in the presence of such structures can be explained by the higher nonlinear optical response of materials caused by the plasmonic properties of NPs.

These NPs were generated during ablation by femtosecond pulses. Their sizes were distributed in a broad range. Most of the inner atoms of these structures, apart from outer atoms on the surface of NPs, do not participate in harmonics generation being screened by the neighboring atoms. Thus one can expect the attractiveness of the smaller-sized aggregates for both THG and HHG. The examples of such already prepared aggregates are the QDs of metal sulfides. One can argue which of multiparticle aggregates is most useful for the frequency conversion of the ultrashort laser pulses? Being qualitatively divided by three groups (clusters with up to 10^3 atoms and dimensions less than 1 nm, QDs with up to 10^5 atoms and dimensions less than 5 nm, and large NPs with up to 10^8 atoms and dimensions up to 100 nm), each of them demonstrates both positive and negative features as the emitters of harmonics. Another important issue is the concentration of active centers of harmonic emission since as mentioned, in the case of large aggregates, only those atoms in the vicinity of NP surface can participate in harmonic emission. The latter issue is especially important in the case of HHG. In this connection, QDs demonstrating strong local field-induced enhancement of the nonlinear optical properties can be considered as the most attractive species for efficient THG and HHG.

The procedure for preparation of the metal sulfide QDs is reported elsewhere [19]. The synthesis of QDs was carried out taking into account a number of criteria for their use to generate harmonics of ultrashort pulses. The first criterion is the formation (as a result of ablation of samples by femtosecond, picosecond, or nanosecond heating pulses) of a plasma cloud consisting of QDs, as well as atoms, ions, and clusters at the time of the passage of the converting femtosecond pulses above the target surface. The second criterion is the presence of the maximal concentration of QDs at the time of transmission of the femtosecond pulses through the plasma. The third criterion is the size characteristics of quantum dots. Previous studies of harmonic generation in multi-particle containing plasmas, as well as qualitative assessments, show that the most optimal sizes of such species should be in the range of 1–4 nm.

The synthesized samples of colloidal solutions of zinc, cadmium, and silver sulfides were further used to produce the solid targets for laser ablation and generation of harmonics. The method of increasing the filling factor of Ag_2S, ZnS, and CdS QDs in the stabilizing polymer (gelatin) was used to develop the solid samples of multi-particle species for laser ablation in vacuum and formation of the plasma torches containing a large amount of QDs required for efficient generation of harmonics.

The mean sizes of ZnS QDs measured using a transmission electron microscope (TEM) were 2 nm, while other species had mean sizes in the range of 1.7–2.0 nm (Ag_2S, Fig. 1.15a), and 3.0–3.2 nm (CdS). The optical spectra of these QDs (Fig. 1.15b) were significantly blue-shifted with regard to the bulk materials of similar elemental consistency.

The atomic mass of the ablated target has a direct effect on the arrival time of the ejected particle at the axis of propagation of the driving pulse. THG in plasmas at

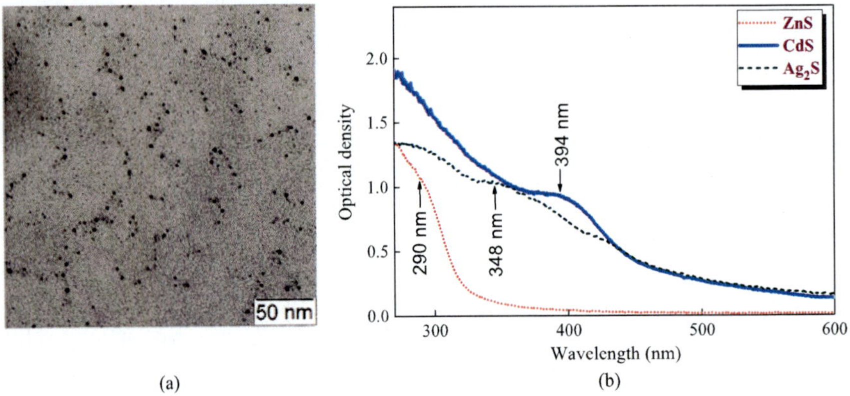

Fig. 1.15 **a** TEM image of Ag_2S QD. **b** Absorption spectra of metal sulfide QDs. Reproduced from [14] with permission from Springer

different delays between heating and driving pulses provides an assessment of the particle size distribution for different materials. The components created during laser ablation can be characterized using their average arrival times, which can be assigned to different cluster sizes governed by conservation of average kinetic energy ($E = mv^2/2$). The delay at which 3H yield reaches its maximum should scale as a square root of atomic weight, e.g., $t_d \propto M^{0.5}$. It reflects that the lighter atoms and ions from the target surface will reach the driving pulse position before the heavier species.

In order to validate this relationship, the 3H intensity measurements during ablation of metal sulfide QDs-containing species at different delays between the heating and driving pulses were performed. Figure 1.16a shows that 3H yield reaches the maximums at the delays of 120–200 ns in the case of Ag_2S QDs. This observation supports the relationship between the arrival times and atomic or molecular masses,

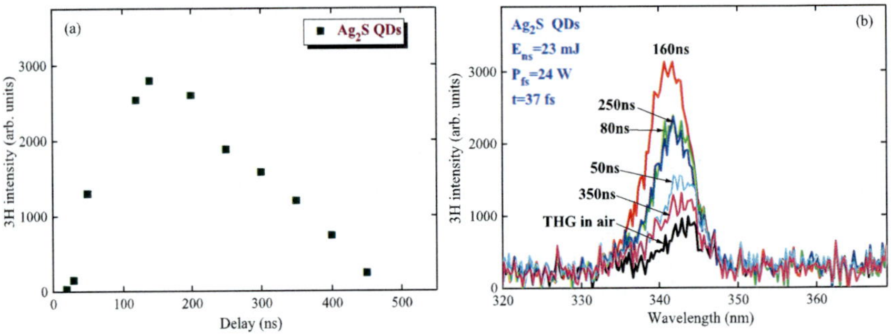

Fig. 1.16 THG in Ag_2S QD LIP. **a** Delay dependence between the beginning of ablation by nanosecond heating pulses and propagation of femtosecond driving pulses through the plasma. **b** Blueshift of 3H spectrum at different delays between pulses. Reproduced from [14] with permission from Springer

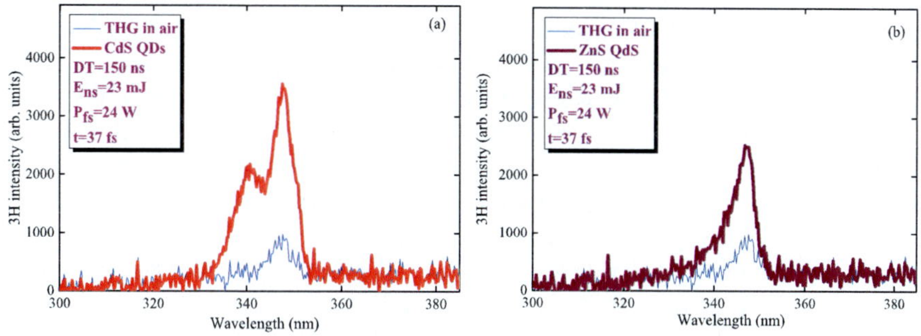

Fig. 1.17 THG in **a** CdS QD LIP and **b** ZnS QD LIP. Reproduced from [14] with permission from Springer

which predicts that the expected arrival time for Ag_2S species (ions or clusters) should be $(M_{Ag2S}:M_{Mn})^{0.5}$ ~1.7 times longer than the arrival time of Mn species (80–100 ns). Thus, one can assume that the above kinetic rule properly explains the dynamics of material spreading out from the ablated target, once one analyzes the particles' movement during laser ablation at relatively moderate fluencies (5–20 J cm^{-2}) of heating nanosecond pulses.

3H from QD-containing LIPs was blue-shifted with regard to the central wavelength of this harmonic (343 nm). The blue shift was larger in the case of denser plasma (Fig. 1.16b). Particularly, the largest blue shift was observed in the case of the optimal delay between heating and driving pulses corresponding to the largest concentration of Ag_2S QDs along the axis of propagation of the driving pulses. Similar dependencies were observed in the case of CdS and ZnS QD LIPs (Fig. 1.17). In each of these cases, the 3H yield was significantly (5–20 times) larger with regard to the 3H generated in the air at similar concentrations of particles.

1.2.4 Discussion

Previously, low-order harmonics from air and LPPs were performed using 10 Hz and 1 kHz class Ti: sapphire lasers. Here we analyze the application of a 100 kHz class ytterbium fiber laser for THG and FHG in air and several plasma formations. Below, we consider a possible explanation for experimental observation of strong 3H and 5H radiation from ablated solids, which is probably related to the cluster formation during laser ablation. Previously, it has been shown that in the case of the plasma plumes containing metal nanoparticles, those species demonstrate increased HHG conversion efficiency compared with the LPPs formed on the surfaces of bulk metal targets when they consist of atoms and ions [20].

Small aggregated particles can be formed in an ablation plume induced by ultrashort laser pulses. The ultrafast laser ablation has been proposed as the simplest method for generating small nanoparticles. Note that these conditions for the formation of NPs are unsuitable for effective plasma HHG since they provide a large number of free electrons, which leads to an increasing phase mismatch between the driving and harmonic waves, especially in the case of higher orders of harmonics.

The absolute measurements of the 343 nm yield from the air were carried out using a power meter, which allowed determining the average power of generated UV radiation (~5 mW) at a 100 kHz pulse repetition rate. By knowing the experimental ratio of 3H yields in air and Mn plasma (1:4), 20 mW was determined as the average power of the 3H generating in the latter medium. By assuming the energy of 343 nm photon (5.8×10^{-19} J) and the whole energy of 343 nm flux during 1 s (20 mJ) one can determine the number of photons generating from Mn plasma during THG (3.4×10^{16}). Correspondingly, the average power of UV photons (3.61 eV, ~20 mW) allows determining the THG conversion efficiency in the case of 20 W driving pulses propagating through the manganese plasma (~10^{-3}). The achievable number of harmonic photons in the case of air and plasma can also be estimated by considering the typical conversion efficiency from these media (~10^{-3} [21]).

Based on the experiments with different targets, one can conclude that overheating of the same spot worsens the conditions of stable plasma formation. The morphology of the heating spot did not remain unchanged during these experiments. At the same time, crater formation in the case of a movable target occurred after a significantly longer period compared to static targets. Once the target begins to move, the previously heated area cools down and can again be used for effective ablation and harmonic generation.

In the case of low-order harmonic generation, two perspectives can be considered: on one hand, it can be useful in the search for new efficient materials that will act as nonlinear media in THG processes. On the other hand, it is suitable as a diagnostic tool to characterize the species presented in LIP. Previous studies of harmonic generation in laser plasma of NPs were conducted only using lasers with a relatively low pulse repetition rate (from 10 Hz to 1 kHz). Since the average harmonic power increases with the repetition rate, it is highly desirable to maintain the concentration of NPs in the plasma during repeated irradiation of the cluster-containing ablation target. The search for methods to ensure a stable concentration of clusters in the plasma will increase not only the output of low- and high-order harmonics, but also the average power of the generated harmonics. At the same time, stable conditions of the nanoparticle plasma allow to investigate the influence of various cluster properties, in particular surface plasmon resonance, on the efficiency of THG. The use of a laser with a high pulse repetition rate (100 kHz and above) is advantageous for the above studies since, with an increase in the average power of the generated harmonics, less time is required to collect all the necessary data to compare the nonlinear optical properties of different plasmas.

1.3 Summary

In this chapter, THG in various plasma targets was analyzed using nanosecond and picosecond heating pulses in both vacuum and air. TH from various plasmas was used as a probe to characterize the components of the plasma. TH intensity shows the cubic dependence on the probe pulse energy in both air and vacuum. It was observed in the case of THG in the air that, at the high intensity of probe pulses, TH intensity decreases, which is an indication of 2free electrons generation in air. TH intensity variations with the distance between the target and the probe pulses reveal the nature of the ejected particles during ablation under vacuum conditions, which was not observed during THG in air. TH intensity measurements with the change of the distance between the target surface and probe pulses position have shown the significant contribution of the smaller-sized particles along with ions and atoms compared to the heavier particles. The modulation of TH intensity with the variation of delay between the heating and probe pulses is attributed to the influence of the masses of ejected particles in accordance with the relation $t_d \propto M^{0.5}$. Among the studied materials under vacuum conditions, the relative TH conversion efficiency is larger for Ag plasma compared to the other species. It was shown that THG in metal plasmas is governed by the atomic or ionic species, contrary to the wide-bandgap semiconductor materials where TH was generated by the small-sized clusters and nanoparticles. TH maxima were observed at the delays in the range of 100–300 ns for metal plasmas as compared to 500–700 ns for semiconducting materials. This difference in optimal delays has revealed that, during LA, the ejected particles from metal targets move faster compared to heavier targets. This study can be generalized to a variety of different materials as the sources of plasma for THG depending on the delay and spatial spreading of the plasma expansion.

The application of a high pulse repetition rate laser allowed increasing the average power of the low-order harmonics generating in air and different plasmas. We analyzed the generation of the third and fifth harmonics of 37 fs pulses using carbon, boron, manganese, and titanium nanoparticle-contained laser-produced plasmas using a 100 kHz class laser. The fluxes of the third harmonic (343 nm) in the plateau region were estimated to be 8×10^{15} and 3×10^{16} photons/s in the case of air and manganese plasma, respectively. We also analyzed third harmonic generation in quantum dot plasmas. We discussed the increase of the average power of coherent ultraviolet harmonics from plasmas by two orders of magnitude compared with previously used 1 kHz lasers. The long-term stability of plasma harmonic generation at this range of pulse repetition rates still remains to be improved. However, the joint application of the rotating target and the up and down movement allows the shift of the heating beam along the target surface thus leading to the improvement of the stability of harmonic yield from 100 kHz class lasers.

References

1. C.-Y. Shih, R. Streubel, J. Heberle, A. Letzel, M.V. Shugaev, C. Wu, M. Schmidt, B. Gokce, S. Barcikowski, L.V. Zhigilei, Nanoscale **10**, 6900 (2018)
2. G.O. Ariunbold, P. Polynkin, J.V. Moloney, Opt. Express **20**, 1662 (2012)
3. S. Mourdikoudis, R.M. Pallares, N.T.K. Thanh, Nanoscale **10**, 12871 (2018)
4. B. Kearton, Y. Mattley, Nat. Phot. **2**, 537 (2008)
5. M. López-Arias, M. Oujja, M. Sanz, R.A. Ganeev, G.S. Boltaev, N.K. Satlikov, R.I. Tugushev, T. Usmanov, M. Castillejo, J. Appl. Phys. **111**, 043111 (2012)
6. R. de Nalda, M. Lopez-Arias, M. Sanz, M. Oujjaa, M. Castillejo, Phys. Chem. Chem. Phys. **13**, 10755 (2011)
7. M. Sanz, M. Lopez-Arias, J.F. Marco, R. de Nalda, S. Amoruso, G. Ausanio, S. Lettieri, R. Bruzzese, X. Wang, M. Castillejo, J. Phys. Chem. C **115**, 3203 (2011)
8. M. Sanz, R. de Nalda, J.F. Marco, J.G. Izquierdo, L. Banares, M. Castillejo, J. Phys. Chem. C **114**, 4864 (2010)
9. R.A. Ganeev, G.S. Boltaev, R.I. Tugushev, T. Usmanov, M. Baba, H. Kuroda, J. Opt. **12**, 055202 (2010)
10. S.K. Maurya, M. Venkatesh, R.A. Ganeev, C. Guo, Opt. Mater. **98**, 109423 (2019)
11. G. Lambert, A. Andreev, J. Gautier, L. Giannessi, V. Malka, A. Petralia, S. Sebban, S. Stremoukhov, F. Tissandier, B. Vodungbo, P. Zeitoun, Sci. Rep. **5**, 7786 (2015)
12. D. Kartashov, S. Ališauskas, A. Pugžlys, A.A. Voronin, A.M. Zheltikov, A. Baltuška, EPJ Web of Conf. **41**, 09007 (2013)
13. D.L. Weerawarne, X. Gao, A.L. Gaeta, B. Shim, Phys. Rev. Lett. **114**, 093901 (2015)
14. G.S. Boltaev, N.A. Abbasi, V.V. Kim, M. Iqbal, S.A. Khan, A.I. Zvyagin, M.S. Smirnov, O.V. Ovchinnikov, R.A. Ganeev, A.S. Alnaser, Appl. Phys. B **126**, 76 (2020)
15. L.B. Elouga Bom, J.C. Kieffer, R.A. Ganeev, M. Suzuki, H. Kuroda, T. Ozaki, Phys. Rev. A 75, 033804 (2007)
16. M. Oujja, I. Lopez Quintas, A. Benitez Canete, R. de Nalda, M. Castillejo, A.B. Canete, Appl. Surf. Sci. 392, 572 (2017)
17. R.A. Ganeev, G.S. Boltaev, K. Zhang, P.V. Redkin, S.K. Maurya, M. Venkatesh, Z. Yu, V.V. Kim, C. Guo, OSA Continuum **2**, 1510 (2019)
18. I. Lopez-Quintas. M. Oujja, M. Sanz, M. Martín, R.A. Ganeev, M. Castillejo, Appl. Surf. Sci. 278, 33 (2013)
19. R.A. Ganeev, G.S. Boltaev, V.V. Kim, K. Zhang, A.I. Zvyagin, M.S. Smirnov, O.V. Ovchinnikov, P.V. Redkin, M. Wöstmann, H. Zacharias, C. Guo, Opt. Express **26**, 35013 (2018)
20. H. Singhal, R.A. Ganeev, P.A. Naik, J.A. Chakera, U. Chakravarty, H.S. Vora, A.K. Srivastava, C. Mukherjee, C.P. Navathe, S.K. Deb, P.D. Gupta, Phys. Rev. A **82**, 043821 (2010)
21. R.A. Ganeev, T. Witting, C. Hutchison, F. Frank, P.V. Redkin, W.A. Okell, D.Y. Lei, T. Roschuk, S.A. Maier, J.P. Marangos, J.W.G. Tisch, Phys. Rev. A **85**, 015807 (2012)

Chapter 2
Different Aspects of High-Order Harmonics Generation in Plasmas

2.1 Incoherent and Coherent Extreme Ultraviolet Emission from Boron Plasma

The analysis of useful sources of coherent emission in the short-wavelength range ($\lambda < 100$ nm) is pursued through the generation of such radiation in synchrotrons and from the optical harmonics of ultrashort laser pulses during their propagation in low-dense media (gases and plasmas) and their specular reflection from surfaces. High-order harmonic generation (HHG) studies in the low-dense plasma notably improved the understanding of the restricting and amending processes, thus allowing to analyze processes such as resonance enhancement of single harmonic, nanoparticle-induced enhancement of harmonics, quasi-phase-matching of the groups of harmonics, etc. [1–30]. Further search for plasma media that is suitable for the frequency conversion of IR laser pulses towards the extreme ultraviolet (XUV) range requires the analysis of the coherent and incoherent properties of such sources.

The analysis of the optical and nonlinear optical properties of laser-induced plasma (LIP) formed on the surface of boron-containing materials can give useful insights about this medium for different applications. Particularly, the studies presented in [31] have shown that the third-order harmonic generation in boron carbide plasma is sensitive to the presence of atoms, molecules, clusters, and nanoparticles in the LIP and can be used as a probe of their density. In [32], 500 fs pulses at $\lambda = 308$ nm were focused on a B_4C plate inserted into the vacuum chamber. They have measured the conversion efficiency for the line emission of H- and He-like B ions, as well as in the range of recombination radiation emitted by He-like B ions (4.86, 6.03, and 4.52 nm, respectively). The feasibility of using thin, freely suspended boron films as a target in femtosecond laser-produced plasma experiments was discussed in [33].

To generate XUV radiation in the vicinity of the water window, laser intensities of 10^{15}–10^{16} W cm^{-2} have to be used, though some schemes of dual laser pumping have demonstrated the application of 10^{14} W cm^{-2} intensities from Nd:YAG class lasers to generate the short-wavelength radiation in this region [34]. The studies of boron emission lines in XUV were reported in [35]. They were generated close to

R. A. Ganeev, *High-Order Harmonics Generation in Plasmas*, Springer Series on Atomic, Optical, and Plasma Physics 122, https://doi.org/10.1007/978-3-031-09040-0_2

solid-state density plasmas with a peak temperature of about 190 eV using 200 μJ, sub-10-fs laser pulses with a high pulse contrast. Time integrated XUV spectroscopy was applied to derive the plasma conditions allowing the generation of 4.86, 5.27, and 6.03 nm lines from the ablated boron carbide target. The lack of the higher lines indicates that there is no significant contribution from the plasma to the XUV emission and that there was no significant expansion of the plasma during the laser pulse. A boron nitride plasma X-ray source produced by irradiation of 100 mJ*Q*-switched YAG laser pulses was also analyzed in [36]. A single line emission at 4.86 nm was selected from the BN plasma using carbon foils.

The interest in boron-containing LIPs is not limited only to the plasma emission and the generation of low-order harmonics. Previous studies have shown the advantages of using this plasma for generating coherent emission in the XUV region [37]. An earlier study has shown that LIP can be used as a nonlinear medium for HHG of ultrashort pulses provided that the limiting factors, such as self-phase modulation and phase mismatch induced by the abundance of free electrons in the plasma plumes, are minimized [38]. There are plenty of specific features of the incoherent XUV radiation and HHG in plasma plumes, the foremost one is the wide range of plasma characteristics available for varying the conditions of irradiation through the above channels. Finding new plasma media and optimizing the best plasma plumes under different conditions of laser ablation, for efficient coherent and incoherent emission, is a way for further enhancement of radiation in the XUV range.

In this section, we analyze boron plasma as a medium for incoherent and coherent radiation in different ranges of XUV [39]. We discuss a method for forming strong incoherent emission from the boron plasma in the spectral region of 6 nm. The optimization of the target ablation allows the generation of XUV emission mainly originated from a single B IV transition ($\lambda = 6.03$ nm). We also analyze the generation of the 806 nm radiation harmonics up to the 55th order ($\lambda = 14.6$ nm) and of tunable near-infrared radiation (1280–1400 nm) up to the 35th order ($\lambda = 38$ nm) during the propagation of femtosecond pulses through the boron plasma.

2.1.1 Formation of Boron Plasma for Generation of Incoherent and Coherent Extreme Ultraviolet Radiation

Ti: sapphire laser was used as the source of uncompressed picosecond pulses for the generation of the LIP on the surface of the bulk boron target. The characteristics of this radiation were as follow: central wavelength 806 nm, pulse duration 370 ps, pulse energy up to $E_{hp} = 6$ mJ, 10 Hz pulse repetition rate. The cylindrical lens (f = 200 mm) was used to form extended LIP with the size of 5×0.08 mm^2 using a heating pulse of intensity up to $I_{hp} = 4 \times 10^9$ W cm^{-2}. This plasma was used for HHG during the propagation of compressed femtosecond driving pulses from the same laser. The parameters of these driving pulses were as follows: pulse duration 35 fs, pulse energy

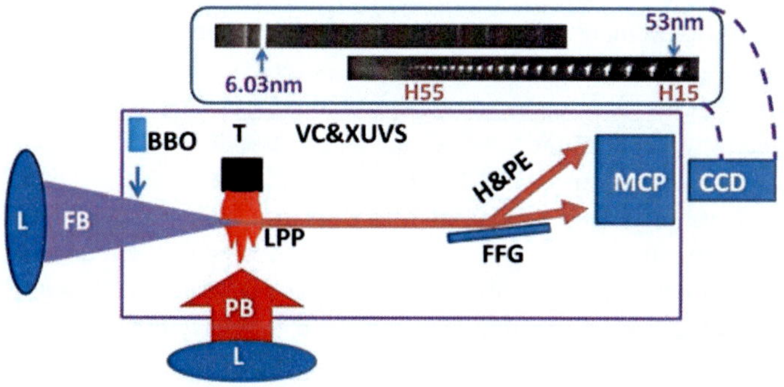

Fig. 2.1 Experimental scheme for the studies of incoherent and coherent emission from the boron plasma. L, focusing lenses for driving and heating pulses; BBO, boron borate crystal; T, boron target; FB, driving femtosecond beam; PB, heating picosecond beam; LIP, laser-produced plasma formed on the surface of boron target; VC&XUVS, vacuum chamber and extreme ultraviolet spectrometer; H&PE, harmonic and plasma emission; FFG, flat field grating; MCP, microchannel plate; CCD, charge-coupled device camera. Inset: images of plasma and harmonic spectra. Upper panel shows the strong incoherent emission of B IV line (6.03 nm) prevailing over other plasma emission components in the XUV range of 2–30 nm. Bottom panel demonstrates the coherent emission spectrum arising from the process of HHG in boron LPP. Harmonic orders between H15 ($\lambda = 53$ nm) and H55 ($\lambda = 14.5$ nm) are shown. Reproduced from [39] with permission from European Physical Journal

up to $E_{dp} = 1$ mJ, the intensity of the driving pulses at the focus area was varied up to 4×10^{14} W cm^{-2}. The optimal conditions for harmonic generation in boron plasma were found at 50 ns delay between the heating and driving pulses and at a distance of ~150 μm from the target surface. These characteristics of the heating and driving pulses were used for the generation of coherent XUV emission through HHG (Fig. 2.1).

The studies of plasma emission in the deep XUV range were performed using 370 ps heating pulses of variable intensity on the target surface in the range of 1×10^{10} W cm^{-2}–3×10^{10} W cm^{-2}. The laser plasma was also formed using nanosecond pulses from Nd:YAG laser. The optical parametric amplifier was also applied to produce the femtosecond tunable IR pulses ($t = 70$ fs, $E = 1.2$ mJ) in the range of 1100–1600 nm, which were used as the driving pulses for HHG. The plasma and harmonic emissions were analyzed using an XUV spectrometer.

As mentioned earlier, laser plasma was formed using picosecond or nanosecond pulses. The latter pulses were used to probe the dynamics of plasma spreading and the formation of the optimum cloud of particles at the moment the femtosecond driving pulses propagate nearby the target surface. A scheme of heating nanosecond pulses and driving femtosecond pulses from different laser sources was used for HHG in plasmas. The ablation of boron-contained targets was carried out by 5 ns, 1064 nm, 10 Hz pulses. An electronic delay between nanosecond and femtosecond pulses allowed applying sufficiently longer delays between two pulses compared with the optical delay technique. The delay between pulses was synchronized and tuned using the delay generator. The synchronization of two laser sources allows analyzing the enhancement of harmonics in different groups of atomic or molecular particles produced during the ablation of bulk materials or targets that initially contain multi-atomic species. The main advantage of this approach is the control of the delay between the heating nanosecond pulses and the driving femtosecond pulses over a wide range between 0 and 10^5 ns, which is sufficient to analyze fast and slow components of ablating material spreading out from the target. The use of the low-pulse repetition rate nanosecond Nd:YAG lasers as the sources of heating pulses may also offer some additional advantages compared with the commonly used picosecond pulses of the same repetition rate and wavelength. The application of nanosecond pulses to ablate the surface of targets allows the formation of less ionized and less excited plasma during longer periods of laser-matter interaction compared with picosecond pulses. This conclusion is based on the analysis of the nanosecond and picosecond ablation-induced plasma emission in the visible and extreme ultraviolet ranges in the case of formation of the optimal plasmas leading to the generation of the highest harmonic yields. Different delays between heating and driving pulses from 5 to 1000 ns were employed in the discussed experiments, with the highest harmonic yield obtained at ~ 90 ns delay. The harmonic yield at the delays larger than 600 ns was notably decreased due to the removal of the largest part of plasma cloud out from the path of the driving beam propagated at the 150 μm distance above the target surface.

To produce multijet plasmas (MJP), a multi-slit mask (MSM) was installed between the cylindrical lens and the Boron target. The application of three MSMs with different slit sizes, two with 0.2 and 0.5 mm slits equidistantly spaced from each other, and one with a slit size of 0.3 mm separated by a 0.8 mm shield between the slits, was analyzed. The plasma formation was accomplished 38 and 75 ns prior to propagation of the femtosecond pulses through the MJP. These delays between the heating and driving pulses were chosen to analyze different properties of the quasi-phase-matching (QPM) between the interacting and generating waves. The maximal used intensity of 800 nm pulses in the plasma area was 5×10^{14} W cm^{-2}. This radiation propagated through imperforated plasma or MJP 75 ns after ablating the targets by picosecond pulses.

The size of the bulk boron target was 5 × 5 × 2 mm^3 (purity 99.6%, Sigma-Aldrich). The B target was mounted on a three-coordinate holder placed inside a vacuum chamber (~10^{-5} mbar) to adjust with respect to the propagating femtosecond pulse. The tablets that were prepared by pressing the boron nanoparticle (NP) powder were also ablated. The mean size of NPs was 120 nm (purity: 99.9%, bulk density: 1.6–1.8 g/cm^3, morphology: spherical particles, appearance: black powder). The diameter of the tablets was 5 mm. In addition, the boron NP-contained targets were formed by gluing NPs on the glass substrates. No XUV emission and harmonic generation were observed in the case of ablation of pure dried glue.

2.1.2 *Incoherent Emission of Boron Plasma at $\lambda = 6.03$ nm*

The formation of a plasma plume should satisfy some specific requirements, which can be essential for the efficient application of such LIP for different tasks. Particularly, as it was shown above, the "optimal" conditions of target ablation allowed the generation of strong 6.03 nm emission, while other emission lines become suppressed. The application of LIP for HHG also requires specific conditions of plasma formation. The over-dense plasmas generated during strong ablation may strongly suppress the efficiency of high-order harmonics due to the destructive influence of the free electrons appearing at these conditions. Previous studies have shown that the "optimal" plasma for HHG is the one that does not demonstrate strong incoherent emission [11, 12, 14, 21]. In that case, one can expect the fulfillment of the phase-matching conditions for as much as possible harmonic orders. In this connection, one can determine the suitable conditions for LIP formation using the analysis of the plasma emission in the visible and UV ranges during the ablation of targets in ambient conditions in the presence of air.

The dynamics of the plasma emission spectrum in the UV range (200–300 nm) during ablation of bulk boron target by 10 ns, 1064 nm pulses are shown in Fig. 2.2a. The emission lines in the UV region were analyzed using laser-induced breakdown spectroscopy (LIBS) during 1000 ns from the beginning of ablation, with the step of 50 ns. The fluence of heating radiation was chosen to be optimal for the 6.03 nm emission being generated at vacuum conditions. One can see the decay of most intense lines in this spectral range [249.7 (B I) and 279.6 (B II) nm], which then almost disappeared after 500 ns (not shown in the picture). The application of time-resolved LIBS for the determination of the best conditions of ablation may be useful for different tasks, like short-wavelength plasma emission, HHG, LIBS characterization, laser ablation deposition, etc.

In those studies, the comparative time-integrated UV spectra of ablated boron NP tablet using the 10 ns and 300 fs heating pulses were analyzed as well (Fig. 2.2b). The

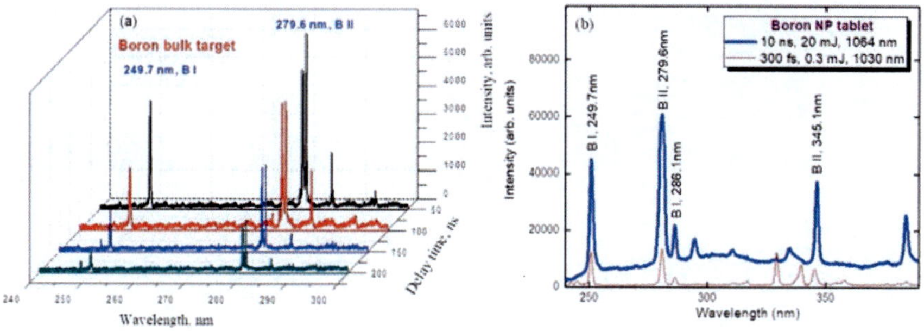

Fig. 2.2 **a** Dynamics of plasma emission spectrum in UV range (200–300 nm) during ablation of bulk boron target by 10 ns, 1064 nm pulses. **b** Time-integrated spectra of plasma emission from boron NP tablet in the case of ablation by 300 fs (thin red curve) and 10 ns (thick blue curve) pulses. Reproduced from [39] with permission from European Physical Journal

application of longer pulses led to the formation of the continuum emission alongside the emission from the B I and B II particles. Femtosecond heating pulses caused the emission of neutral and singly charged boron particles without the continuum component.

Prior to the XUV studies of boron plasma, the emission from different ablated materials was analyzed. Plasma emission from different ablating targets demonstrated a notable difference in the intensities of deep XUV (i.e., in the range of >15 nm). In this regard, boron plasma can provide strongly distinguished emission in the 4.9–6.0 nm range, without the notable emission in other ranges of shorter-wavelength part of XUV using the specially chosen pulse duration and fluence of the heating radiation. The discussed experiments with other metal targets did not show such a difference in the emission lines in the shortest wavelength range of XUV.

Figure 2.3a shows two images of plasma emission from ablated solid boron target in the case of ablation by 370 ps heating pulses of different fluence. In these spectra, the B V (B Ly$_{\alpha}$, 4.86 nm), B IV (B He$_{\alpha}$, 6.03 nm), and B IV (B He$_{\beta}$, 5.27 nm) lines are identified. The upper panel shows that there are only two emission lines from B IV ions (6.03 nm, 2p-1 s, and 5.27 nm, 3p-1 s), which dominate in the XUV range of 3–20 nm. Two other weak lines (12.06 and 18.09 nm) shown by arrows correspond to the second and third orders of diffraction from the 2400 grooves/mm flat field grating of the XUV spectrometer. This spectrum was obtained during ablation of bulk target by picosecond pulses at a moderate fluence (2 J cm^{-2}). The increase of the fluence of the heating pulses up to 6 J cm^{-2} (Fig. 2.3a, bottom image) led to a further increase of the intensity of 6.03 nm emission. Alongside the increase of this

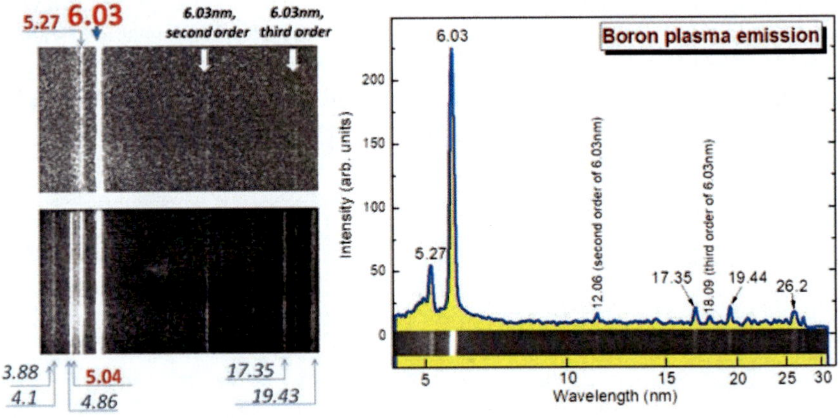

(a) (b)

Fig. 2.3 **a** Images of plasma emission from ablated bulk boron target using the 200 ps heating pulses of different fluence (upper panel: 2 J cm^{-2}; bottom panel: 6 J cm^{-2}). **b** Intensity distribution of B NPs plasma emission in the case of ablation by 50 fs pulses. Reproduced from [39] with permission from European Physical Journal emission, the spectral lines from B V shown by italic letters (3.88, 4.1, 4.86, 17.35, 19.43 nm) appeared in these spectra, though their intensities, except 4.86 nm line, were incomparable with the B IV emission.

The boron NP tablet was also ablated by the same fluence of femtosecond pulses as in the latter case. Plasma emission from B IV at $\lambda = 6.03$ nm was 10–20 times stronger than other emission lines from B V ions (Fig. 2.3b). 50 fs, 800 nm pulses at the fluence of 1 J cm^{-2} on the NP tablet surface were used in those experiments. The ablation of NPs by smaller fluence of heating radiation compared with the case of picosecond pulses led to an even larger ratio between the B IV emission and other emission lines. Similar features were observed in the case of ablation of the glued boron NPs.

The quantitative comparison of 6.03 nm emission was carried out during the ablation of bulk boron target and B NPs-contained target at similar experimental conditions (fluence, pulse duration, and energy, focusing conditions). The emission from the NP target was 3–5 times stronger compared to the ablation of the bulk target. The main obstacle in using these efficient NP-based sources of strong B IV emission ($\lambda = 6.03$ nm) is the difficulty in maintaining the stability of this emission due to the degradation of the NP-contained target. The rotation of the NP target may, to some extent, diminish this problem.

Due to the high pulse intensities, incoherent short-wavelength radiation can be generated if laser radiation interacts with solid and nanostructured matter. In cases of those materials processing, the emission rate decisively depends on the laser parameters and the physical properties of the target material. The interaction of intense short-pulse laser with atomic clusters [40, 41] has become an important area of research. The main characteristic of these small-sized species is the near-solid density inside them that results in enhanced absorption of laser energy via collisional processes and consequently in an increase of X-ray conversion efficiency. A theoretical model of strong soft X-ray emission from laser-irradiated clusters was developed in [42]. In this model, an intense short-pulse laser of Gaussian radial and temporal profiles impinged on a clustered target heats the cluster electrons and leads to Bremsstrahlung emission of X-rays. The plasma frequency of the cluster electrons decreases with the expansion of the clusters under hydrodynamic pressure. It has been suggested that the strong light-NP interaction can cause the efficient blast of clusters leading to strong emission covering the 1–10 nm spectral region. The larger efficiency of these sources can be attributed to the confinement of a larger portion of heating light in a small area, which causes the transfer of this energy into the energy of short-wavelength photons.

2.1.3 Characterization of Plasma for Efficient HHG

The analysis of plasma dynamics was carried out by studying the delay dependence of the harmonic yield from the B plasma. For these purposes, the nanosecond pulses were used for plasma formation. The radiation of Ti: sapphire laser ($\lambda = 806$ nm, $\tau = 35$ fs) was used for the interaction with the plasma produced by Nd:YAG laser ($\lambda = 1064$ nm, $\tau = 5$ ns) to generate the high-order harmonics. The use of a light target such as B (Z = 5) allows maximal yield of HHG at a short delay between the heating pulses producing LIP and the driving femtosecond pulses propagating through the different stages of spreading plasma, prior to the interaction of bulk amount of the ablated material with the driving beam.

Figure 2.4 shows the dynamics of boron plasma formation and spreading along with a broad range of delays between the heating nanosecond pulses and driving femtosecond pulses. It was shown that the optimization of high-order harmonic yield strongly depends on the delay between the heating and driving lasers. The maximal harmonic yield in the discussed case was observed at 50 ns from the beginning of ablation. The novelty of the discussed study stems from using nanostructured targets and nanosecond heating pulses, which allowed achieving the highest yield of harmonic emission from the boron plasma in different spectral ranges. Firstly, harmonics from B plasma have been obtained earlier using ablation by picosecond pulses [37, 43]. Secondly, the targets, which were used in previous studies, comprised of species in the bulk state. The combination of nanostructured targets and nanosecond heating produced the highest harmonic yield for the conditions of the discussed studies. The analysis of the components of plasma ablation was carried out by studying the debris deposited on a nearby silicon substrate during the ablation of bulk boron and B NPs. Deposited B NPs were observed in the case of B NP tablet ablation, as well as during ablation of the bulk target (see the insets in Figs. 2.4 and 2.6). Optimal fluencies for ablation of boron bulk target, boron NPs tablet, and glued boron NPs were chosen by achieving the strongest harmonic yield (Fig. 2.5).

The quantitative determination of plasma concentration is yet resolved properly, though different methods were offered to determine this parameter. One has to determine the concentration of particles at the moment of the driving pulse propagation leading to the highest harmonic conversion efficiency. Obviously, the pressure of

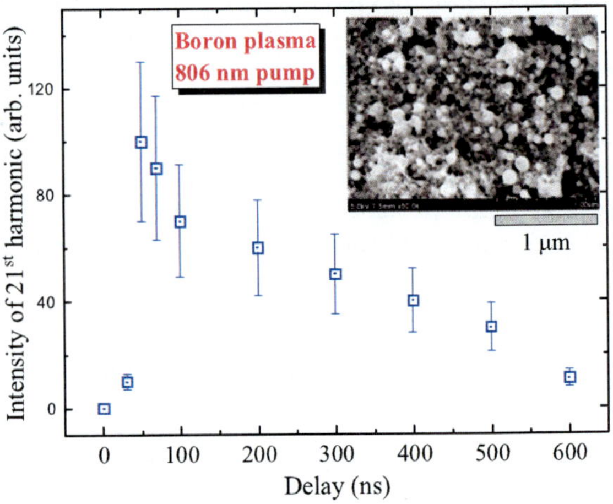

Fig. 2.4 Variation of H21 generated in boron plasma at different delays between the heating 5 ns pulse and driving 35 fs pulse. Inset shows the SEM image of deposited debris from ablated bulk boron. Reproduced from [39] with permission from European Physical Journal

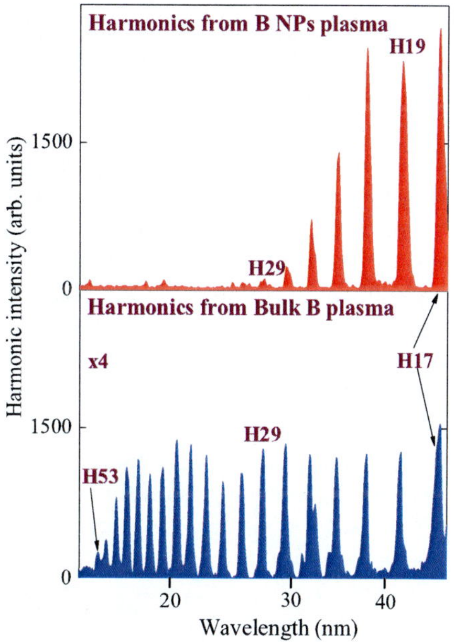

Fig. 2.5 Comparison of lower- and higher-order harmonics generated in the NP-contained plasma (upper panel) and the plasma produced on the bulk boron target (bottom panel). Bottom spectrum was magnified with a factor of four for comparison with upper spectrum. Reproduced from [39] with permission from European Physical Journal

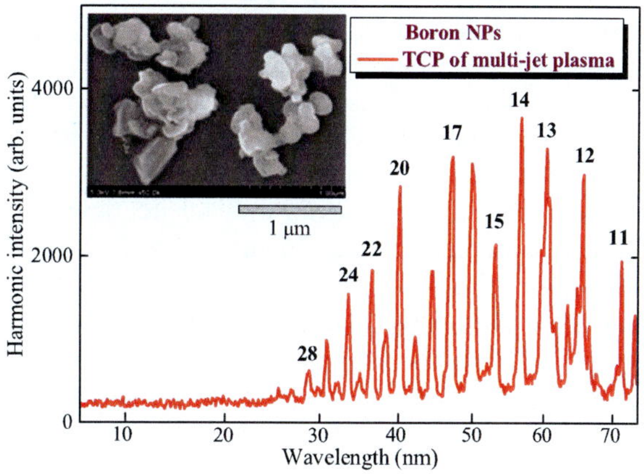

Fig. 2.6 Two-color pump of multi-jet boron plasma. Inset shows the SEM image of deposited debris from ablated NPs. Reproduced from [39] with permission from European Physical Journal

particles initially grows from zero level and then gradually decreases. No measurements of the dynamics of plasma concentration were carried out but rather estimated based on the principles developed earlier in [44] for the laser ablation of solids. Regarding the plasma concentration, one has to determine (or at least, estimate) the concentration of neutrals, ions, and electrons to define their influence on the process of HHG. To understand this phenomenon, the simulations using the hydrodynamic code HYADES have been reported [45]. The expansion of a silver slab interacting with a laser pulse was simulated and then the electron density, ionization level, and ion density were determined as the functions of the heating pulse intensity at a fixed distance from the target surface.

Another approach for determining the plasma characteristics from the ablation of carbon, at the conditions that led to the largest harmonic yield, was reported in [46]. A three-dimensional molecular dynamical simulation of laser ablation of graphite was performed using the molecular dynamics code ITAP IMD [47]. In those studies, the concentrations of carbon plasma were calculated under the experimental conditions of target ablation allowing efficient harmonic generation. The corresponding concentrations were found to be exceeding 10^{17} cm^{-3} in the case of ablation by 8 ps pulses. Notice that the concentrations of plasma particles above the graphite target surface in the area of the femtosecond pulse propagation were calculated to vary between 1.1×10^{17} and 4×10^{17} cm^{-3} depending on the conditions of optimal HHG at two different energies and fluencies of heating pulses. This difference highlights the important role of pulse energy, which is responsible for the difference in the concentrations of the plasma plumes.

Obviously, the difference in the experimental arrangements makes it difficult to determine with high accuracy the plasma concentration, as well as the partial ratio of its components, in another set of experiments. Meanwhile, the similarity in the dynamics of HHG variations with the delay between heating and driving pulses, fluence of heating pulses, distance from the target, etc. may lead to the conclusion on the approximate similarity of the plasma characteristics at the conditions of the maximal harmonic yield. Based on the above assumptions and taking into account the studies reported in [45, 46], it was concluded about the estimations of the plasma characterization in the case of ablated solid boron at the moment of propagation of the driving beam through the plasma cloud at the conditions of highest HHG efficiency (2×10^{17} cm^{-3}).

Meanwhile, no estimates are available for the case of ablated powdered materials. Different absorbance characteristics, uncertainty in the distribution of absorbed heat among the NPs, a difference of their thermodynamic characteristics (i.e., melting and evaporation) with regard to the solid boron, and other peculiarities did not allow determining the concentration of NPs at the moment of the driving beam propagation.

2.1.4 Analysis of Harmonic Cutoffs from Different Plasma Consistencies

Different cutoffs were observed in the case of bulk (fifties orders) and nanopowder (twenties orders) boron targets. The former results were reported in earlier studies of ablated bulk boron targets [37, 43], while the lower orders of harmonics from the nanoparticle-contained plasmas became a frequently reported feature once the comparison of two LIP media (i.e., those contained mono-particles and multi-particles of the same elemental consistency) was discussed earlier (see for example [48]).

Figure 2.5 shows that lower-order harmonics from NP plasma were approximately eight times stronger compared to those produced from monomer-containing plasma. The ablation conditions with respect to pulse energy and fluence of the ablation pulse as well as of the driving fundamental pulse remained identical for these two samples. The same holds for the other experimental parameters, like the distance between the target surface and the driving beam, the pulse duration of the ablation pulses, and the delay between the ablation and driving pulses. The intensity ratio strongly decreases for higher harmonics, until at the 29rd order the harmonic from the nanoparticle target shows the intensity of only 3% of that at the 17th harmonic and becomes smaller by a factor of five compared to the B bulk harmonic.

The harmonics generated from the nanoparticle plasma target showed a larger divergence than those from the bulk target. The observation of only relatively low-order harmonics from the B nanoparticle-containing plasma could be related to the origin of nanoparticle-induced HHG, which is assumed to occur from neutral particles. Meanwhile, the higher-order harmonics from ablated bulk boron target were attributed to the involvement of singly charged ions in the HHG.

The plasma produced on the bulk boron target allowed the generation of the harmonics above the fifties orders. The other two plasmas showed lower harmonic cut-offs (H25 and H21 in the case of B NP tablet and glued B NPs, respectively), while the intensities of the harmonics in the plateau region were notably stronger compared with those generated from the ablated bulk boron target. The comparison of these plasma formations, from the point of view of the stability of harmonic emission, showed that plasma formed on the bulk target was the best choice among the three ablated species. The most probable reason for the difference of harmonic emission from the ablated bulk boron and the two other targets is the difference in the constituencies of plasmas in these three cases at the optimal conditions of laser ablation. Probably, the neutral atoms of boron play important role in harmonic generation from NP-containing plasmas. Correspondingly, their involvement in HHG may explain the low harmonic cut-off observed in the case of these plasmas. Similarly, the lower conversion efficiency in bulk B ablation compared with two boron NP plasmas could be also explained by the preferable involvement of ions in harmonic generation in the former case, while, in the case of ablated NP targets, one can expect the appearance of the boron clusters alongside the NPs, which may enhance the harmonic emission [11, 13].

When comparing harmonics from NP and bulk targets, an essential parameter that needs to be taken into account is the heating pulse intensity. The heating pulse intensity cannot be maintained the same for the three targets used in these experiments. The comparison of harmonic yields was provided at the best conditions of the bulk boron plasma and NP boron plasma formation. In the case of 370 ps heating pulses the intensity $I_{hp} = 4 \times 10^9$ W cm^{-2} was maintained along with the whole extended bulk target. This intensity corresponded to 1.5 J cm^{-2} fluence. Previous studies of HHG in different plasmas have shown that the fluence rather than the intensity of heating pulses is the most relevant parameter that defines the best conditions of plasma formation. Particularly, the comparative studies were carried out in [49] on the plasma formation using pulses of different duration (12 ps and 40 fs).

Note that in these two cases approximately the same optimal fluence of the heating pulses was determined (~1 J cm^{-2}) by using the same focusing conditions and heating pulse energy, while the intensities of the heating 12 ps and 40 fs pulses were 9 × 10^{10} W cm^{-2} and 2.5 × 10^{13} W cm^{-2}, respectively.

One of the main goals of discussed studies was to define which parameter, fluence, or intensity of the heating pulse, plays the dominant role in the formation of suitable plasma for efficient HHG. HHG experiments using 5 ns heating pulses also determined the optimal fluence (~8 J cm^{-2}) of the same order on the bulk boron target at which the maximal harmonic yield was achieved, while the intensity of these pulses on the ablating surface (~1.5 × 10^9 W cm^{-2}) was notably smaller with regard to the case of ablation by 370 ps pulses.

2.1.5 *Two-Color Pump and Quasi-Phase-Matching Studies*

Previous studies have shown that enhanced harmonics can be achieved at the QPM conditions of generating coherent XUV photons once multi-jet LIP are used instead of extended imperforated plasma formations [26, 50–52], which is similar to earlier reported achievements in QPM using gas multi-jet gas targets [53–56]. The QPM during ablation of targets is based on the division of extended plasma into a set of equidistantly separated plasma jets by inserting the MSM in front of the ablating surface or by using the perforated targets.

This mechanism is useful for producing efficient sources of short-wavelength ultra-short pulses for various applications and studies of the properties of harmonic emitters. The dephasing between the propagating harmonic wave and laser-induced polarization is mainly caused by the dispersion of the medium and becomes essential in the presence of a significant number of free electrons. Due to the difference in the velocity of the waves, at some distance from the beginning of the medium, the phase shift becomes close to π. Beyond this distance, called the coherence length, the constructive accumulation of harmonic photons is reversed and the harmonic energy starts to decrease. The implementation of this concept for the case of plasma harmonics widely broadens the field of studies due to the overwhelming prevalence of the ablated solid species used in the latter case over a few gases routinely exploited in conventional HHG. One also should note that the gases used in HHG at QPM conditions are usually denser than LIP (~10^{18} and ~ 10^{17} cm^{-3} respectively), which can restrict the conversion of plateau harmonics due to absorption in noble gases, while for plasmas absorption is insignificant. In the meantime, the formation of QPM conditions in the laser plasmas containing free electrons is more complicated due to the low concentration of plasma for intensity-based optimization and general problems related to precise control of the concentration of ions and electrons in the interaction region.

A two-color pump (TCP) scheme can be used to further improve the HHG yield. The commonly used way is the application of Ti: sapphire laser and its second-order harmonic (H2) leading to doubling the number of generated harmonics from odd orders, in the case of the single-color pump (SCP), to odd and even orders, in the case of TCP. TCP with gases and plasmas targets allowed the analysis of various microscopic and macroscopic effects and polarization-related processes [43, 57, 58]. In the discussed studies, the TCP scheme comprising 806 nm pulses and its second harmonic radiation (403 nm) was used. The 0.5-mm-thick boron borate crystal (BBO, type I, $\theta = 21^0$) was installed inside the vacuum chamber on the path of the focused driving pulse (Fig. 2.1). The conversion efficiency of H2 pulses was 4%. In those studies, MJPs produced by installing the multi-slit mask in front of the extended target were used to analyze the role of quasi-phase-matching in different ranges of the XUV spectrum. The target was irradiated by the heating pulse focused using a cylindrical lens. The MJP configuration was comprised of 12 plasma jets with sizes of 0.2 mm each.

Figure 2.6 shows the TCP-induced harmonic spectrum from the plasma produced on the B NPs pressed tablet. Though the ratio of the second field (403 nm) compared with the fundamental field was notably small (1:25), the application of TCP allowed the four-fold growth of the conversion efficiency of odd harmonics compared with the case of SCP. The stronger even harmonics compared with the odd ones were also achieved. Meanwhile, in the case of TCP, no enhancement of the group of harmonics for which the quasi-phase-matching conditions were fulfilled was achieved (Fig. 2.6).

Another pattern of the harmonic spectrum from MJP was obtained in the case of SCP of boron plasma (Fig. 2.7a, dotted curve). In that case, the plasma contained four 0.5 mm long jets. One can see that the shorter wavelength harmonics became stronger compared to the case of extended homogeneous plasma produced on the boron bulk target (Fig. 2.7a, solid curve).

It was possible to further modify the harmonic distribution from the multi-jet plasma by using the 10-jet plasma produced during B NP ablation. In the case of extended homogeneous plasma produced by picosecond pulses on the B NP pressed tablet, the ordinary decrease of harmonics up to H25 was observed (Fig. 2.7b, solid

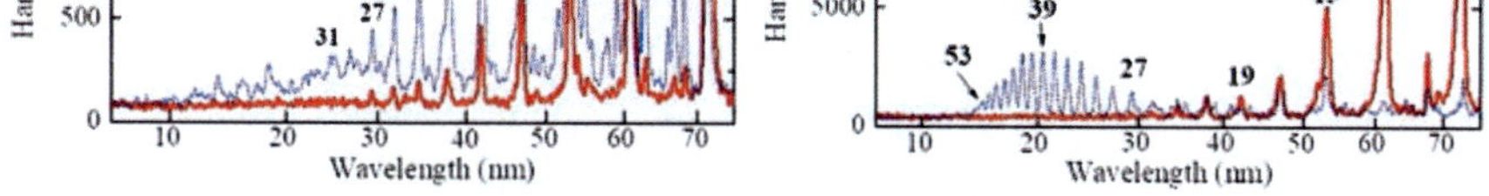

Fig. 2.7 **a** Harmonic spectra from ablated bulk boron using the extended (solid curve) and multi-jet (dotted curve) plasmas. Inset: images of extended and 4-jet plasmas. **b** The same in the case of the ablation of B NP-contained tablet. Inset: images of extended and 10-jet plasmas. Reproduced from [39] with permission from European Physical Journal

curve). The formation of multi-jet plasma using the multi-slit mask containing 0.2 mm slits led to a drastic change in the harmonic distribution (Fig. 2.7b, dotted curve). The group of enhanced harmonics centered around H39 dominated the HHG spectrum, while the lower-order harmonics were suppressed.

The mechanism of this group of harmonics enhancement is well understood and determined by the coherent addition of the harmonic yield in each plasma jet at the conditions of coincidence of the coherence length of harmonic and the sizes of the jet. The discussed studies show the applicability of this approach for the NP-containing plasmas. The comparison of HHG in boron-containing plasmas using different SCP and TCP schemes was studied. The lower conversion efficiency in bulk B plasma compared with B NP plasmas could be explained by the efficient participation of B clusters in HHG, contrary to the atoms and ions of boron. A similar conclusion was drawn during lower-order harmonic generation studies of the plasmas produced by femtosecond and picosecond pulses [59, 60]. Thus the most probable reason for the difference in the harmonic emission properties of bulk B and two B NP ablated targets is the difference in the constituencies of plasmas in these three cases.

The steep decrease in the intensity of low-order harmonics was followed by a plateau pattern. The harmonic conversion efficiency was estimated by considering the typical conversion efficiency from carbon plasma (8×10^{-6} [61]). Similar or even higher values of this parameter were reported by different groups studying carbon plasma as the efficient medium for harmonics generation [11, 46, 62]. The conversion efficiencies from boron and carbon plasmas were compared. From this comparison, one can deduce the ratio between the HHG conversion efficiencies in the plateau range of harmonics distribution in these two cases and correspondingly determined the conversion efficiency in the case of ablation of the bulk boron target. The conversion efficiencies were measured to be between 2×10^{-4} (for third harmonic) to 10^{-7} (along the plateau region).

The HHG appeared to be efficient in the case of neutral and singly ionized plasma when the multiply charged ion and electron concentrations were negligible. It was also observed that the coincidence of harmonics and plasma emission wavelengths does not lead to the enhancement or decrease of the conversion efficiency of these harmonics. The process of harmonic enhancement in the vicinity of some ionic lines requires the strong oscillator strength of the involved transitions, as well as the fulfillment of the suitable phase-matching conditions between the driving and harmonic waves. Meanwhile, being earlier demonstrated only in a few LIPs (In, Mn, Cr, Sn), this process was not observed in the discussed studies in the case of boron plasma (Fig. 2.8a). One can see the coincidence or closeness of H11 and H15 with the strong emission lines of boron, which did not cause the enhancement of those harmonics.

The spectral tuning of driving radiation was carried out using the optical parametric amplifier pumped by an 800 nm laser. This radiation was tuned between 1100 and 1600 nm. Figure 2.8b shows the tuning of harmonic spectra along the strong ionic lines of boron (67.7 and 72.5 nm, see upper panel) using the TCP configuration. No influence of strong B III transition (67.7 nm) on the tunable H20 was

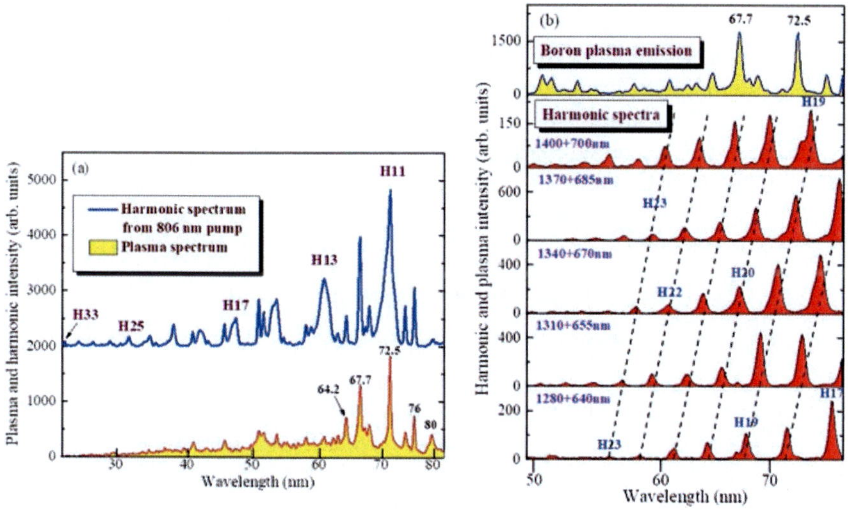

Fig. 2.8 **a** Harmonic spectrum along with some plasma emission lines from bulk boron at optimal conditions of ablation leading to maximal harmonic yield (upper panel) and plasma emission spectrum at stronger irradiation of target, without the propagation of driving pulses through LIP (bottom panel). **b** Plasma emission (upper panel) and harmonic spectra (other panels) using tunable near infrared radiation (1280–1400 nm) and its second harmonic from the optical parametric oscillator. Reproduced from [39] with permission from European Physical Journal

observed, similarly to the absence of the enhancement of H18 tuned along the strong B IV transition (72.5 nm). All spectra of generated coherent radiation represented a plateau-like distribution of harmonics.

The first-ever observation of a plateau in the high-order harmonics generated by an ablation plume at a solid surface was reported in [37]. Boron plasma has accidentally demonstrated advanced properties as a medium that demonstrates the harmonic orders far exceeding those reported in previous studies. Further optimization of this process was reported in [43]. The analysis of the harmonics generation from low-excited boron plasma was presented in the case when the confocal parameter of the focused radiation was close to the plasma sizes (1.2 and 0.6 mm, respectively). The motivation of these studies was to reveal the restricting factors for conversion efficiency and cutoff energy. The conversion efficiencies for the low- and high-order harmonics reported in [63] focused on the longitudinal arrangement of heating and driving pulses propagation through the hole in the boron target, while the discussed studies employed the orthogonal scheme of two pulses.

New relevant knowledge of these studies is related to the application of suitable ablating material using the optimized experimental conditions of plasma formation, which allows the generation of either stable high-order harmonics or strong incoherent emission in the short-wavelength range of XUV. This knowledge is applicable for the practical use of the plasma emission and HHG technique, as well as for the fundamental understanding of the harmonic and incoherent emission generation in the plasmas. The analysis of the components of plasma ablation was carried out by studying the debris deposited on the nearby silicon substrate during the ablation of bulk boron and B NPs. Deposited B NPs were observed in the case of B NP tablet ablation, as well as during ablation of the bulk target. Optimal fluencies for ablation of boron bulk target, boron NPs tablet, and glued boron NPs were chosen by achieving the strongest harmonic yield.

2.2 Calculation of High-Order Harmonic Generation in Laser-Produced Lithium Plasma

High harmonics generation is one of the fundamental phenomena that take place during strong laser field interactions with atomic or molecular systems. One of the promising possibilities is the generation of ultrashort pulses in the form of single attosecond pulses [64, 65], or in the form of attosecond pulse trains [66]. HHG spectroscopy is another important application that allows analyzing the atomic and molecular structures of the source targets using the HHG spectra [9, 67–74]. Targets of high ionization potential have been considered promising sources of short-wavelength HHG. This is attributed to the three-step model of HHG [75, 76], where the cutoff energy is given by $E_{cutoff} = I_p + 3.17U_p$. Here, I_p is the ionization potential of the atom, and U_p is the ponderomotive energy of the electron in the laser field. One method to extend the cutoff energy is to use target materials that have high ionization potential (such as helium or neon) to reduce the effect of ionization. An alternative method to extend the HHG cutoff has been demonstrated by using alkali ions from a laser-plasma source [38, 77] as the nonlinear medium. Their higher ionization potential will allow the increase of I_s, which should also contribute to increasing the cutoff energy.

Suzuki et al. [77] measured the 43rd harmonic of a Ti: sapphire laser ($\lambda = 795$ nm) at a wavelength of 18.49 nm (photon energy of 67.05 eV) from laser-ablated lithium. The analyzed experimental HHG spectra and the cutoff position showed a strong mismatch with the expected contribution from Li^+ ions (ionization potential of 75.6 eV), which hindered a proper interpretation of the observed extended cutoff position. In this section, we analyze and compare the calculated spectrum with the experiment reported in [77] in order to explain the aforementioned mismatch [78]. We discuss detailed numerical calculations using the strong-field approximation (SFA) approach, commonly referred to as the Lewenstein HHG model [76]. The discussed calculations show good agreement with the experimental findings reported in [77]. Additionally, the comparison of HHG spectra for Li^+ ion and He atom is analyzed, since they both have closed 1 s atomic orbital.

For the numerical calculations, the SFA approach [76] and atomic units were used throughout the paper. To characterize the interaction of the atomic system with the strong laser the Keldysh parameter $\gamma_k = (I_p/2U_p)^{1/2}$ [79] was used, where U_p is ponderomotive potential or quiver energy of the electron, which is given by:

$$U_p = \frac{I_n}{4\omega^2} = 3.4 \times 10^{-7} I_n \lambda^2 \tag{2.1}$$

Here I_n is the intensity of the laser field, λ and ω are the wavelength and frequency, correspondingly. For $\gamma_k > 1$, the ionization process is described using the multiphoton picture, i.e. resulting quantities are proportional to $(I_n)^N$, where N is the number of absorbed photons. For $\gamma_k < 1$, the tunneling model becomes valid and the so-called plateau-like features are observed in HHG spectra. HHG spectrum can be calculated from the time-dependent induced dipole moment [76]:

$$\begin{aligned} D(t) = i \int\limits_0^\infty d\tau \left(\frac{\pi}{\varepsilon + \frac{i\tau}{2}}\right)^{\frac{3}{2}} d^*(p_{st}(t,\tau) - A(t)) d_i(p_{st}(t,\tau) - A(t-\tau)) \\ E_i(t-\tau)\exp(-iS_{st}(p_{st}, t, \tau)) + c.c. \end{aligned} \tag{2.2}$$

and depends of its Fourier component as

$$P_i(\Omega) \propto (|D_i(\Omega)|)^2 = \left| \frac{1}{2\pi} \int\limits_{-\infty}^{\infty} dt \exp(i\Omega t) D_i(t) \right|^2 \tag{2.3}$$

The function $d(p)$ in Eq. (2.2) denotes the atomic dipole matrix element of bound to continuum transitions $d(p) = \langle p|r|0\rangle$ The initial bound state is approximated by the atomic orbital with energy ε_0 ($I_p = -\varepsilon_0$) and presented by decomposition into a series of Slater-type basis functions (STO) [80]. The coefficients of decomposition were calculated by the Hartree–Fock procedure [81]. These calculations provide the analytical Hartree–Fock wave functions that accurately reproduce the ionization potentials. Atomic orbitals (AO) used in discussed work are well constrained with the electron-nuclear cusp conditions. Each atomic orbital for ground states of Li, Li^+, and He is represented as the expansion in the series of STO. For Li 2 s AO expanded in a series of 8 basis functions, Li^+ 1 s AO and He 1 s AO represented by 5 basic STO functions. The Fourier transform of $\chi_{nlm}(\alpha)$ STO functions can be represented as

$$\langle p|\chi_{nlm}(a)\rangle = \frac{l!(n-l)!}{2\pi^2} \frac{(2\alpha)^{n+\frac{l}{2}}}{[(2n)!]^{\frac{l}{2}}} \frac{(2ip)^l}{\left[p^2+\alpha^2\right]^{1+\frac{n+l}{2}}} C_{n-l}^{l+1}\left(\frac{\alpha}{\left[p^2+\alpha^2\right]^{\frac{1}{2}}}\right) Y_{lm}(p) \tag{2.4}$$

Here $C_n^k(x)$ Gegenbauer polynomial and Y_{lm} are spherical harmonics. The Fourier transform of the basis functions [Eq. (2.4)] is connected with dipole matrix elements via simple relation

$$\langle p|r|\chi_{nlm}(\alpha)\rangle = i\frac{\partial}{\partial p}\langle p|\chi_{nlm}(\alpha)\rangle \tag{2.5}$$

Initial state was always the ground state of each modeled atomic species. For these model calculations of Li, Li^+ and He atoms one need only 1 s and 2 s Slater-type basis functions, which correspond to χ_{100} and χ_{200}. Taking in account Eqs. (2.4) and (2.5) one can obtain:

$$\begin{aligned}\langle \boldsymbol{p}|\boldsymbol{r}|\chi_{100}(\alpha)\rangle &= \frac{4\alpha^{5/2}}{i\pi^{5/2}}\frac{\boldsymbol{p}}{(p^2+\alpha^2)^3} \text{ and } \langle \boldsymbol{p}|\boldsymbol{r}|\chi_{200}(\alpha)\rangle \\ &= \frac{i\boldsymbol{p}}{\sqrt{3}\pi^{\frac{5}{2}}}\left(\frac{4\alpha^{\frac{5}{2}}}{(p^2+\alpha^2)^3} - \frac{24\alpha^{9/2}}{(p^2+\alpha^2)^4}\right)\end{aligned} \tag{2.6}$$

The laser field in Eq. (2.2) is modeled with $E(t) = E_0 f(t)\cos(\omega t)$, where $f(t)$ is equal to 1 for t from 0 to T_{pulse} (T_{pulse} is laser pulse duration), and 0 for rest of the time interval. $E_0 = (I_n)^{1/2}$, where I_n is the intensity of laser pulse. The vector potential of the applied field is given by $A(t) = \{-(E_0/\omega)\sin(\omega t),0,0\}$. Further integrations by t and τ in Eqs. (2.2) and (2.3) were performed numerically.

For the 2 s atomic orbital of Li atom, with ionization energy 5.39 eV, an expansion in a set of 8 basis functions was used [81]. The HHG spectra were calculated for parameters of laser field close to those used in the experiment [77] with the wavelength equal to $\lambda = 795$ nm and pulse duration $T_p = 150$ fs. The results for $P_i(\Omega)$ [Eq. (2.3)] of 2 s atomic orbital of Li atom are presented in Fig. 2.9 for intensities (a) $I_n = 3 \times 10^{14}$ W/cm^2 and $\gamma_k = 0.39$; (b) $I_n = 3.5 \times 10^{14}$ W/cm^2 and $\gamma_k = 0.36$; (c) $I_n = 4 \times 10^{14}$ W/cm^2 and $\gamma_k = 0.33$ and (d) $I_n = 4.5 \times 10^{14}$ W/cm^2 and $\gamma_k =$ 0.31. For the range of laser intensities used, γ_k is less than 1 ensuring the validity of the tunneling model leading to the well-defined plateau region.

The position of the cutoff harmonic that is predicted by the phenomenological expression [75]:

$$N_{cutoff} = \left(I_p + 3.17U_p\right)/\omega \tag{2.7}$$

is marked with an arrow and labeled by the notation H[N], where N is the harmonic order. It is hard to say unambiguously from Fig. 2.9, which harmonic order corresponds to the cutoff position, so one has to compare it not only with Eq. (2.7) but also has to use more physically clear quantum orbits (QO) theory [76, 82].

The resulting QO cutoff position is labeled as H[N](SP) in Fig. 2.9, where [N] is the harmonic order. As seen in Fig. 2.9, there is a slight difference between the position of the cutoff obtained from phenomenological formula Eq. (2.7) and that obtained by QO calculations. For lithium atoms, the QO cutoff position corresponds to the last harmonic in the plateau region and agrees well with the numerical calculations. In the experiment [77], the cutoff position was identified as the 43rd order of 795 nm radiation for intensity estimated to be 4×10^{14} W/cm^2. The discussed calculations

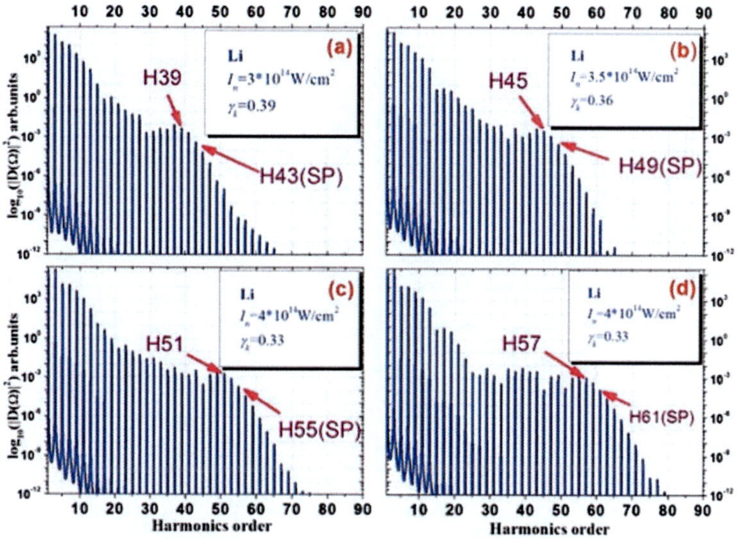

Fig. 2.9 HHG spectra $P_i(\Omega)$ of 2 s atomic orbital of Li atom for different laser intensities I_n and γ_k. Cutoff position according to Eq. (2.7) and QO theory indicated on each figure with H[N] and H[N](SP) notations, where N is a number of harmonic. **a** $I_n = 3 \times 10^{14}$ W/cm^2 and $\gamma_k = 0.39$, H39 according to Eq. (2.7) and H43 (SP) is the result of QO theory; **b** $I_n = 3.5 \times 10^{14}$ W/cm^2 and $\gamma_k = 0.36$, H45 and H49(SP) **c** $I_n = 4 \times 10^{14}$ W/cm^2, $\gamma_k = 0.33$, H51 and H55(SP); **d** $I_n = 4.5 \times 10^{14}$ W/cm^2 and $\gamma_k = 0.31$, H57 and H61(SP). Reproduced from [78] with permission from The Optical Society of America

demonstrated that the cutoff position at 43rd order corresponds to a laser intensity of 3.5×10^{14} W/cm^2, which is close to the experimental value of the driving pulse intensity.

In [58] the shift of the cutoff position to higher orders was attributed to contributions from the singly charged Li^+ ions, which possess a high second ionization potential (75.64 eV). The barrier-suppression intensity I_{BSI} [83] for the singly-charged lithium ion is estimated to be 3.2×10^{16} W/cm^2, while $I_{BSI} < 10^{12}$ W/cm^2 for neutral Li atom. Therefore it is expected that with such a small I_{BSI} for neutral Li, most of the atoms will be ionized already in the initial part of the 150-fs laser pulse. According to Eq. (2.7), for Li^+ the cutoff would be around the 97-th order for laser intensity $I_n = 4 \times 10^{14}$ W/cm^2. However, the experiment had shown no influence on HHG cutoff position from singly charged ions.

To further support the conclusion that singly charged ions do not contribute to the HHG cutoff, the interactions using the Keldysh parameter for $I_p = 75.64$ eV and $I_n = 4 \times 10^{14}$ W/cm^2 were analyzed. At these values of the above parameters, γ_k is equal to 1.26, which means that the process is of multiphoton nature. Actually, at these parameters the applicability of Lewenstein's model and SP approximation become questionable. Nevertheless, the calculations were performed for Li^+ 1 s atomic orbital

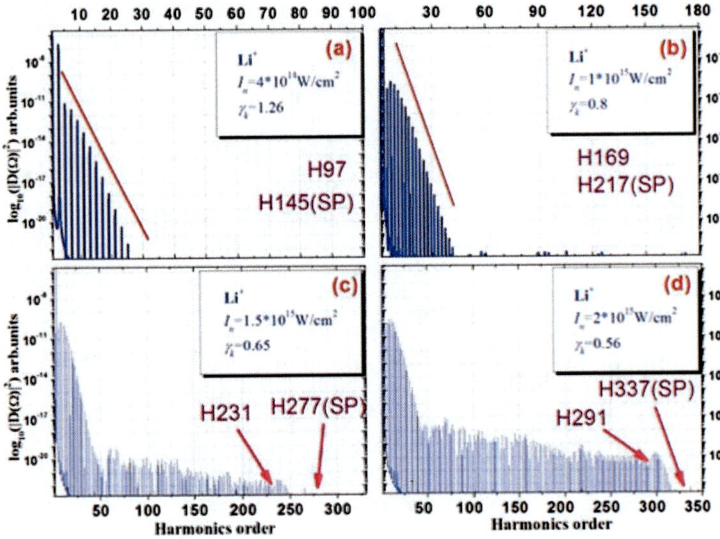

Fig. 2.10 HHG spectra 1 s atomic orbital of Li^+ ion for different intensities of laser field In and γ_k. Cutoff position according to Eq. (2.7) and QO theory indicated on each figure with H[N] and H[N](SP) notations, where N is a number of harmonic. **a** $I_n = 4 \times 10^{14}$ W/cm^2 and $\gamma_k = 1.26$, H97 according to Eq. (2.7) and H145 (SP) is result of QO theory; **b** $I_n = 1 \times 10^{15}$ W/cm^2 and $\gamma_k = 0.8$, H169 and H217(SP); **c** $I_n = 1.5 \times 10^{15}$ W/cm^2 and $\gamma_k = 0.65$, H231 and H277(SP); **d** $I_n = 2 \times 10^{15}$ W/cm^2 and $\gamma_k = 0.56$, H291 and H337(SP). Reproduced from [78] with permission from The Optical Society of America

[81], and the results are presented in Fig. 2.10 for (a) $I_n = 4 \times 10^{14}$ W/cm^2 and $\gamma_k = 1.26$, (b) $I_n = 1 \times 10^{15}$ W/cm^2 and $\gamma_k = 0.8$, (c) $I_n = 1.5 \times 10^{15}$ W/cm^2 and $\gamma_k = 0.65$, and (d) $I_n = 2 \times 10^{15}$ W/cm^2 and $\gamma_k = 0.56$. The QO estimation of the cutoff position at these conditions was also carried out.

In the case of Fig. 2.10a and b, HHG spectra have a multiphoton nature. The slope function (red line), corresponds to the dependence of the form of ~ $C_0(C_1 \times I_n)^k$, where k is the order of the harmonic. Plateau harmonics start to appear only at very high driving pulse intensities, as can be seen in Fig. 2.10c and d. It is clear that for $\gamma_k > 1$ and $\gamma_k \approx 1$ in the case of Li^+ ions the QO approach fails to describe the position of the cutoff, as well as it does not fit with the phenomenological formula [Eq. (2.7)]. One can see it in Fig. 2.10b, for $\gamma_k = 0.8$, when again no plateau region and no harmonics at 169th and therefore no harmonics at 217th order appear in the HHG spectra. The exact cutoff positions in Fig. 2.10c and d are placed between Eq. (2.7) and QO results. Thus these calculations demonstrate that with parameters close to those used in [77] there is no cutoff extension.

Li^+ ion and He atom were also compared since they both have the closed 1 s atomic orbital and He also possesses a high ionization potential that is equal to 24.58 eV. In terms of the SFA model, there is no difference between ions and neutral atoms since the Coulomb potential is neglected in the final state for both systems. The results for

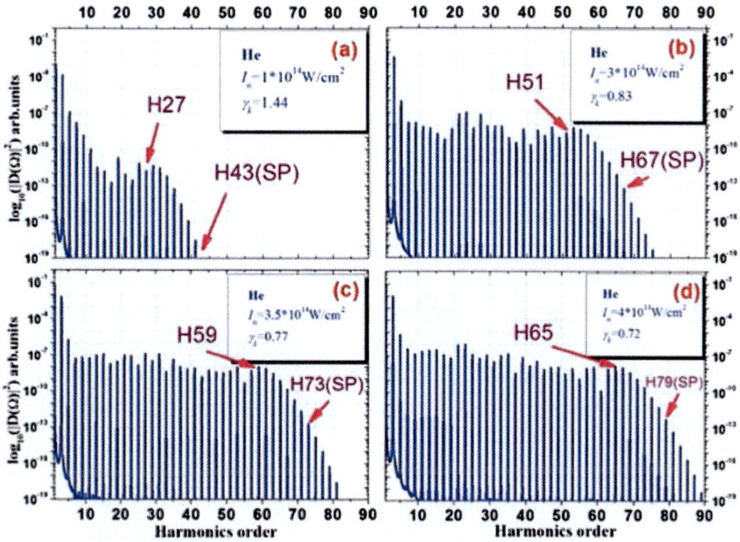

Fig. 2.11 HHG spectra 1 s atomic orbital of He atom for different intensities of laser field I_n and γ_k. Cutoff position according to Eq. (2.7) and QO theory indicated on each figure with H[N] and H[N](SP) notation, where N is a number of harmonic. **a** $I_n = 1 \times 10^{14}$ W/cm^2 and $\gamma_k = 1.44$, H27 according to Eq. (2.7) and H43 (SP) is result of QO theory; **b** $I_n = 3 \times 10^{14}$ W/cm^2 and $\gamma_k = 0.83$, H51 and H67(SP); **c** $I_n = 3.5 \times 10^{14}$ W/cm^2 and $\gamma_k = 0.77$, H59 and H73(SP); and **d** $I_n = 4 \times 10^{14}$ W/cm^2 and $\gamma_k = 0.72$, H65 and H79(SP). Reproduced from [78] with permission from The Optical Society of America

the He 1 s atomic orbital [81] are presented in Fig. 2.11 for (a) $I_n = 1 \times 10^{14}$ W/cm^2 and $\gamma_k = 1.44$, (b) $I_n = 3 \times 10^{14}$ W/cm^2 and $\gamma_k = 0.83$, (c) $I_n = 3.5 \times 10^{14}$ W/cm^2 and $\gamma_k = 0.77$, and (d) $I_n = 4 \times 10^{14}$ W/cm^2 and $\gamma_k = 0.72$. Resulted QO cutoff position is pointed in Fig. 2.11 for HHG spectra in form of H[N](SP). Those calculations show that, for helium, in spite of larger γ_k than a similar parameter for lithium ion, there is no multiphoton-like picture. For helium one can see the plateau-like spectrum of harmonics, while it should be similar to that from lithium ions provided the character of HHG spectrum is defined by Keldysh parameter.

Moreover, Eq. (2.7) gives the cutoff order quite close to the actual positions in Fig. 2.11. Again, the QO cutoff is strongly overestimated. Also for higher intensities and when $\gamma_k < 1$ (Fig. 2.11b–d), one can see that harmonics of the plateau region are less suppressed compared with the low-order harmonics and plateau region of Li^+ ion. The calculations demonstrate quite a different character of the high-order harmonics response of the He atom at comparable values of γ_k.

The Lewenstein model includes all three stages of a three-step model of HHG: ionization or, more correctly, the transition of an active electron to the continuum, acceleration in a laser field, and subsequent recombination with the emission of a harmonic photon. The first step, the transition to the continuum (ionization), is represented by the corresponding amplitude of the dipole transition. This is exactly, in the length gauge, the amplitude, which was used in the Keldysh model [79]. For the case of the $\gamma_k > 1$, it clearly gives the probability of multiphoton ionization. In that case, to overcome the barrier, the main role plays the multiphoton ionization, which in the case of a very deep potential of Li^+ ion (more than 2 a.u.) gives a very small probability of the appearance of an electron in the continuum. This may be the reason for the observed features of the spectra of higher harmonics.

The ionization in a strong laser field is a complex problem. The entire account of ionization would seriously complicate the calculations. The discussed calculations were focused on the role of different emitters contributing to the HHG plateau. Those calculations showed that, under the considered conditions, the contribution from ions plays no role. On the other hand, barrier suppression intensity is a rather rough estimate, obtained from consideration of the suppression of the Coulomb barrier in a static electrical field and verified in case of notably long pulses of the order of picoseconds.

In principle, it is possible to include in calculations different initial states of the same material except for ground state, for example, one or more excited states could be included [72]. In particular, the enhancement of single harmonics in HHG spectra is commonly associated with the presence of multiphoton resonances. One can include the ionization rates in the calculations [72] by accounting for the depletion of the ground state. It is also possible to estimate the population of the ground state of Li if one calculates the saturation intensity. This calculation gives an estimate of the magnitude for the saturation intensity much lower than the intensities used in the experiment when the lithium atoms turn out to be ionized already at the beginning of a rather long (150 fs) pulse. However, the experiment showed that the extension of the HHG plateau is not achieved. Comparison of the HHG spectra using Li and Li^+ at similar intensities taking into account the tunnel character of ionization in the case of Li and the multiphoton pertrubative-like process in the case of Li^+ allows concluding about the contribution of only neutral lithium atoms, even without directly including the decay of the ground state.

The multi-photon ionization and tunnel ionization may happen at the same time when the Keldysh parameter is close to 1. In the Keldysh model for ionization [79], the expressions for the ionization amplitudes were obtained in the cases when the $\gamma_k > 1$ and when the $\gamma_k < 1$. Formally, the same amplitude of the transition (or ionization) from the ground state to the Volkov continuum enters Lewenstein's model as the first step of the three-step model. Meantime, the discussed calculations show a substantially different character of the spectra of He atoms for $\gamma_k = 1.44$ and Li^+ for $\gamma_k = 1.26$, while one can expect to obtain similar spectra in the cases of He and Li^+ with a predominance of a multiphoton process.

In general, the calculations of HHG in plasma should take into account the presence of neutral components, positively charged ions, and free electrons. If one considers the application of the SFA method to ions, then from this point of view there is no difference between a neutral atom and its ions, since in the final Volkov state the effect of the atomic core on the active electron is not taken into account. In order to take into account the influence of the residual ion on the continuum state of an active electron, several techniques can be used, in particular, the Coulomb-Volkov approximation, but this leads to a substantial growth of calculation time.

The discussed calculations do not explicitly take into account the dependence of the cut-off position on the delay between the heating pulse and the main pulse. Taking into account the estimation of the propagation velocity of particles in the plume of the order of 10^5 cm/sec and the fact that the driving pulse is focused at the distance of 100–200 μm from the target surface, the main point of optimizing delay is to synchronize the plasma plume with the driving pulse arrival time.

2.3 Spectral Modification of Converting Radiation and High-Order Harmonics Through Filamentation in Argon and Propagation in Laser-Produced Plasmas

High-order harmonic generation of laser radiation in various media is a widely used technique for the generation of coherent radiation in the extreme ultraviolet range. Different methods allow modification of the characteristics of this radiation over a broad range. One of them is the phase modulation of driving radiation, which leads to rearrangement of the spectrum and intensity of generated harmonics. The self-phase modulation (SPM) of laser pulses is widely used to generate additional frequencies in the spectrum of laser radiation. In particular, the introduction of an optical medium, e.g., a glass plate, into the propagation path of strong laser pulses leads to spectral broadening due to the SPM. In this relation, one can expect that the passage of the focused radiation through the gases at higher pressures than those commonly used for HHG can cause a considerably larger broadening of the spectrum of the laser pulse, which in turn makes it possible to significantly change the spectral pattern of the harmonics induced by this radiation upon its passage through the converting medium. Additionally, in the case of the pulses that underwent a significant SPM due to the filamentation of the medium through which it propagated, a considerable heterogeneous broadening of the spectral distribution can be achieved.

HHG can be obtained in gases and laser-induced plasma. Meanwhile, the efficiency of HHG upon passage through the gases and LIP still remains too low to use them as real sources of coherent short-wavelength radiation for photolithography, biomedical applications, plasma diagnostics, microscopy, etc. One of the reasons for this restriction is the spectral variation of converting pulse during propagation through the medium. The phase modulation of laser pulses largely limits HHG conversion efficiency. The LIP possesses various attractive features allowing the efficient generation of coherent short-wavelength radiation. The high efficiencies obtained in a number of nanoparticle-containing plasma formations, the implementation of resonant amplification of individual harmonics, a considerable increase of harmonic intensity in different spectral ranges using the quasi-phase-matching, etc. [84] have shown the attractiveness of the HHG in LIP. The advantages of HHG upon propagation of the ultrashort pulses through the LIP could be realized at the conditions of the weakly excited and weakly ionized medium. The impeding processes that restrict the conversion efficiency of laser radiation towards XUV play a limited role in these conditions.

Several studies of the spectrally broadened harmonics have been performed in gases and LIP [85, 86]. Those studies were carried out under the conditions of harmonic spectra broadening at a comparatively "weak" influence of the SPM on the spectral properties of harmonics. This term refers to the conditions under which the spectral and spatial parameters of radiation to be converted change insignificantly upon passage through the nonlinear medium. Particularly, the spectra of harmonics generating in LIP using narrowband and broadband laser pulses were analyzed in [86]. Those experiments were performed at weak variations of the characteristics of laser radiation using different techniques (the SPM in plasma and in thick glass plates, the variation of radiation chirp by changing the spacing between the diffraction gratings in the pulse compressor, the use of a two-color pump, etc.), which caused relatively small changes in the spectral and spatial components of the radiation to be converted.

In this section, we analyze the harmonic spectra generated in the LIP using the 806 and 1310 nm pulses that result in broadening and blueshift of the XUV components [87]. We discuss the possibility of the substantial spectral variations of harmonics in laser-plasma upon the transformation of the spectrum of laser radiation passed through the filaments formed during propagation of femtosecond pulses through the additional chamber filled in with the argon gas. We also analyze the spectral variations of harmonics during the propagation of laser pulses through the over-dense plasmas. The modifications of harmonic spectra occurring in these two cases are analyzed and compared.

2.3.1 Experimental Conditions of HHG in LIP Using 806 nm and 1310 nm Pulses

The mode-locked Ti: sapphire laser pumped by diode-pumped CW laser as the source of 803 nm, 55 fs, 82 MHz, 450 mW pulses was used for injection in the pulsed Ti: sapphire regenerative amplifier with pulse stretcher and additional double passed linear amplifier. The output characteristics from this laser were as follows: central wavelength 806 nm, pulse duration 370 ps, pulse energy 5 mJ, 10 Hz pulse repetition rate. This radiation was further amplified in homemade three-passed Ti: sapphire linear amplifier. Part of this radiation was separated from a whole beam and used as a heating pulse for homogeneous extended plasma formation using the 200-mm focal length cylindrical lens installed in front of the extended solid target placed in the vacuum chamber (Fig. 2.12). The intensity of the heating pulse on the target surface was varied up to 6×10^9 W cm^{-2}. The ablation sizes were 5×0.08 mm^2.

The remaining part of amplified radiation was delayed with regard to the heating pulse in such a way that the compressed pulse propagated through the preformed

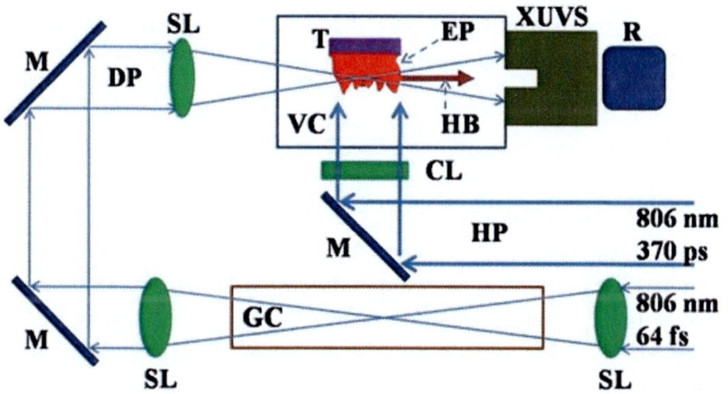

Fig. 2.12 HHG scheme using Ti: sapphire laser radiation. DP, driving pulse; HP, heating pulse; SL, spherical lenses; GC, gas chamber; M, mirrors; CL, cylindrical lens; VC, vacuum chamber; T, target; EP, extended plasma; HB, harmonic beam; XUVS, extreme ultraviolet spectrometer; R, CCD registrar. Reproduced from [87] with permission from Springer

plasma plume 35 ns from the beginning of target ablation. After propagation of compressor stage the output characteristics of Ti: sapphire laser were as follows: pulse energy 8 mJ, pulse duration 64 fs, 10 Hz pulse repetition rate, pulse bandwidth 17 nm, central wavelength 806 nm. Most HHG experiments were carried out using the 806 nm radiation, while the 1310 nm pulses from optical parametric amplifier [88] were also used in these studies. The intensities of the 806 and 1310 nm pulses focused by 400 mm focal length lens inside the extended plasma were varied up to 1×10^{15} and 2×10^{14} W cm^{-2} respectively. The driving pulses were focused onto the extended plasma from the orthogonal direction, at a distance of ~100 μm above the target surface. The plasma and harmonic emissions were analyzed using an XUV spectrometer containing a cylindrical mirror and a 1200 grooves/mm flat field grating with variable line spacing. The XUV spectrum was detected by a microchannel plate detector with the phosphor screen and recorded by a CCD camera.

Silver, zinc, manganese, and graphite plasmas were analyzed as the media for a harmonic generation. The 5-mm-long samples were installed in the vacuum chamber for laser ablation. The interest in these LIPs was related to some attractive peculiarities shown in previous studies of HHG in the plasmas using both fixed (800 nm) and tuneable (1150–2400 nm) laser pulses. Earlier, some of those plasmas (Ag, Mn) allowed the generation of the highest harmonic cut-off energy (up to 170 eV) among ablated media, while other plasmas showed the strongest lower-order harmonic yield (C, Zn) and resonance-induced enhancement of single harmonics (Zn, Mn) [84]. Application of these plasmas for HHG using the spectrally modulated pulses propagating through filaments, as well as the introduction of the phase modulation during propagation of chirp-free pulses, allow the definition of the best conditions for frequency conversion of ultrashort pulses.

2.3.2 Harmonics Induced by the Pulses Propagated Through Gas Filaments

The filamentation of 806 nm pulses was performed in the 1500-mm long chamber filled with argon at different pressures varying between vacuum conditions and 600 mbar. The driving radiation was focused using a 1000 mm focal length lens inside the gas chamber (Fig. 2.12). The propagated pulses were then defocused by a similar lens and again focused in the vacuum chamber containing LIP for a harmonic generation. The spectral characteristics of driving pulses before and after the gas chamber were analyzed using a fiber spectrometer. The spectral characteristics of 806 nm pulses were also recorded upon passage through the LIP.

In the case of focusing of the 806 nm, 64 fs, 3 mJ pulses in the gas chamber, the 250-mm-long filaments and the variation of the output spectrum of propagated pulses starting from 40 mbar of argon pressure were observed. The spectrum of driving radiation became blueshifted and spectrally broadened upon the increase of gas pressure up to 600 mbar. Further growth of gas pressure led to strong white light generation, modification of the spatial shape of the laser beam, and a notable decrease of HHG conversion efficiency during experiments using extended plasmas. After passage through the telescope, the blueshifted and spectrally broadened radiation was directed into a chamber with a plasma target. Under these conditions, the harmonics were observed, which experienced a blueshift compared with the case of an almost empty gas chamber (see the raw spectra of harmonics generated in silver plasma at 10 and 540 mbar in a gas chamber, Fig. 2.13).

Raw spectra of harmonics are shown for the better visual distinction of the blueshifts of harmonics at similar conditions of experiments and different pressures of Ar. Two series of experiments at 10 and 540 mbar of Ar in the gas chamber were performed under similar conditions of the formation of LIP. Furthermore, the pulses to be converted with the same energy and similar conditions to record the XUV spectra were used. Note the growth of harmonic conversion efficiency and an extension of harmonic cut-off in the case of SPM pulses (Fig. 2.13, bottom panel).

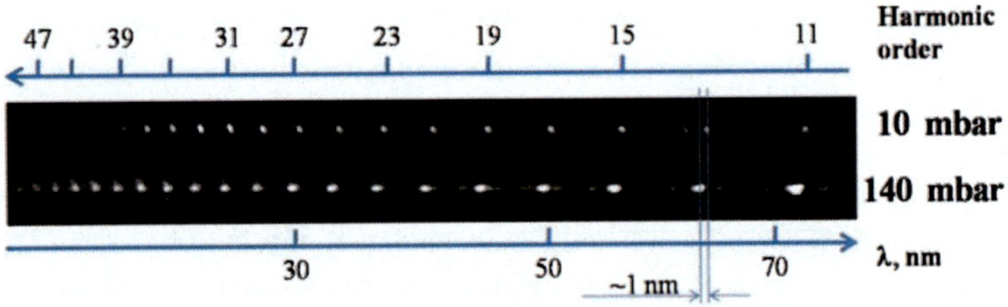

Fig. 2.13 Raw images of harmonic spectra generated in silver plasma at different pressures in gas chamber (upper panel–10 mbar, bottom panel–140 mbar). The blueshift of 13th harmonic was ~1 nm. The spatial components seen in both sides of lower-order harmonics correspond to the second-order diffraction of higher-order harmonics. Reproduced from [87] with permission from Springer

A comparison of these spectra shows that the SPM and blueshift of driving radiation led not only to a notable blueshift of generated harmonics but also to their spectral broadening. Similar blueshifts were observed in the case of the LIP formed on the surfaces of other targets.

Below we discuss the spectral characteristics of the laser radiation that passed through the telescope and created filaments in the argon and compare them with the spectral parameters of the laser radiation propagated through the empty chamber (Fig. 2.14). The SPM in the filamentation region of the laser beam led to a variation of the spectral distribution of laser radiation. Whilst the spectral shape of laser radiation in front of the gas chamber showed a nearly smooth distribution centered at 806 nm (spectral bandwidth of 17 nm), the pulses propagated through argon were blueshifted depending on gas pressure (Fig. 2.14). Particularly at 570 mbar of argon in the filamentation chamber, the central wavelength of broadened radiation was shifted from 806 to 792 nm. The spectral width of blueshifted pulses was 22 nm. Correspondingly, the 13th harmonic at these conditions was ~1.1 nm blue-shifted compared with the gas-free case. Notice that these experiments with Ag plasma were conducted at the conditions when plasma did not alter the harmonic spectrum due to SPM in the converting medium.

The analysis of these studies shows the link between the measured properties of the HHG spectra and the properties of the driver pulse after propagation in the argon gas cell. The variation of driving pulse in the latter case causes the changes in the spectral properties of harmonics, which is natural, since the characteristics of converted radiation are directly related to the characteristics of pump radiation.

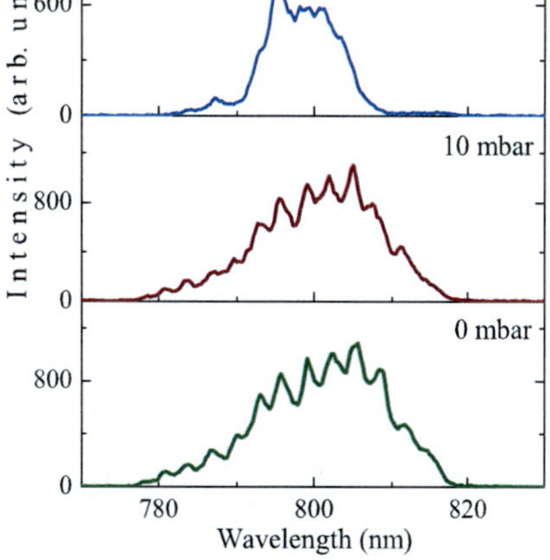

Fig. 2.14 Spectra of driving radiation upon passage of gas chamber at different pressures of argon. Reproduced from [87] with permission from Springer

2.3.3 SPM in LIP

The above-described studies were carried out by analyzing the spectral characteristics of the harmonics produced by already spectrally broadened and blueshifted laser pulses. Meanwhile, the plasma itself is a nonlinear medium, and any laser pulse passing at appropriate intensity through the relatively dense plasma may undergo SPM, spectral broadening, and blueshift. The information about the SPM of a laser pulse inside a plasma plume may be retrieved from the bandwidth of harmonics. The SPM of a laser pulse should increase with the increase of laser intensity or the particle density in the plasma plume.

The widths and blueshifts of the spectra of generated harmonics in the case of the free of chirp pulses (i.e. during propagation through the gas-free chamber) were studied at the conditions of over-dense LIP ($\geq 6 \times 10^{17}$ cm^{-3}) and strong laser radiation. The SPM of laser radiation at these conditions occurred due to the relatively high density of plasma particles and driving pulse intensity (1×10^{15} W cm^{-2}). One can note that the maximum yield of harmonics from most plasmas was achieved when no plasma emission appeared in the XUV spectra.

The use of denser plasma in the case of a single-color (1310 nm) pump at similar energy of pulses (1 mJ) led to significantly smaller conversion efficiency and larger bandwidths of harmonics compared with a two-color pump, though the cut-off energy, in that case, was larger (Fig. 2.15a). Plasma concentration, in that case, was estimated to be 6×10^{17} cm^{-3} using the quasi-phase-matching technique

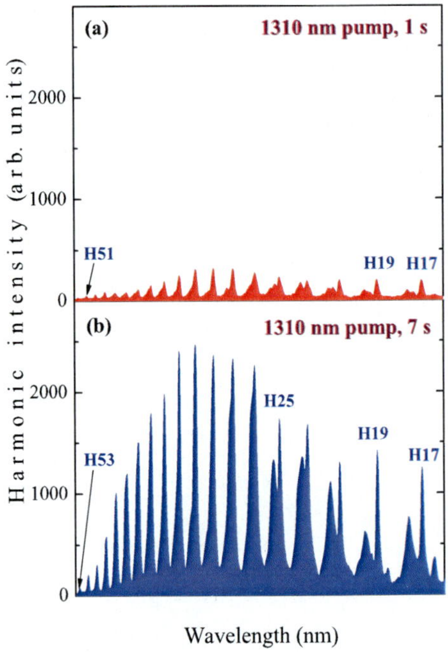

Fig. 2.15 Harmonic generation in over-dense (6×10^{17} cm^{-3}) graphite plasma using SPM-free pulses. **a** Single-colour (1310 nm) pump. **b** The same as **a** but using longer collection time for better visualization of the peculiarities of harmonic spectra. **a** and **b** spectra were captured using 1 s and 7 s collection time respectively. Reproduced from [87] with permission from Springer

[89], which was approximately three times larger than the one commonly used for efficient HHG. The widths of harmonics were larger than one could expect for the driving pulses with a spectral width of 17 nm.

One can see both broadening and blueshift of harmonics, alongside the redistribution of harmonic emission in the XUV range. To make the visual impression better, the collection time of the registrar of harmonic spectra was increased from 1 to 7 s. Figure 2.15b shows the same HHG spectrum as the one presented in Fig. 2.15a, while underlining the peculiarities of spectral variations of harmonics using longer collection time. One can see the strong blue components of 19th–25th harmonics alongside the weaker red-shifted components. The harmonic cut-off was extended up to the 53rd order. The maximum yield was obtained for the 33rd harmonic. The unusual shape of harmonic distribution was attributed to better phase-matching conditions in the 5-mm-long LIP for the harmonic orders centered at the wavelength of 40 nm.

The spectrum of driving pulses was affected during propagation through such LPP. The spectral analysis of the 806 nm pulses propagated through the plasma in the case of 2×10^{17} and 8×10^{17} cm^{-3} graphite plasma densities is shown in Fig. 2.16. One can see the unaffected spectral shape in the former case, while other spectra show dominantly blueshifted radiation. The instability of plasma formation led to the variable shape of the spectrum of propagated pulses (compare three bottom spectra).

The peculiarities of SPM-transformed spectra of harmonics were distinguished from each other once different targets were used. Figure 6 shows the blueshifts of harmonics in the case of (a) Zn and (b) Mn plasmas. In the case of weak 806 nm

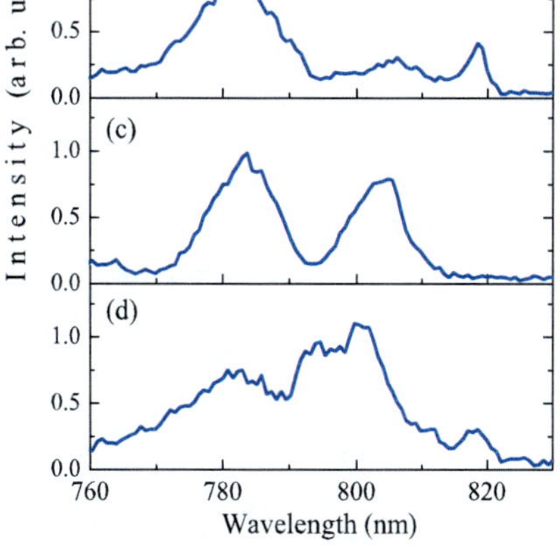

Fig. 2.16 Spectra of 806 nm pulses measured after propagation through the LIP at **a** weak and **b–d** strong excitation of graphite target. Reproduced from [87] with permission from Springer

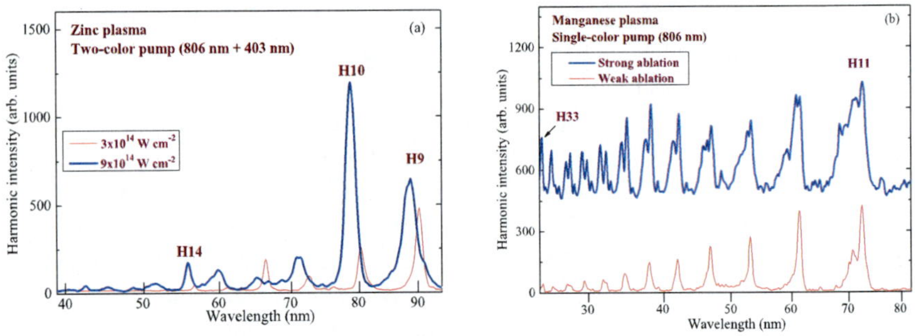

Fig. 2.17 **a** Harmonic spectra from Zn plasma in the case of weak (3×10^{14} W cm^{-2}, thin curve) and strong (9×10^{14} W cm^{-2}, thick curve) intensities of driving pulses. **b** Harmonic spectra from Mn plasma in the case of weak (2×10^{9} W cm^{-2}, thin curve) and strong (4×10^{9} W cm^{-2}, thick curve) ablation. Two curves were separated from each other by vertical axis for better viewing of the difference between these spectra. Reproduced from [87] with permission from Springer

pulses, no significant broadening and absorption were observed in both plasmas (see thin curves in Fig. 2.17). The growth of pulse intensity (from 3×10^{14} to 9×10^{14} W cm^{-2}) led to the blueshift of odd and even harmonics in the case of the two-color pump of Zn plasma. In the case of Mn plasma, rather sophisticated blueshift and broadening of harmonics were observed by increasing the intensity of heating pulse (from 2×10^{9} to 4×10^{9} W cm^{-2}) and plasma density (from 2×10^{17} to 5×10^{17} cm^{-3}). The choice of plasma significantly alters the broadening and blueshift of harmonics at similar conditions of experiments (i.e., at the same laser intensity and plasma density). Particularly, the studies of carbon-containing targets (fullerenes, diamond nanoparticles, carbon nanotubes, and so on) have previously shown that they are very useful in generating broadband and blueshifted HHG [87].

2.3.4 Discussion

The variations of HHG spectra in these studies were achieved using the driving pulses, which were phase modulated (i) by preliminary focusing in gas, and (ii) inside the plasma medium itself. Filamentation in gases appears as a nonlinear self-guided propagation regime sustained by a dynamic balance between nonlinear self-focusing and defocusing effects [90]. Depending on experimental conditions, the latter can include higher-order Kerr terms and free electrons with respective weights depending on the propagation medium considered. Meanwhile, the laser intensity at the plasma region was kept low enough ($\leq 8 \times 10^{14}$ W cm^{-2}) to avoid any breakdown, white light generation, or filamentation but high enough to induce the SPM in propagated pulse. At relatively higher intensities of femtosecond pulse ($>1 \times 10^{15}$ W cm^{-2}), the intensity of generated harmonics was decreased due to additional impeding processes, such as self-defocusing, ionization, and disintegration of plasma particles, phase mismatch, and filamentation.

The blueshift of harmonics can be explained in terms of the tuning of some parts of harmonic spectra due to the redistribution of laser spectrum along the main pulse caused by SPM and following preferential harmonic generation induced by the leading part of the pulse [91]. The harmonics produced in the trailing portion also get dephased due to propagation in the already-produced high-electron-density plasma; however, their weight in an overall yield of harmonics is insignificant due to growing phase-mismatch between the interacting waves. Since the laser has more bandwidth, the harmonics generated with these pulses have relatively higher bandwidths.

One has to underline the role of the variable redistribution of pulse spectrum along the time scale. In the case of chirp-free pulses, each component of the spectrum, which possesses a sufficient intensity, participates in the formation of the spectrum of harmonic. It means that both long- and short-wavelength components of the driving pulse spectrum become converted, to some extent, into the harmonic components. In other words, the harmonic spectrum contains both short- and long-wavelength spectral components, which can be attributed to the corresponding parts of the spectrum of driving pulse. In the case of SPM pulses, for instance, those possessing positive chirp, the spectrum of driving radiation is redistributed along the laser pulse in such a manner that the leading part of the pulse contains predominantly red components, while the trailing part contains predominantly blue components. In that case, the harmonics will contain the red-shifted part, alongside the main unshifted spectrum. And vice versa, the negatively chirped pulses will form the harmonic spectrum containing the blue-shifted component (which is the case reported in this paper). The reasons in the prevailing translation of the spectral components of the leading part of chirped pulse towards the harmonic spectrum are related to the nature of harmonic generation. This process realizes in gases and plasmas predominantly through the interaction of the leading part of the laser pulse, while at the peak of pulse some impeding processes such as the additional tunnel ionization of parent particles leading to the appearance of a large number of free electrons and the phase mismatch induced by spatio-temporal variations of the driving pulse in the focal area. In other words, the trailing part of the laser pulse does not actively participate in the formation of the harmonic spectrum. At real experimental conditions (i.e., ionized plasma, some excess of electron density, heterogeneous blueshift of the driving pulse, etc.) the "sufficient" phase matching conditions become depending on the redistribution of the driving spectrum along the pulse.

Figure 2.17b shows the sub-peaks at the blue sides of the harmonics generating in the Mn plasma, which are absent in the case of Zn plasma. They become even stronger than the main peaks (see the 25th and 27th harmonics). Probably, the phase-matching conditions for these components become optimal, though it is not clear why the following harmonics (e.g., H31 and H33) were entirely "cleaned" from the blue-shifted components. Probably, the difference between the conditions of the formation of two plasmas (Zn and Mn) at which one can achieve the highest harmonic yield can be assumed as a reason for the difference of these harmonic spectra.

Figures 2.13 and 2.14 show that the wavelength of the driving pulse was blue-shifted by 14 nm after SPM, while the blue shift of the 13th harmonic was only ~ 1 nm. The blue shift of harmonic depends on the order of this harmonic and blueshift of the driving radiation. There is also the transient blueshift, which obviously depends on the driving field intensity. The blueshift of the 13th harmonic could be roughly estimated as the blueshift of the driving field divided on the harmonic order (14 nm: 13 $\approx$ 1 nm). Different spectral components appearing during SPM convert into the harmonics since their intensity is large enough to participate in the generation of high-order harmonics. Moreover, in some cases, the energy contained in the blueshifted parts of harmonic becomes comparable with the main spectral component of the harmonic (Figs. 2.15b and 2.17b).

At peak laser intensity close to 10^{15} W cm^{-2}, the nonlinear addition to the refractive index of the plasma at 806 nm could be estimated as -2×10^{-3}. This can produce a strong refractive index gradient resulting in SPM and even self-defocusing of the femtosecond laser beam at high intensities. Next, at the given density of plasma with ~ 10% ionization, the coherence length for the 27th harmonic was estimated to be ~ 0.6 mm. Thus any increase in ionization beyond the optimal plasma condition would severely degrade the phase relations between interacting waves and lower the generation of high-order harmonics.

As it was mentioned, there are a few HHG studies using gases and plasmas, which have analyzed the variation of harmonic bandwidth due to spectral broadening of the driving radiation [85, 86]. However, most previous studies (excluding Ref. [92]) were addressed the conditions when laser broadening became induced by the relatively "mild" influence of the SPM on the harmonic properties. In discussed work, the observation and analysis of harmonic's characteristics were presented at the conditions when laser radiation significantly changes due to propagation in argon, filamentation in this medium and remarkable growth of spectral and spatial frequency components prior to and during the interaction with the nonlinear medium.

2.4 Summary

We discussed the strong incoherent emission from boron plasma at ~6 nm and coherent emission in the 15–53 nm spectral range. Particularly, we analyzed generation of the 806 nm radiation harmonics up to the 55th order ($\lambda = 14.6$ nm) and of tunable near-infrared radiation (1280–1400 nm) up to the 35th order ($\lambda = 38$ nm) during the propagation of femtosecond pulses through the boron plasma. The incoherent and coherent emissions from this target were compared and analyzed by considering the influence of ionic transitions on the harmonic conversion efficiency. We discussed different schemes of laser-plasma interaction (two-color pump of B plasma, generation of tunable harmonics in XUV range, formation of structured extended plasma, application of boron nanoparticles, laser-induced breakdown spectroscopy, overheating of boron targets, quasi-phase-matching of plasma harmonics, etc.) and we found major characteristics that would prevail during laser ablation and HHG. In particular cases (i.e., the plasma produced on the three boron-containing targets) we compared three characteristics: (a) stability of harmonic emission, (b) extension of harmonic cutoff, and (c) harmonic yield. The plasma produced on bulk boron can be considered as the best choice among the three ablated targets when it comes to harmonic emission stability and cut-off extension. On the other hand, the B NP plasma showed the best characteristics from the point of view of the highest yield. The fluencies of heating pulses for the HHG experiments were found different from those used in the plasma emission studies. In each case, the "optimal" fluence of heating pulses was chosen. This term refers to the maximal efficiency of harmonic generation or incoherent emission in the studied plasmas plumes. Furthermore, it was found that the conditions of efficient emission of 6 nm radiation from the over-ionized plasma were unsuitable for efficient HHG. We analyzed the plasma during ablation of boron-containing targets and found the optimal conditions allowing the generation of almost a single emission line ($\lambda = 6.03$ nm, B IV) in the 2–20 nm spectral range.

We discussed the detailed calculations of HHG spectra for Li, Li^+, and He using the Lewenstein model. It was found that the results from the experiment reported in [77], which describes HHG in Li plasma, are in good agreement with the chosen model estimations. For neutral lithium atom, the cutoff position at 43rd harmonic of 795 nm wave corresponded to simulated intensity 3×10^{14}–3.5×10^{14} W/cm^2. The difference between the theoretical calculations and the experimentally estimated value can be well explained by the measurement errors of laser intensity in the focal plane. Lithium ions will start to contribute only at very high intensities, and in general, it is not always beneficial to look for the material with higher ionization potentials for the extension of the HHG cutoff.

Finally, the spectral features of the harmonic generation using the laser pulses subjected to a considerable phase modulation were analyzed. The modulation was induced by transmitting laser pulses to be converted through the argon under the conditions of the formation of filaments. The filamentation was followed by the change in the spectral characteristics of laser pulses. Correspondingly, the harmonic spectra followed those of driving pulses. The blueshift and spectral broadening of fundamental and harmonic radiation also occurred during propagation through the over-dense laser-produced plasmas. The use of the spectrally blueshifted and broadened pulses for the HHG led to an increase in the spectral width of generated harmonics and their blueshift.

References

1. R.A. Ganeev, H. Singhal, P.A. Naik, U. Chakravarty, V. Arora, J.A. Chakera, R.A. Khan, M. Raghuramaiah, S.R. Kumbhare, R.P. Kushwaha, P.D. Gupta, Appl. Phys. B **87**, 243 (2007)
2. D.B. Milošević, J. Phys. B **40**, 3367 (2007)
3. R.A. Ganeev, M. Suzuki, M. Baba, H. Kuroda, Phys. Rev. A **76**, 023805 (2007)
4. L.B. Elouga Bom, F. Bouzid, F. Vidal, J.-C. Kieffer, T. Ozaki, J. Phys. B 41, 215401 (2008)
5. H. Singhal, V. Arora, B.S. Rao, P.A. Naik, U. Chakravarty, R.A. Khan, P.D. Gupta, Phys. Rev. A **79**, 023807 (2009)
6. I.A. Kulagin, T. Usmanov, Opt. Lett. **34**, 2616 (2009)
7. D. Cricchio, P.P. Corso, E. Fiordilino, G. Orlando, F. Persico, J. Phys. B **42**, 085404 (2009)
8. V. Strelkov, Phys. Rev. Lett. **104**, 123901 (2010)

9. D.B. Milošević, Phys. Rev. A **81**, 023802 (2010)
10. M.V. Frolov, N.L. Manakov, A.F. Starace, Phys. Rev. A **82**, 023424 (2010)
11. L.B. Elouga Bom, Y. Pertot, V.R. Bhardwaj, T. Ozaki, Opt. Express 19, 3077 (2011)
12. R.A. Ganeev, P.A. Naik, H. Singhal, J.A. Chakera, M. Kumar, M.P. Joshi, A.K. Srivastava, P.D. Gupta, Phys. Rev. A **83**, 013820 (2011)
13. Y. Pertot, L.B. Elouga Bom, V.R. Bhardwaj, T. Ozaki, Appl. Phys. Lett. 98, 101104 (2011)
14. L.B. Elouga Bom, Y. Pertot, V.R. Bhardwaj, T. Ozaki, Opt. Express 19, 3677 (2011)
15. M. Tudorovskaya, M. Lein, Phys. Rev. A **84**, 013430 (2011)
16. Y. Pertot, S. Chen, S.D. Khan, L.B. Elouga Bom, T. Ozaki, Z. Chang, J. Phys. B 45, 074017 (2012)
17. R.A. Ganeev, V.V. Strelkov, C. Hutchison, A. Zaïr, D. Kilbane, M.A. Khokhlova, J.P. Marangos, Phys. Rev. A **85**, 023832 (2012)
18. S. Haessler, L.B. Elouga Bom, O. Gobert, J.-F. Hergott, F. Lepetit, M. Perdrix, B. Carré, T. Ozaki, P. Salières, J. Phys. B 45, 074012 (2012)
19. S. Haessler, V. Strelkov, L.B. Elouga Bom, M. Khokhlova, O. Gobert, J.-F. Hergott, F. Lepetit, M. Perdrix, T. Ozaki, P. Salières, New J. Phys. 15, 013051 (2013)
20. M. Kumar, H. Singhal, J.A. Chakera, P.A. Naik, R.A. Khan, P.D. Gupta, J. Appl. Phys. **114**, 033112 (2013)
21. H. Singhal, P.A. Naik, M. Kumar, J.A. Chakera, P.D. Gupta, J. Appl. Phys. **115**, 033104 (2014)
22. N. Rosenthal, G. Marcus, Phys. Rev. Lett. **115**, 133901 (2015)
23. M.A. Fareed, N. Thiré, S. Mondal, B.E. Schmidt, F. Légaré, T. Ozaki, Appl. Phys. Lett. **108**, 124104 (2016)
24. M.A. Fareed, S. Mondal, Y. Pertot, T. Ozaki, J. Phys. B At. Mol. Opt. Phys. **49**, 035604 (2016)
25. M.A. Fareed, V.V. Strelkov, N. Thiré, S. Mondal, B.E. Schmidt, F. Légaré, T. Ozaki, Nature Commun. **8**, 16061 (2017)
26. M. Wöstmann, L. Splitthoff, H. Zacharias, Opt. Express **26**, 14524 (2018)
27. Z. Abdelrahman, M.A. Khohlova, D.J. Walke, T. Witting, A. Zair, V.V. Strelkov, J.P. Marangos, J.W.G. Tisch, Opt. Express **26**, 15745 (2018)
28. M.A. Fareed, V.V. Strelkov, M. Singh, N. Thiré, S. Mondal, B.E. Schmidt, F. Légaré, T. Ozaki, Phys. Rev. Lett. **121**, 023201 (2018)
29. I.S. Wahyutama, T. Sato, K.L. Ishikawa, Phys. Rev. A **99**, 063420 (2019)
30. M. Kumar, H. Singhal, J.A. Chakera, J. Appl. Phys. **125**, 155902 (2019)
31. M. Oujja, A. Benítez-Canete, M. Sanz, I. Lopez-Quintas, M. Martín, R. de Nalda, M. Castillejo, Appl. Surf. Sci. **336**, 53 (2015)
32. S.A. Akhmanov, I.M. Bayanov, S.V. Gaponov, V.M. Gordienko, M.S. Dzhidzhoev, V.V. Ivanov, S.V. Krayushkin, S.A. Magnitsky, V.T. Platonenko, Y.Y. Platonov, Y.V. Ponomarev, A.B. Savel'ev, N.N. Salastchenko, E.V. Slobodchikov, A.P. Tarasevich, Bull. Russ. Acad. Sci. Phys. 56, 1372 (1992)
33. M.S. Dzhidzhoev, V.M. Gordienko, V.V. Kolchin, S.A. Magnitsky, V.T. Platonenko, A.B. Savel'ev, A.P. Tarasevitch, J. Opt. Soc. Am. B 13, 193 (1996)
34. G. Arai, H. Hara, T. Hatano, T. Ejima, W. Jiang, H. Ohashi, S. Namba, A. Sunahara, A. Sasaki, M. Nishikino, G. O'Sullivan, T. Higashiguchi, Opt. Express **26**, 27748 (2018)
35. J. Osterholz, F. Brandl, T. Fischer, D. Hemmers, M. Cerchez, G. Pretzler, O. Willi, S.J. Rose, Phys. Rev. Lett. **96**, 085002 (2006)
36. H. Kondo, T. Tomie, H. Shimizu, Appl. Phys. Lett. **72**, 2668 (1998)
37. R. Ganeev, M. Suzuki, M. Baba, H. Kuroda, T. Ozaki, Opt. Lett. **30**, 768 (2005)
38. Y. Akiyama, K. Midorikawa, Y. Matsunawa, Y. Nagata, M. Obara, H. Tashiro, K. Toyoda, Phys. Rev. Lett. **69**, 2176 (1992)
39. M. Iqbal, R.A. Ganeev, G.S. Boltaev, V.V. Kim, A.S. Alnaser, Eur. Phys. J. D **74**, 28 (2020)
40. H.M. Milchberg, S.J. Mcnaught, E. Parra, Phys. Rev. E **64**, 056402 (2001)
41. V.P. Krainov, M.B. Smirnov, Phys. Rep. **370**, 237 (2002)
42. M. Kumar, R. Singh, U. Verma, Laser Part. Beams **32**, 9 (2014)
43. R.A. Ganeev, M. Baba, M. Suzuki, H. Kuroda, J. Appl. Phys. **99**, 103303 (2006)
44. A.M. Rubenchik, M.D. Feit, M.D. Perry, J.T. Larsen, Appl. Surf. Sci. **129**, 193 (1998)

45. L.B. Elouga Bom, J.-C. Kieffer, R.A. Ganeev, M. Suzuki, H. Kuroda, T. Ozaki, Phys. Rev. A 75, 033804 (2007)
46. R.A. Ganeev, C. Hutchison, T. Witting, F. Frank, W.A. Okell, A. Zaïr, S. Weber, P.V. Redkin, D.Y. Lei, T. Roschuk, S.A. Maier, I. López-Quintás, M. Martín, M. Castillejo, J.W.G. Tisch, J.P. Marangos, J. Phys. B **45**, 165402 (2012)
47. J. Roth, F. Géahler, H.-R. Trebin, Int. J. Mod. Phys. C **11**, 317 (2000)
48. M. Wöstmann, P.V. Redkin, J. Zheng, H. Witte, R.A. Ganeev, H. Zacharias, Appl. Phys. B **120**(1), 17 (2015)
49. R.A. Ganeev, J. Zheng, M. Wöstmann, H. Witte, P.V. Redkin, H. Zacharias, Eur. Phys. J. D **68**, 325 (2014)
50. R.A. Ganeev, M. Suzuki, H. Kuroda, Phys. Rev. A **89**, 033821 (2014)
51. R.A. Ganeev, V. Tosa, K. Kovács, M. Suzuki, S. Yoneya, H. Kuroda, Phys. Rev. A **91**, 043823 (2015)
52. V.V. Strelkov, R.A. Ganeev, Opt. Express **25**, 21068 (2017)
53. J. Seres, V.S. Yakovlev, E. Seres, C.H. Streli, P. Wobrauschek, C.H. Spielmann, F. Krausz, Nature Phys. **3**, 878 (2007)
54. A. Pirri, C. Corsi, M. Bellini, Phys. Rev. A **78**, 011801 (2008)
55. V. Tosa, V.S. Yakovlev, F. Krausz, New J. Phys. **10**, 025016 (2008)
56. T. Fok, Ł Węgrzyński, M. Kozlova, J. Nejdl, P.W. Wachulak, R. Jarocki, A. Bartnik, H. Fiedorovicz, Photonics Lett. Pol. **6**, 14 (2014)
57. M. Negro, C. Vozzi, K. Kovacs, C. Altucci, R. Velotta, F. Frassetto, L. Poletto, P. Villoresi, S. De Silvestri, V. Tosa, Laser Phys. Lett. **8**, 875 (2011)
58. E. Takahashi, P. Lan, O.D. Mücke, Y. Nabekawa, K. Midorikawa, Nature Commun. **4**, 2691 (2013)
59. M. Oujja, I. Lopez Quintas, A. Benitez Canete, R. de Nalda, M. Castillejo, A.Benítez Cañete, Appl. Surf. Sci. 392, 572 (2017)
60. M. Oujja, J.G. Izquierdo, L. Banares, R. de Nalda, M. Castillejo, Phys. Chem. Chem. Phys. 20, 16956 (2018)
61. R.A. Ganeev, T. Witting, C. Hutchison, F. Frank, P.V. Redkin, W.A. Okell, D.Y. Lei, T. Roschuk, S.A. Maier, J.P. Marangos, J.W.G. Tisch, Phys. Rev. A **85**, 015807 (2012)
62. R.A. Ganeev, P.A. Naik, H. Singhal, J.A. Chakera, M. Kumar, U. Chakravarty, P.D. Gupta, Opt. Commun. **285**, 2934 (2012)
63. R.A. Ganeev, M. Suzuki, M. Baba, H. Kuroda, Eur. Phys. J. D **37**, 255 (2006)
64. E. Goulielmakis, M. Schultze, M. Hofstetter, V.S. Yakovlev, J. Gagnon, M. Uiberacker, A.L. Aquila, E.M. Gullikson, D.T. Attwood, R. Kienberger, F. Krausz, U. Kleineberg, Science **320**, 1614 (2008)
65. K. Zhao, Q. Zhang, M. Chini, Y. Wu, X. Wang, Z. Chang, Opt. Lett. **37**, 3891 (2012)
66. R. López-Martens, K. Varjú, P. Johnsson, J. Mauritsson, Y. Mairesse, P. Salières, M.B. Gaarde, K.J. Schafer, A. Persson, S. Svanberg, Claes-GöranWahlström and Anne L'Huillier. Phys. Rev. Lett. **94**, 033001 (2005)
67. M. Lein, N. Hay, R. Velotta, J.P. Marangos, P.L. Knight, Phys. Rev. Lett. **88**, 183903 (2002)
68. M.V. Frolov, N.L. Manakov, T.S. Sarantseva, A.F. Starace, J. Phys. B **42**, 035601 (2009)
69. X. Liu, X. Zhu, P. Lan, X. Zhang, D. Wang, Q. Zhang, P. Lu, Phys. Rev. A **95**, 063419 (2017)
70. S. Odžak, D.B. Milošević, Phys. Rev. A **79**, 023414 (2009)
71. T.S. Sarantseva, M.V. Frolov, N.L. Manakov, M.Y. Ivanov, A.F. Starace, J. Phys. B **46**, 231001 (2013)
72. D.B. Milosevic, J. Opt. Soc. Am. B **23**, 308 (2006)
73. F. Wang, L. He, C. Zhai, W. Shi, Q. Zhang, P. Lan, P. Lu, Phys. Rev. A **92**, 063839 (2015)
74. M.V. Frolov, N.L. Manakov, T.S. Sarantseva, A.F. Starace, Phys. Rev. A **86**, 063406 (2012)
75. P.B. Corkum, Phys. Rev. Lett. **71**, 1994 (1993)
76. M. Lewenstein, P. Balcou, M.Y. Ivanov, A. L'Huillier, P.B. Corkum, Phys. Rev. A **49**, 2117 (1994)
77. M. Suzuki, M. Baba, R.A. Ganeev, L.B. Elouga Bom, H. Kuroda, T. Ozaki, J. Phys. B 45, 065601 (2012)

78. V.V. Kim, R.A. Ganeev, G.S. Boltaev, M. Iqbal, A.S. Alnaser, Opt. Lett. **44**, 3693 (2019)
79. L.V. Keldysh, Sov. Phys. JETP **20**, 1307 (1965)
80. B. Dźevad, S.T. Howard, Phys. Scr. **39**, 226 (1989)
81. T. Koga, K. Kanayama, S. Watanabe, A.J. Thakkar, Int. J. Quantum Chem. **71**, 491 (1999)
82. W. Becker, F. Grasbon, R. Kopold, D. Milo, D. Milošević, G. Paulus, H. Walther, Adv. At. Mol. Opt. Phys. **48**, 35 (2002)
83. S. Augst, D. Strickland, D.D. Meyerhofer, S.L. Chin, J.H. Eberly, Phys. Rev. Lett. **63**, 2212 (1989)
84. R.A. Ganeev, J. Opt. Soc. Am. B **31**, 2221 (2014)
85. V. Tosa, H.T. Kim, I.J. Kim, C.H. Nam, Phys. Rev. A 71, 063 808 (2005)
86. R.A. Ganeev, H. Singhal, P.A. Naik, J.A. Chakera, M. Tayyab, M. Baba, H. Kuroda, P.D. Gupta, J. Opt. Soc. Am. B **26**, 2143 (2009)
87. R.A. Ganeev, Appl. Phys. A **123**, 294 (2017)
88. R.A. Ganeev, Z. Wang, P. Lan, P. Lu, M. Suzuki, H. Kuroda, Phys. Rev. A **93**, 043848 (2016)
89. R.A. Ganeev, M. Suzuki, S. Yoneya, H. Kuroda, Appl. Phys. B **121**, 307 (2015)
90. P. Béjot, J. Kasparian, S. Henin, V. Loriot, T. Vieillard, E. Hertz, O. Faucher, B. Lavorel, J.-P. Wolf, Phys. Rev. Lett. **104**, 103903 (2010)
91. R.A. Ganeev, P.A. Naik, H. Singhal, J.A. Chakera, P.D. Gupta, Opt. Lett. **32**, 65 (2007)
92. R.A. Ganeev, H. Kuroda, Appl. Phys. Lett. **95**, 201117 (2009)

Chapter 3
High-Order Harmonic Generation in Plasmas Using High-Pulse Repetition Rate Lasers

3.1 Time-Dependent Optimization of Laser-Produced Molecular Plasmas Through High-Order Harmonic Generation

High-order harmonic generation (HHG) of ultra-short laser pulses in an isotropic medium is a nonlinear optical method for the generation of coherent extreme ultraviolet (XUV) radiation. Various theoretical and experimental studies have been devoted to the increase of HHG conversion efficiency [1, 2]. HHG in gases, laser-induced plasmas (LIP), and solid surfaces demonstrate the possibility of attosecond pulses generation in the XUV range [3]. Early studies of HHG in molecular species were considered as a spectroscopy tool [4]. The analysis of HHG spectra can be useful for indication of the alignment of molecules at different angles which show the signatures of the molecular structure. HHG spectroscopy can also be considered as a method for the analysis of complex biomolecules. The ablation of the molecules is also important for the enhancement of HHG efficiency [5].

The low-excited and low-ionized laser plasma plumes can be considered as the suitable media for efficient generation of high-order harmonics (HH) allowing demonstration of the resonance enhancement of single harmonic, quasi-phase-matching in multi-jet plasmas, and nanoparticle-enhanced harmonics [6–15]. Various LIPs were considered as the nonlinear optical media for the generation of the low- and high-order harmonics of picosecond and femtosecond driving pulses [16]. The harmonic cutoff in the case of the targets with different atomic numbers shows a strong dependence on the delay between the heating and driving pulses. In those experiments, the optimization of plasma parameters and delay between ablating and driving pulses led to the growth of harmonic yield [17]. The correlation between those results and the appearance of a second component of the third harmonic (TH) at longer delays confirmed the presence of aggregates in the graphite plasma and established the analysis of their presence in LPPs with excellent spatiotemporal resolution through the harmonic generation [18].

R. A. Ganeev, *High-Order Harmonics Generation in Plasmas*, Springer Series on Atomic, Optical, and Plasma Physics 122, https://doi.org/10.1007/978-3-031-09040-0_3

There are a few techniques, which were used for the analysis of the composition and dynamics of LIPs. The optical emission spectroscopy (OES) and time-of-flight mass spectrometry (TOF–MS) stand as some of the most used methods of plasma characterization [19]. The laser-induced breakdown spectroscopy (LIBS) allows indicating and distinguishing different plasma species by their emission spectra during the excitation of targets by nanosecond pulses [20]. The information retrieved by OES from the spontaneous emission of LIPs allows determining important parameters of plasma formation, such as the electron temperature and density, while TOF–MS provides the information about the presence of ions in LIP [21].

In this section, we analyze the dynamics of plasma formation using the molecular bulk targets possessing different molar masses at the conditions of maximal HH yield along a broad range of delays between the heating nanosecond and driving 1 kHz femtosecond pulses [22]. We discuss the optimization of HH yield, which strongly depends on the delay between heating and driving pulses and correspondingly depends on the molar weights of different species presented in LIP. Different applications of this method of the nonlinear optical characterization of LIPs are described. The experimental setup of HHG in LIPs using a two-laser configuration is shown in Fig. 3.1. The key feature of the two-laser configuration is the possibility to control the delay time between heating nanosecond pulses and driving femtosecond pulses over long-range using the digital generator. The delay between the heating and driving pulses was monitored on an oscilloscope using a fast photodiode. The radiation of Ti:Sapphire laser ($\lambda = 800$ nm, $\tau = 40$ fs, 1 kHz) was used for interaction with the laser-plasma produced by Nd:YAG laser (1064 nm, 5 ns, 10 Hz) leading to the generation of harmonics.

The laser-plasma plumes were created on the surfaces of different targets using the nanosecond Nd:YAG laser. The diameter of the ablation zone was 0.5 mm. The fluence of 10 mJ, 5 ns heating pulses on the target surface was 4 J/cm^2. The position of the target with regard to the optical axis of propagation of the driving femtosecond pulses was controlled by the micro-translating stage to achieve the maximal efficiency in the conversion of the driving pulse energy to the harmonics yield. The step of the micro-translating stage was equal to 0.1 mm. The femtosecond driving pulses were

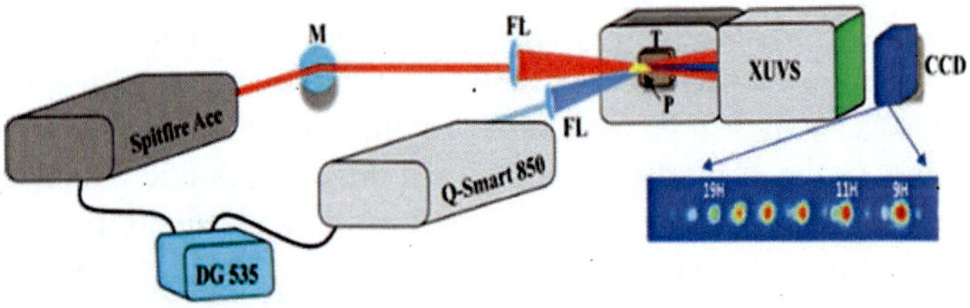

Fig. 3.1 Setup for high-order harmonic generation using a synchronized two-laser system. Spitfire Ace, femtosecond laser; Q-Smart 850, nanosecond laser; DG535, delay generator; M, mirror; FL; focusing lenses; T, targets; P, plasmas; XUVS, XUV spectrometer; CCD, CCD camera. Inset shows the raw image of HHG spectrum generated in ZnO plasma. Reproduced from [22] with permission from American Institute of Physics

focused by a 300 mm focal length lens at a distance of ~1 mm above the surface of solid targets which was optimal to achieve a relatively high yield of HH generated in the plasma plumes. An electronic delay between the heating (nanosecond) and driving (femtosecond) pulses was varied between 0 and 5000 ns though it can be also extended up to 50 μs to determine the influence of large aggregates in LIPs on HHG. The harmonic spectra were analyzed using an extreme ultraviolet spectrometer.

B_4C, ZnO, GaP, ZnS, GaAs, CdS, Ag_2S, MoS_2, and $CH_3NH_3PbBr_3$ samples with different molar weights were used as the targets for ablation and plasma formation for HHG. The sizes of all targets were $5 \times 5 \times 2$ cm^3. The ablation of targets was carried out in the vacuum chamber at 2.5×10^{-4} Pa. The targets were chosen by the difference of their molar weight to analyze the influence of different components on HHG efficiency.

The velocity of plasma species depends on the atomic or molecular weight of the particles. Figure 3.2 shows spectra of the harmonics generated in the plasma plumes produced on the B_4C and ZnO targets ablated using 5 ns, 1064 nm pulses. In the case of ZnO plasma (84 g/mol), the harmonics up to the 23rd order of 800 nm, 40 fs

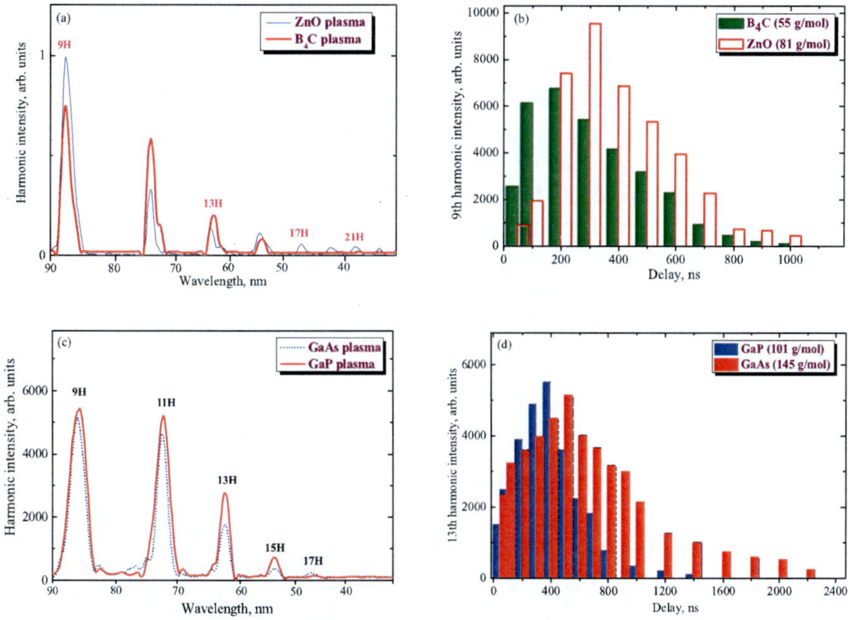

Fig. 3.2 **a** Spectra of the harmonics generated in B_4C and ZnO plasmas. **b** Variations of intensity of the 9th harmonic generated in B_4C and ZnO plasmas at different delays between the heating 1064 nm, 5 ns pulses and driving 800 nm, 40 pulses. The maximal yields of the 9th harmonic were obtained at 200 and 300 ns delays for B_4C and ZnO plasmas, respectively. **c** Harmonic spectra obtained from GaP and GaAs plasmas. **d** Dependences of the intensity of 13th harmonic on the delay between the heating and driving pulses in the cases of GaP and GaAs plasmas. In these cases, the longer optimal delay time corresponded to heavier molecules (GaAs). Reproduced from [22] with permission from American Institute of Physics

pulses were achieved. The plasma containing B_4C (55 g/mol) allowed observing the harmonics up to the 15th order at the same experimental conditions of ablation and driving beam propagation (Fig. 3.2a).

The variation of harmonic intensity depends on the delay between the heating nanosecond and driving femtosecond pulses, as shown in Fig. 3.2b. The optimal delays allowing the generation of the strongest 9th harmonic were found to be 300 and 200 ns for ZnO and B_4C plasmas. One can see the shift of the maximal yield of the 9th harmonic towards the longer delay in the case of the plasma containing heavier particles (ZnO) with regards to lighter ones (B_4C). Other harmonics from those plasmas also showed analogous dependence.

Similar dependences of the optimal delay on the molecular weight of the components of plasmas were observed in the cases of different groups of ablating targets. Figure 3.2c and d show the comparative harmonic spectra and delay dependences of the 13th harmonic for GaP (101 g/mol) and GaAs (145 g/mol) plasma plumes. The same delay-dependent optimization of harmonics was obtained in the case of ZnS (97 g/mol) and CdS (144 g/mol) plasmas. The delay dependences of the lower-order harmonics generating in the plasmas produced during ablation of ZnS were also analyzed in [23]. For the 3rd–7th orders, the maximum harmonic yields were obtained at around 250 ns, which means the generation of harmonics from the ZnS molecules.

Figure 3.3a shows the harmonic spectrum generated in the silver sulfide LIP. This plasma allowed the generation of the harmonics up to the 39th order. Here we also analyzed the variation of the intensity of the 11th harmonic generated in this plasma by changing the delay between heating and driving pulses (Fig. 3.3b). The maximum yields of the 11th harmonic were observed at 140 and 700 ns delays. These delays correspond to the contributions of Ag (108 g/mol) ions and neutrals and Ag_2S (248 g/mol) ions and molecules to the harmonic yield. During ablation of Ag_2S target at relatively large fluencies of heating pulses, the molecules can be

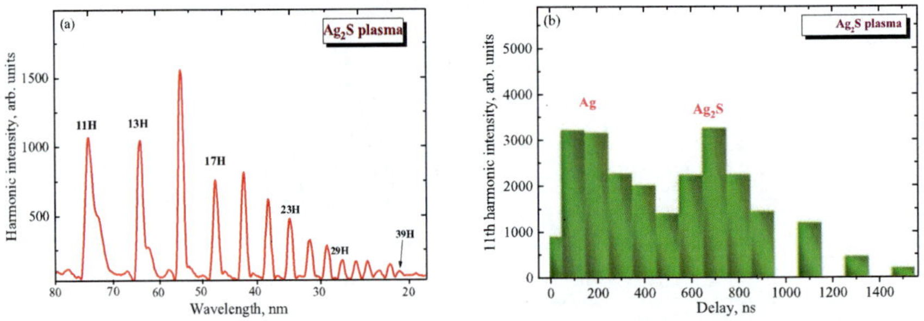

Fig. 3.3 **a** Harmonic spectrum generated in LIP produced on the Ag_2S target by 1064 nm, 5 ns pulses. **b** Variations of 11th harmonic intensity at different delays between the heating and driving pulses in the Ag_2S LIP. Two peaks of the maximal intensity of 11th harmonic were revealed in the case of HHG in Ag_2S target ablation (at ~ 140 and 700 ns delays). Reproduced from [22] with permission from American Institute of Physics

partially disintegrated to form the groups of Ag and Ag_2S plasma clouds, which arrived at different times at the area of interaction with the driving pulses.

The optimization of HHG strongly depends on the position of the sample with regard to the optical axis of the driving pulses. The delay dependence for different positions of targets has been discussed in [5]. The intensity of the harmonics generated from atoms rapidly decreased with the growth of the distance from the optical axis to the target. For each distance, the optimum delay between ablating laser pulse and fundamental pulse allowing the generation of strongest harmonics was found. Meanwhile, clusters and nanoaggregates can contribute to HH efficiency at notably longer delays (up to 40 μs).

Here we discuss the variation of the optimal delay at the same distance for different targets. The analysis of HHG allows determining the optimal delay during laser ablation using the nanosecond heating pulses. The studies of the ablated plasma expansion using pulses of different duration have shown the variation of the sizes of plasma in the case of ablation by the nanosecond pulses [24]. In the case of femtosecond pulses, the dimensions of plasma plumes remain unchanged. Typical spectral- and time-integrated two-dimensional images of the nanosecond and femtosecond induced LIPs recorded using ICCD have clearly demonstrated the difference in the hydrodynamics of these two plasma plumes. Under similar ablation conditions, the spatial extension of the femtosecond LIP in the direction along the normal to the target was found to be longer compared with the nanosecond pulses induced LIP. This behavior can influence the conversion efficiency of the HH, which directly depends on the concentration of the ablated plasma species.

The velocities of the components of the ablated plume can be estimated as *d/t*, where *t* is the optimal delay corresponding to the maximal harmonic yield and *d* is the distance between the target surface and the axis of femtosecond beam propagation. During the ablation using stronger pulses the larger amount of the components of disintegrated molecules leaves the target surface. The produced atoms faster eject from surfaces of the target and reach the driving beam at a shorter time (Fig. 3.3b). The variable fluence of heating pulses can change the plasma dynamics of targets [25]. This behavior has earlier been analyzed for ejected Sn particles, which possessed 10 km/s velocities in the case of plasma formation using the nanosecond pulses [25]. In the discussed case, 1064 nm, 5 ns heating pulses were also used as heating pulses and the plasma plumes mainly consisted of the molecules, neutral atoms, and single charged particles.

Based on the kinetic theory, the velocity of particles can be described as [26]:

$$v = (2E/m)^{0.5} \tag{3.1}$$

Here *E* is the kinetic energy of particles and *m* is the mass of particles (atoms or molecules). Correspondingly, the arrival time of the plasma at the distance *d* above the target surface is described by:

$$t \propto d/v \propto (m)^{0.5} \tag{3.2}$$

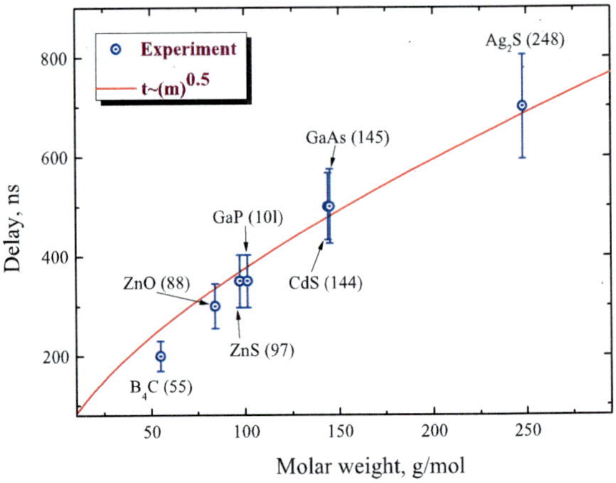

Fig. 3.4 Variation of the optimal delay between the heating and driving pulses on the molar weight of ablated molecules (empty circles). Theoretical fit was plotted based on the kinetic theory of particles distribution by velocity (solid line). Reproduced from [22] with permission from American Institute of Physics

Figure 3.4 shows the dependence of the optimal delay between the heating and driving pulses on the molecular weight of the studied molecular species. Here the theoretical $t \propto (m)^{0.5}$ dependence is also presented, which shows the reasonable coincidence with the experimental data taking into account the accuracy in the determination of the optimal delay (30%). The fitting data were plotted for ≤ 1 mm distance between the target surface and the optical axis of the driving femtosecond laser pulses.

Some deviation from the theoretical dependence could be attributed to the larger laser fluence required for evaporation of the species possessing smaller molar weight [26–28]. The plasma components possessing higher velocity reach the interaction zone with driving pulses during a shorter period. In some cases, the formation of clusters can also influence the velocity of plasma species during laser ablation using nanosecond pulses.

The developed technique is based on the two-laser approach when two lasers emit the pulses synchronized with each other. In that case, the application of a digital delay generator allowed the tuning of the delay between heating and driving pulses by the electronic method rather than the optical one. Moreover, this method allows significantly broadening the range of delay variations (up to tens of microseconds). This notably larger range of delay variations becomes extremely useful once one considers the analysis of the movement of not only molecular species but also the larger-sized species (clusters, quantum dots, and even large nanoparticles) being ablated on the surfaces of targets. The optical method of delay variations used in previous studies of HHG in laser-produced plasmas is useless in that case. The additional advantage of the proposed method is the maintenance of the same characteristics of the driving beam entering the plasma volume.

The advancement presented in this study is the development of the above approach, which is not restricted only with the technical amendment yet demonstrated by other researchers but rather uses for the analysis of the dynamics of different morphological components during laser ablation and their role in the conversion of IR femtosecond pulses towards the XUV. The timeliness of this approach is related to the recent studies of the role of different multiparticle species during the enhanced harmonics generation in plasma [22]. Notice the absence of studies on the microsecond timescale of delays assuming the hypothesis of the slower velocities for larger-sized species. The only way to investigate this optimization is the application of the electronically-driven delay variations. The importance of the suggested approach can be further emphasized in the case of multi-particle systems.

Apart from this novelty, the analysis of ablated molecular species for HHG was demonstrated. The example is presented in Fig. 3.3b, which shows that, at relatively strong ablation, the disintegration of molecular species leads to optimization of HHG at different delays attributed to the specific components of those molecules.

The well-developed three-step model of HHG in gases and plasmas [29–31] can describe some peculiarities of the discussed studies. The recombination time of excited electrons depends on the wavelength of the driving femtosecond pulses. The resonance enhancement of a single high-order harmonic relies on the transitions of ions or atoms and molecules possessing strong oscillator strength close to the wavelength of some harmonics. This approach has been earlier summarized as a four-step model for the analysis of the observed peculiarities of single harmonic enhancement in different plasmas [12, 32, 33]. Meanwhile, in discussed case, the optimization of HHG was achieved through the analysis of the dynamics of plasma spreading rather than using the fundamental processes of HHG and resonance enhancement.

The discussed approach allowing the manipulation of the delay between pulses by electronic rather than optical methods opens up new opportunities in the analysis of the spreading of different components of plasma. It also gives clues on how and when large aggregates, like clusters and quantum dots, can be involved in HHG. The two-laser approach may open the possibility to electronically control the delay time between ablating and driving pulses in the case of the large nanoparticles spreading out from the ablated surfaces. Notice that the details of the theoretical studies of velocities variations of the species ablated from the materials possessing different molar weights were reported in [24–28]. Thus the developed experimental approach allows the broadening of the area of plasma components studies through the nonlinear optical methods at longer timescales from the beginning of plasma formation during laser ablation.

The importance of the new technique is justified with its application for the studies of, for instance, the resonance processes in molecular plasmas, as well as the analysis of perovskite quantum dots and large molecules. Since the developed approach allows the consideration of slowly moving heavy plasma species, alongside the fast-moving atoms, the comparative studies of light and heavy plasma components offer the analysis of different physical processes in plasmas. Among them, one can consider the resonance enhancement of single harmonic from variable plasma constituents, comparison of the efficiency of quantum dot harmonic emitters possessing different masses, etc.

The first such example is related to the resonance processes during HHG leading to enhancement of the single harmonic over the neighboring ones. The mechanism of resonance-induced enhancement of single harmonic described by four-step model [32] allowed determining the conditions at which this process occurs in the metal

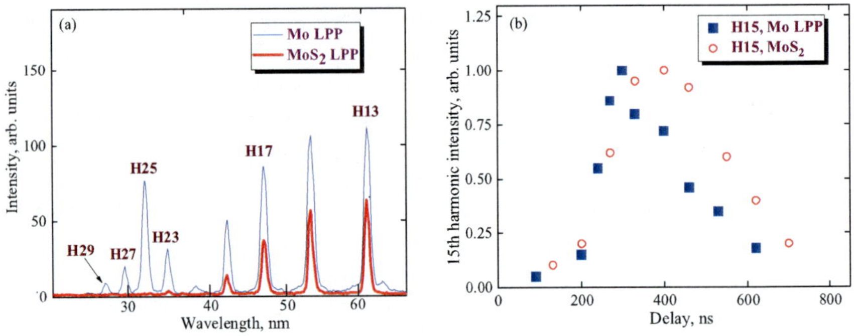

Fig. 3.5 **a** Harmonic spectra from Mo (thin curve) and MoS_2 (thick curve) LIPs. **b** Normalized 15th harmonic yields from Mo (filled squares) and MoS_2 (empty circles) at different delays between heating and driving pulses. Reproduced from [22] with permission from American Institute of Physics

plasmas such as In, Cr, Mn, etc. [16]. Earlier, only GaAs molecular plasma has demonstrated a similar process when the 21st harmonic of the 800 nm pump showed a stronger yield than the neighboring harmonics [34]. The discussed experiments with MoS_2 plasma did not show the resonance enhancement of a single harmonic during the formation of plasma at a relatively small fluence of heating pulses (2 J cm^{-2}; Fig. 3.5a, thick curve). Meanwhile, similar experiments using atomic target (Mo) allowed observation of resonantly enhanced 25th harmonic, which was notably stronger than 21st, 23rd, 27th, and 29th harmonics (Fig. 3.5a, thin curve).

This process was attributed to the closeness of the wavelength of the 25th harmonic and the wavelength of Mo ionic transition possessing large oscillator strength. The absence of a similar effect in MoS_2 plasma was probably caused by the shift of the wavelength of above transition in this molecule out from the wavelength of 25th harmonic ($\lambda = 32$ nm). The application of a two-laser scheme allowed achieving maximal harmonic yields from Mo and MoS_2 plasmas at the delays of femtosecond pulses ~300 and 400 ns from the beginning of the ablation of those materials (Fig. 3.5b). The square root of the ratio of molar and atomic masses of MoS_2 and Mo (~1.3) approximately corresponds to the ratio of the optimal delays for these plasmas, thus confirming the above consideration of this process using different species. The ablation of the molecular target at a larger fluence of heating pulses (4 J cm^{-2}) led to broadening of the delay-dependent curve for MoS_2, thus indicating the partial disintegration of this molecule on Mo and S components. The latter component does not notably influence the harmonic yield, while the appearance of Mo ions allowed observation of the enhanced 25th harmonic, similarly to the ablation of bulk Mo target [Fig. 3.5a, thick curve]. Thus the electronically driven variations of the delay allowed analyzing the processes attributed to the modification of the ionic transitions of plasma components leading to the enhancement of some harmonics, contrary to the plateau-like shapes of harmonic distribution commonly observed during plasma HHG studies.

The second example of the application of the developed two-laser HHG scheme is related to the analysis of the suitability of very large aggregates for a harmonic generation. Below we analyze the delay dependence of HHG in the plasmas containing the multi-component $CH_3NH_3PbBr_3$ perovskite quantum dots produced on the surface of thin perovskite films by 1064 nm, 5 ns heating pulses and appeared on the path of femtosecond driving pulses at notably longer delays from the beginning of ablation ($\geq$5 μs). The 150–nm thick perovskite film was prepared using the two-step method [35]. Figure 3.6a shows spectra of the harmonics generated in the plasmas containing $CH_3NH_3PbBr_3$ perovskite species appearing in the area of femtosecond pulses propagation at different delays. The analysis of plasma components appearing during ablation of thin films was reported elsewhere [36]. In the discussed case, the two-laser system was used for the analysis of the contribution of different multi-atomic perovskite species to the harmonic yield. The strongest harmonics were observed at 800 ns delay (thick blue curve), while the relatively weaker yield was obtained at 6000 ns delay time (thin red curve). Figure 3.6b shows the variations of 11th harmonic yield from perovskite LIP at different delays between two laser pulses. The increase of delay between heating and driving pulses led to a decrease of harmonic yield, which was observed up to 20 μs. No harmonics were observed at the delays exceeding 25 μs, probably due to the notable decrease of the concentration of LIP.

In previous studies, the restriction of optically modified delays did not allow analyzing the temporally modified contribution of the molecules and their components to the high-order harmonic spectra, particularly, the variations of resonance enhancement for different plasma components. The demonstrated examples of the variable contribution of molecules to high-order harmonic spectra reveal new capabilities of the electronic delay technique.

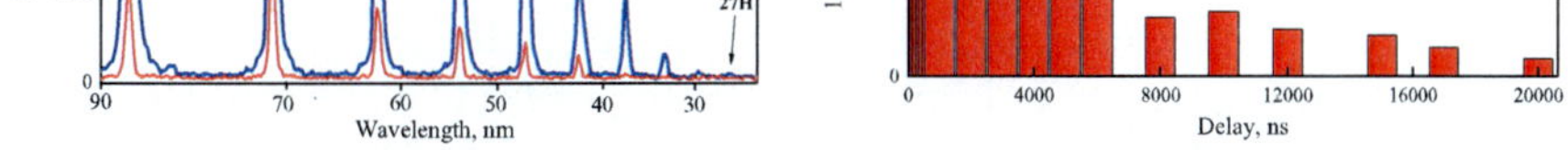

Fig. 3.6 **a** Harmonic spectra generated during propagation of 40 fs pulses through $CH_3NH_3PbBr_3$ plasma produced on the surface of thin film. Thick and thin curves correspond to 800 and 6000 ns delays. **b** Variations of 11th harmonic intensity at different delays between the heating and driving pulses in the $CH_3NH_3PbBr_3$ LIP. The maximal intensity of 11th harmonic was obtained at 800 ns delay, while the emission of this harmonic was observed along a broad time scale (>20 μs) due to contribution from the large molecular aggregates of perovskites. Reproduced from [22] with permission from American Institute of Physics

3.2 High-Order Harmonics Generation in the Plasmas Produced on Different Rotating Targets During Ablation Using 1 kHz and 100 kHz Lasers

High-order harmonic generation during propagation of ultrashort laser pulses through gases [37] and laser plasmas [38] is a well-developed method for the formation of coherent extreme ultraviolet sources. LIPs are considered as media that allow the observation of different physical and optical processes during HHG, which one can hardly realize in the gaseous media. Among them are the resonance-induced enhancement of single harmonic [39], nanoparticle-induced growth of HHG conversion efficiency [40], quasi-phase-matching (QPM) in the spatially modulated plasma allowing the enhancement of groups of harmonics in different XUV ranges [41], etc.

Despite these advantages of HHG in LIPs, the application of high-pulse repetition rate lasers gave rise to issues that pertain to the insufficient stability of the formed plasma plumes, which causes decay of harmonic yield after a few hundred thousand shots on the target surface. This problem does not exist with 10 Hz class lasers. However, when using kHz class driving lasers the formation of craters on the surface of the studied materials, during ablation of the same spot of the target, becomes inevitable. The up-and-down movement of the target can, to some extent, be used for resolving this issue of the instability of plasma characteristics, but cannot alone fully resolve it.

The rotation of the target can notably diminish this problem in the case of 1 kHz lasers. In such a case, the formation of stable weakly ionized plasmas from laser ablation of solid targets using a 1 kHz pulse repetition rate lasers can be used for stable high-order harmonic generation from plasma plumes [42]. The developed technique allowed achieving the long-term stability of harmonic yield.

The justification in finding optimal procedures for improving the stability of the harmonics generating in LIPs is related to the observations of the advanced properties of plasma harmonics over gas harmonics [43]. Particularly, the HHG conversion efficiency in some plasma plumes was one order of magnitude stronger compared with in-gas HHG efficiency [44]. Earlier reported analysis of plasma formation by pulses of different duration [45] also pointed out the importance of determining the optimal conditions of laser-target interaction in the case of high-pulse repetition rate sources of coherent IR radiation. All these requirements for the stability of plasma play an exceptionally crucial role in the case of the application of a new generation of lasers operating at two orders of magnitude larger pulse repetition rates (100 kHz and higher).

Here we discuss the application of rotating targets for LIP formation and HHG at two ranges of pulse repetition rates (1 and 100 kHz) [46]. The latter case has never been studied. We analyze the high average flux of short-wavelength radiation using 100 kHz pulses while comparing the plasma formation conditions at different pulse repetition rates. The advantages of shorter pulses induced ablation for the improvement of harmonic generation stability are discussed. The formation of modulated multi-jet plasma on the plane and curved surfaces during ablation by 100 kHz pulses is analyzed.

3.2.1 Experimental Arrangements for Rotating Target

The experimental setup used in the presented investigations is shown in Fig. 3.7. The experiments were carried out using two lasers: Ti:sapphire laser and Yb-doped fiber laser operated at 1 and 100 kHz pulse repetition rates, respectively. In the case of the first laser (L1, $\lambda = 800$ nm), ablation of targets to create the plasma plume was carried out by pulses with durations of 200 ps and 35 fs. In the first case, the heating pulse was created by splitting off a portion (2 mJ) of the uncompressed 200 ps laser pulses and focused by a spherical lens on the target surface for plasma formation. The remaining driving pulse was compressed in a compressor and then focused into plasma plumes with a 500 mm focusing lens. The driving pulse propagating through LIP was delayed with respect to the ablation pulse by 50 ns to allow the plasma to move away from the target surface to such extent, which allowed converting pulse to pass at the distance of 0.2 mm from the target surface through the densest part of plasma without being clipped by the target. In the second case, the 35 fs pulse was split into two parts, one of which was delayed with regard to the other by 50 ns. This combination of two pulses allowed the formation of plasma by the first 35 fs pulse with the second 35 fs pulse being propagated through the LIP.

In the case of the second laser (L2), two outputs were used that the system delivers through two ports. The first port provided 250 fs, 1030 nm, 50—150 kHz, 0.3 mJ pulses that were used for LIP formation. The second port provided 37 fs pulses (1030 nm, 50—150 kHz, ~ 1 mJ), which were used as the driving radiation for harmonics generation. The experiments with this laser were carried out at a 100 kHz pulse repetition rate. The target movement and rotation system consisted of three linear stages driven by stepper motors. The target was attached to an axis of the

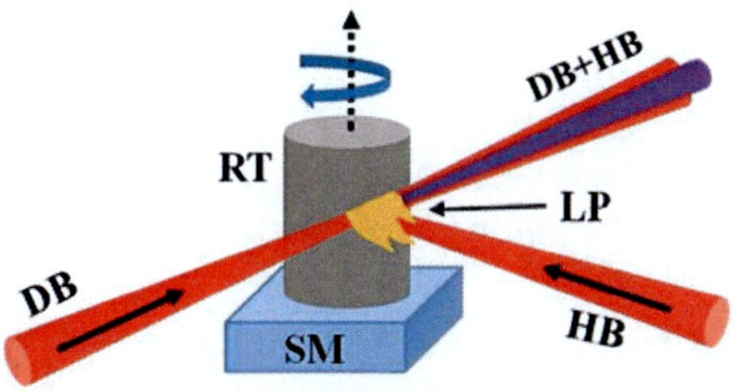

Fig. 3.7 **a** Experimental setup for HHG using rotating target. DB: femtosecond driving beam, SM: motor, RT: rotating target, HB: picosecond (or femtosecond) heating beam, LP: laser plasma, DB + HB: driving beam and harmonic beam. Reproduced from [46] with permission from The Optical Society of America

fourth motor, which provided rotation with a variable velocity of rotation (from a few rotations per minute (rpm) up to 150 rpm).

The 250 fs pulses from the first port were directed towards the vacuum chamber and focused using a 300 mm focal length spherical or cylindrical lens on the surface of the target. The optimal fluence and intensity of 250 fs heating radiation on the target surface were $F = 0.5$ J cm^{-2} and $I_{hp} = 2 \times 10^{12}$ W cm^{-2}, respectively. At these conditions of ablation, the maximal yield of harmonics was observed from different plasmas. The nanosecond pulses from Nd: YAG laser (5 ns, 1064 nm, 10 Hz, 5 mJ) were also used for ablation of targets using approximately the same fluence to determine the optimal delay between heating and driving pulses to increase the harmonic yield, as well as to compare the harmonic yield with regard to the ablation using femtosecond pulses.

HHG from gases and plasma was compared as well. The gas jet (diameter of nozzle 0.25 mm) was placed at the focal plane of the focused radiation and operated with a continuous Ar flow at a pressure of 70 mbar.

The harmonic radiation was analyzed by an XUV spectrometer consisting of a cylindrical (or flat) gold-coated mirror, 1200 lines/mm flat-field grating, and an imaging microchannel plate detector with phosphor screen imaged onto a CCD camera. The difference in harmonic yields in the case of different repetition rates of driving and heating pulses was adjusted by varying the collection time of the CCD camera. Commonly, this parameter was tuned for comparative analysis of the harmonic yields generated from low and high repetition rate pulses. In the case of the 1 kHz repetition rate experiments, 10 ms integration time of CCD was used, while for 100 kHz case this value was equal to 60 μs. Thus in the case of a 1 kHz laser, 10 shots were collected for the HHG spectrum, while in the case of a 100 kHz laser six shots were collected for a single spectrum of harmonics.

The Al, In, Ag, C, and Au rods were used for plasma formation on the rotating targets. The targets (cylindrical rods with a diameter of 15 mm and a length of 30 mm) were installed on the rotating motor as shown in Fig. 3.7, with the driving pulse propagating at a distance of 100–200 μm above the target surface. Silver thin foil wrapped around the stainless steel rod was also used to form the homogeneously ablated areas during the rotation of this target.

3.2.2 *Analysis of Harmonics Stability from the 1 kHz and 100 kHz Plasmas*

The irradiation of the fresh surface of the target by each next pulse is not a necessary requirement for achieving harmonic stability. The heating beams can overlap on the target surface during its rotation until the overheating notably changes the conditions of plasma formation on the corrupted surface. During these studies, the rotating targets were used to exclude or diminish the instability of harmonic yield caused by the degradation of the surface during laser ablation. In the case of a 1 kHz repetition rate laser, the commonly used rotating speed was varied between 20 and 30 rpm for different targets. This rotation speed allowed achieving the stable harmonic output. However, even at 5 rpm, the stability of harmonic yield was significantly better compared with the unmovable target. In the case of the 20 rpm speed of targets, approximately 30 pulses overlapped on the same spot.

In the case of ablation using 100 kHz pulses, the rotating speeds up to 150 rpm were used for the generation of stable harmonics. In that case, approximately 400 laser shots were overlapped on the same spot. At these conditions, the harmonic yield started to gradually decrease after some time from the beginning of ablation. The increase of rotation speed up to 600 rpm did not allow the improvement of harmonic stability. Thus in the case of 100 kHz ablation, the main factor of instability was the insufficient flow of heat out from the surface of the heated spot. However, the stability improved once one dragged the rotating target up and down. The optimization of this vertical movement of the target allows ensuring the reliable continuous generation of harmonics from LIP.

Different harmonic orders can be used for the analysis of the stability of harmonic yield. In discussed studies, the harmonic order in the mid of the whole spectrum of generated XUV radiation in the plateau-like range (15th to 55th orders) was chosen. Figure 3.8a shows the variation of the 27th harmonic from the Ag plasma produced on the rotating target using 1 kHz, 200 ps, 800 nm heating pulses. The 35 fs pulses from L1 were used as the driving radiation for harmonics generation. The target was rotated at a velocity of 10 rpm. The reasonably good stability, with variations of harmonic yield not exceeding 5%, was maintained for a long time. In the case of the abrupt stop of the rotating motor, the harmonic was significantly decreased. The

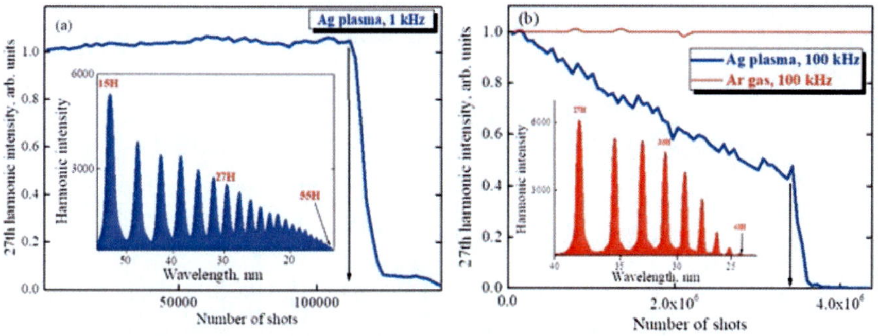

Fig. 3.8 **a** Stability measurements of the 27th harmonic generating in the silver plasma produced on the target rotating at 5 rpm during 110 s from ablation using 1 kHz, 800 nm, 200 ps pulses (laser L1). The harmonic was abruptly decreased once the motor stopped after 110 s from ablation (shown by the arrow). Inset: HHG spectrum from Ag plasma. **b** Stability measurements of the 27th harmonic generating in Ar gas (red thin curve) and Ag plasma (blue thick curve) produced on the target rotating at 100 rpm during 35 s from the beginning of ablation using 100 kHz, 1030 nm, 250 fs pulses (laser L2). The harmonic was almost disappeared at the moment when the target was stopped (shown by the arrow). Inset: HHG spectrum from Ag plasma. Reproduced from [46] with permission from The Optical Society of America

moment of motor stop is shown in Fig. 3.8a by an arrow after 110 s corresponding to 0.11×10^6 laser shots from the beginning of ablation. After that, the harmonic yield was decreased within 10 to 15 s down to ~7% compared with the case of the rotating target. The change in the target rotation velocity did not lead to a decrease in the harmonic yield as far as this velocity was maintained above 3 rpm. The inset in Fig. 3.8a shows the harmonic spectrum from stable plasma produced on the rotating silver target in the XUV range. The harmonic spectrum was extended up to 55th order (i.e., up to the wavelength of 14.5 nm and photon energy $E_{ph} = 85.2$ eV).

In the case of L2, the stability of harmonics from LIP was worsened. Figure 3.8b shows the decay of 27th harmonic yield from the plasma produced on the rotating Ag target by 100 kHz pulses within the first 35 s from the beginning of ablation (blue thick curve) corresponding to 3.5×10^6 shots. When the motor stopped, almost all harmonics disappeared except for a few lowest orders. These experiments were carried out at 100 rpm. The increase of rotation velocity (up to 150 rpm) did not lead to improvement of the stability of harmonic yield, probably, due to the overheating of the target area and craters formation. Meanwhile, the harmonic generation from the gas target was stable within the whole set of similar experiments using 100 kHz pulses (red thin curve). The inset in this figure shows the harmonic spectrum from Ag plasma extended up to the 43rd order ($\lambda = 23.9$ nm, $E_{ph} = 51.8$ eV). Notice that higher conversion efficiency of lower-order harmonics in LIPs compared with HHG in argon gas was achieved even at the conditions when the concentration of particles in plasma was lower than in gas (3×10^{17} and 3×10^{18} cm^{-3}, respectively).

The application of a 1 kHz laser (L1) for HHG in other LIPs (Au, Al) showed similar features to those presented in Fig. 3.8a. The difference between the dynamics of harmonic decay after motor stopping was observed once the picosecond or femtosecond pulses were used as the heating radiation. Figure 3.9a shows the decay of the 19th harmonic generated in the Al LIP produced by 35 fs (blue thick curve)

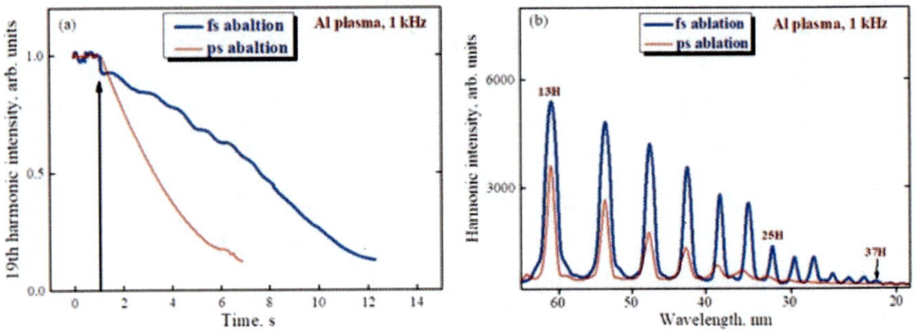

Fig. 3.9 **a** Decay of the 19th harmonic generated from the Al LIPs produced by 35 fs, 1 kHz pulses (blue thick curve) and 200 ps, 1 kHz pulses (red thin curve) after stopping the motor rotating aluminum rod (shown by arrow). **b** Spectral distribution of the harmonics generated by 800 nm, 35 fs driving pulses in the stable LIP produced on the rotating Al rod using femtosecond (thick blue curve) and picosecond (thin red curve) heating pulses. Reproduced from [46] with permission from The Optical Society of America

and 200 ps (red thin curve) pulses. The decay of the 19th harmonic started at the moment when the motor with an aluminum rod was stopped (black arrow). One can see that the decay down to 10% of the stable level of harmonic yield was achieved within 7 s from stopping the motor in the case of 200 ps heating pulses, while in the case of 35 fs heating pulses this decay was less abrupt (12 s).

The purpose of stopping the rotation of the motor was to show how fast the impeding factors originated on the overheated target surface destroy the maintenance of "optimal" plasma in the case of ablation by 200 ps or 35 fs pulses. The motivation of those experiments using 1 kHz heating pulses of different duration was to show that femtosecond pulses lesser modify target surface compared with picosecond pulses at an approximately similar fluence of both pulses on the target surface (~1 J cm^{-2}). These findings are even more important for the case of 100 kHz pulses. An earlier report on this decay of harmonics from 1 kHz LIP [42] was aimed at demonstrating the worsening plasma formation conditions in the case of the picosecond ablation of a stopped target. During the discussed studies, the difference between the dynamics of harmonic decay was observed once either picosecond or femtosecond pulses were used as the heating radiation from a 1 kHz laser. Thus one has to underline the preference in ablation of rotating target by shorter (35 fs) pulses rather than longer ones (200 ps) at a 1 kHz repetition rate. This finding was used during the following experiments with a 100 kHz laser when the 250 fs pulses were applied for plasma formation on the target surface.

The observed difference of the harmonic intensity decay after stopping the rotating target can be explained by the depth of the crater formed by different heating pulses, which depends on the duration of the ablating pulses [47]. The total heat penetration depth (l) in materials depends on the optical penetration depth (l_a) and thermal penetration depth l_t in metals ($l = l_a + l_t$) and in metals actually equal to l_t, which is mostly determined by pulse duration, $l_t = 2(K_e\tau_p)^{0.5}$ [11]. Here K_e is the heat conductivity due to electrons and τ_p is the pulse duration. The width of the heat-affected zone in metals can be estimated from this relation, as long as laser pulse duration exceeds electron–phonon relaxation time. Shorter pulses reduce this zone and make the ablation process more uniform and stable. Meanwhile, the deeper crater in the case of heating picosecond pulses compared to femtosecond pulses causes worsened conditions for the optimal LIP formation. The results of discussed studies corroborated with the theoretical consideration of surface modification by different ablating pulses described by the above relation.

In Fig. 3.9b, the spectra of the harmonic spectra generated in Al LIP using different 1 kHz heating pulses are shown. In the case of ablation by 35 fs pulses, the extension in the harmonic cutoff energy up to 57.3 eV (λ = 21.6 nm, 37H) was achieved, which was notably larger than the harmonic cutoff observed in aluminum plasma produced by 200 ps pulses (E_{ph} = 38.7 eV, λ = 32 nm, 25H). This extension of cutoff can be attributed to the involvement of double-charged Al ions with higher ionization potential. A similar extension has earlier been reported in the case of manganese plasma [48].

3.2.3 Two-Color Pump of the Plasmas Produced on the Rotating Targets by 1 and 100 kHz Pulses

The two-color pump (TCP) scheme of HHG in LIPs was performed using two orthogonally polarized 800 nm and 400 nm driving pulses (L1) and 1030 nm and 515 nm driving pulses (L2), respectively. In the case of L1, the second harmonic generation (SHG) at $\lambda = 400$ nm was achieved using a 0.2 mm thick barium borate (BBO) crystal, which was inserted inside the vacuum chamber on the path of the 800 nm fundamental radiation. The SHG conversion efficiency was ~3%, so, in the plasma area, the ratio of fundamental and second harmonic intensities was 33:1, which corresponded to the 5.7:1 ratio of the amplitudes of fundamental and second harmonic waves. Even this small part of the second field allowed a generation of almost equal odd and even harmonics.

The example of TCP (800 nm + 400 nm) of the plasma produced on the rotating indium target is shown in Fig. 3.10a (blue dashed curve). The generation of odd and even harmonics up to the 21st harmonic was observed. In the meantime, the application of a single-color pump (SCP, 800 nm) at similar conditions (i.e., by removal of BBO from the path of 800 nm beam) led to the generation of weaker harmonics (red solid curve). The application of TCP allowed the generation of even harmonics. Some of them (12H, 14H) were close to the ionic transition of indium possessing large oscillator strength, which led to the enhancement of those even harmonics, alongside H13. Application of Al rotating target for TCP at similar conditions demonstrated the generation of equal gradually decreasing odd and even harmonics (Fig. 3.10b, blue dashed curve). HHG using the SCP scheme also showed a similar gradually

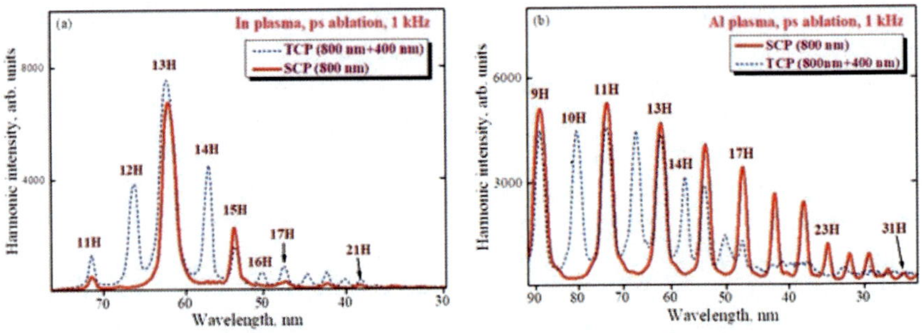

Fig. 3.10 **a** Spectral distributions of the harmonics generated in the indium plasma using SCP (800 nm) and TCP (800 nm + 400 nm) in the case of 1 kHz laser. Blue dashed curve shows the odd and even harmonics obtained in the case of TCP and red solid curve shows the odd harmonics obtained in the case of SCP. The SHG conversion efficiency in TCP scheme was equal to 3%. **b** Spectral distributions of the harmonics generated in the aluminum plasma using SCP and TCP. Blue dashed curve shows the odd and even harmonics obtained in the case of TCP and red solid curve shows the odd harmonics obtained in the case of SCP. Reproduced from [46] with permission from The Optical Society of America

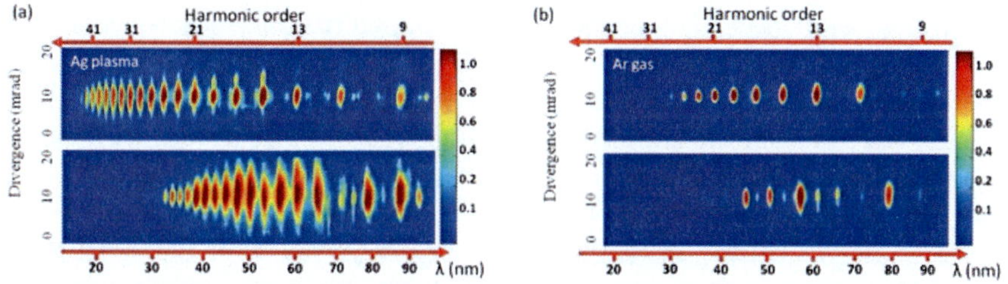

Fig. 3.11 Raw images of harmonic spectra from LIP and gas using 1 kHz pulse repetition rate. **a** HHG in silver plasma using single color (upper panel) and two-color (bottom panel) pump schemes. **b** HHG in Ar gas using single color (upper panel) and two-color (bottom panel) pump schemes. Reproduced from [46] with permission from The Optical Society of America

decreased odd harmonics (red solid curve). In both cases (In and Al rod targets), the picosecond pulses were used as heating radiation.

Figure 3.11 shows the raw images of harmonic spectra generated in (a) silver plasma and (b) argon gas in the case of SCP and TCP at a 1 kHz pulse repetition rate. The presentation of the images of HHG spectra collected by the CCD camera can visually demonstrate various features of this process. These spectra were collected using a flat gold-coated mirror in an XUV spectrometer, which allowed the analysis of the divergence of high-order harmonics, alongside the observation of the variations of phase-matched conditions along the XUV spectra range.

The harmonic spectra at two regimes of pumping (SCP, TCP) of the silver plasma produced on the rotating Ag rod are presented in Fig. 3.11a. One can see the enhancement of harmonic yield during SCP of Ag LIP in the case of higher orders of harmonics compared to the lower orders (upper panel). Also notice larger divergence of the harmonics belonging to the shorter wavelength range, contrary to the common rule of wavelength-dependent divergence (θ) and sizes (d) of the harmonic beam ($\theta(d) \propto \lambda$). This deviation is related to the prevailing influence of the collective effect of phase matching for the shorter-wavelength harmonics, which caused their stronger yield.

Another pattern of harmonic distribution appeared in the case of TCP (bottom panel). One can see the gradually decreased harmonic yield for both odd and even harmonics, alongside the decrease of divergence for each next harmonic order. One can also admit the significantly larger harmonic yield in the case of TCP compared with SCP, as well as similar intensities of neighboring odd and even harmonics. This advantage of using TCP has already been demonstrated during HHG studies in gases and plasmas (for example [49, 50]). The difference in harmonic divergence can be attributed to the recombination process of ionized electrons through the long or short trajectories.

In the case of TCP of silver plasma (bottom panel of Fig. 3.11a), stronger lower order odd and even harmonics gradually decreased towards the shorter wavelength range. These results also underline the significant increase of harmonic yield in the case of the presence of very weak second harmonic field (2–6% of the intensity of fundamental wave, depending on the thickness of used BBO crystal), which, in the case of gas HHG, was attributed to the formation of a quasi-linear field and higher ionization rate compared with SCP [49].

These spectra from LIP were compared with those from Ar gas. In the case of SCP (800 nm, Fig. 3.11b, upper panel), a gradual decrease of harmonic yield and divergence characterized the behavior of HHG in a gaseous medium. The comparison of SCP in plasma and gas media showed the notable prevalence of harmonic yield from the former medium. Approximately similar concentrations of plasma and gas (~3×10^{17} cm^{-3}) were maintained during these experiments at similar intensities (2×10^{14} W cm^{-2}). The application of TCP of Ar led to the appearance of even harmonics, which were stronger than the odd ones (bottom panel).

Application of 100 kHz laser for TCP in plasma also led to significant growth of the harmonic yield compared with SCP. Below we analyze the role of the second field in the variations of the harmonic intensity generated in the plasmas produced on the rotating targets. Different thicknesses of BBO and velocities of target rotation were used to optimize the harmonic yield. Figure 3.12 demonstrates different conditions of HHG in the silver plasma produced by 250 fs, 1030 nm, 100 kHz, 0.25 mJ pulses. Two silver samples were used: silver bulk rod and thin silver film wrapped around the stainless rod to analyze whether heat transfer inside the target depends on the thermal conductivity of samples. SCP of the plasma produced on the bulk rod rotated at 30 rpm led to the generation of harmonics only up to the 17th order (Fig. 3.12a, upper panel). These conditions of harmonic generation using 1030 nm radiation were notably worse compared with the above-studied case (compare with Fig. 3.11a). The application of 0.1 mm thick BBO allowed the generation of a few even orders during TCP of silver plasma (up to 14H, Fig. 3.12a, second panel from the top). Though the overlap between 1030 and 515 nm pulses in the area of LIP was perfect in the case of using this thin crystal, the small ratio of the amplitudes of overlapped 515 and 1030 nm pulses did not allow the generation of strong odd and even harmonics.

Two bottom panels of Fig. 3.12a were obtained using 0.4 mm thick BBO at 30 and 15 rpm of bulk Ag rod. The third panel demonstrates a significant improvement of odd and even harmonics compared with the application of 0.1 mm thick BBO. SHG efficiencies in these comparative experiments were 1.5% (in the case of 0.1 mm thick BBO) and 4.7% (in the case of 0.4 mm thick BBO). Harmonics up to the 28th order were routinely obtained during this set of studies. The application of thicker crystal for SHG can cause a temporal separation of two pumps (fundamental and second harmonic waves) due to group velocity dispersion in BBO. This separation did not allow efficient generation of even harmonics in previous studies using 800 nm fundamental radiation from Ti:sapphire lasers. Meanwhile, in the case of longer wavelength driving sources, like Yb-doped fiber laser, the difference in group velocity dispersion of ω and 2ω waves was smaller, which allowed the application of thicker crystal to produce a larger amount of second harmonic photons while maintaining sufficient overlap with the fundamental radiation in the area of LIP.

Here we discuss the comparison of HHG at two regimes of rotation velocities in the ablating silver rod (30 and 15 rpm; two bottom panels of Fig. 3.12a). The decrease of rotation speed did not cause the worsening of harmonic stability, but

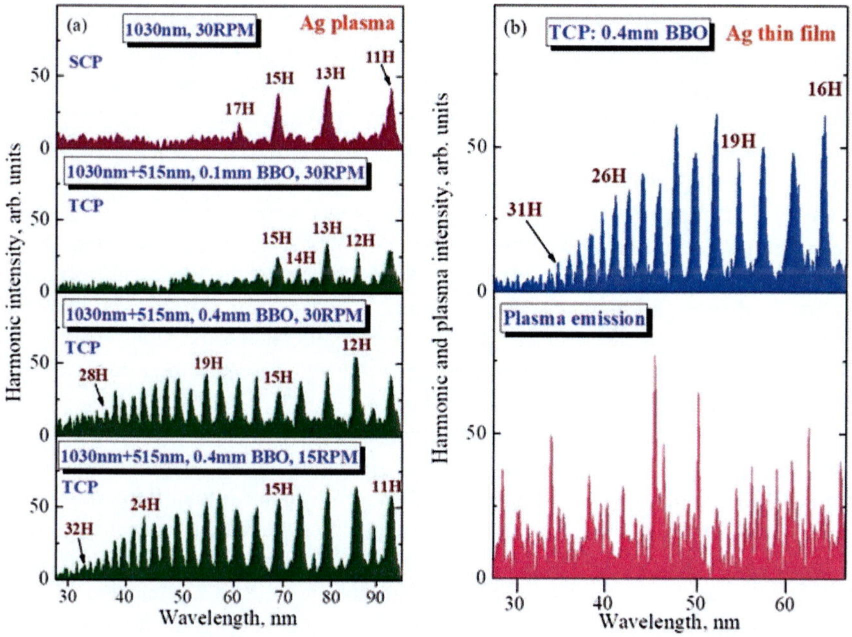

Fig. 3.12 TCP of silver plasma produced at 100 kHz pulse repetition rate. **a** HHG in the plasma produced on the rotating bulk silver rod. Upper panel corresponds to SCP and 30 rpm of target, while other panels correspond to TCP. Second panel corresponds to TCP using 0.1 mm thick BBO. Two other panels were obtained using 0.4 mm thick BBO at 30 and 15 rpm of bulk Ag rod. **b** Upper panel: HHG using TCP of the plasma produced at a fluence $F = 1$ J cm^{-2} on the silver thin film rounded the stainless steel rod. Approximately similar spectrum was obtained with regard to the Ag bulk rod. Bottom panel shows the plasma spectrum without harmonic emission once stronger excitation of thin film ($F = 4$ J cm^{-2}) leads to significant phase mismatch between the driving (1030 nm and 515 nm) waves and harmonics. Reproduced from [46] with permission from The Optical Society of America

rather improved the harmonic yield and cutoff (up to the 32nd order; bottom panel). To analyze the role of heat removal, thin (0.1 mm) silver film was used. The upper panel in Fig. 3.12b shows the HHG spectrum using TCP of the plasma produced on such film. Notice that the thin Ag film was wrapped around a stainless steel rod, but the harmonic yield was the same as for bulk silver (compare a bottom panel of Fig. 3.6a and an upper panel of Fig. 3.12b), while steel has a much worse thermal conductivity than silver. This observation allowed concluding that the thermal penetration depth has no significant influence on heat dissipation and correspondingly HHG yield.

This assumption can be qualitatively considered when the heat distribution is calculated. A previous study [51] allows assessing the surface temperature after applying the heating pulse. It was pointed out that for the short pulses the widely used surface generation model can give erroneous results compared to the more accurate volume generation model. Nevertheless, both models predict the relatively slow flow of heat inside the volume, which has been proven in the case of the tungsten heated by 30 ps pulses. In the discussed case, the application of 250 fs or 200 ps pulses should also lead to the slow decrease of temperature along the depth of material. That is why the expected similarity in surface temperature maps for the bulk and 0.1 mm thick silver may point out the similarity in the plasma formation and harmonic generation, which was proven in the studies of these silver samples. However, the preferable flow of heat along the surface will cause its accumulation in the case of high pulse repetition ablation of the same or nearby place of rotated target and faster modification of the surface compared with 1 kHz laser. One has to note that these estimates seem insufficient to deal with another problem occurring during overheating of the same spot by multiple shots differently overlapped with each other at variable velocities of the rotating target. The issue, which can play a decisive role in that case, is the appearance of free electrons in the plasma cloud, which can destroy the phase-matching conditions.

The conditions of plasma formation were chosen such that maximal yield of odd and even harmonics, as well as longevity of stable harmonics generation, were achieved. The optimal fluence of heating 250 fs pulses in these conditions was found to be $F = 1$ J cm^{-2}. The increase of the fluence led to a decrease in both the stability of harmonic yield and the conversion efficiency. The formation of denser plasma did not compensate for the growing influence of free electrons that appeared at these conditions. The free electrons are responsible for the abrupt deviation from the optimal phase-matching conditions for efficient HHG in LIP. The bottom panel in Fig. 3.12b shows only plasma emission spectrum at the conditions when stronger excitation of thin-film ($F = 4$ J cm^{-2}) led to significant phase mismatch between the driving (1030 nm and 515 nm) waves and the harmonics.

Similar results were obtained using other targets. Figure 3.13a shows the harmonic spectra generated during SCP and TCP of Au LIP using a 100 kHz laser. The upper panel demonstrated the harmonic spectrum obtained during SCP of the plasma produced by 250 fs pulses on the surface of a gold ring. Two other panels were obtained using TCP of Au LIP when 0.1 and 0.4 mm thick crystals were used for SHG. One can see the similarities of the harmonics generated in gold (Fig. 3.13a) and silver (Fig. 3.12a) plasmas.

The application of soft material (graphite) for HHG worsened the stability of harmonics, while the harmonic yield at these conditions was higher compared to the case of ablated metals. Figure 3.13b presents the SCP- and TCP-induced HHG spectra in the carbon plasma produced on the graphite rod. 0.1 mm thick crystal was used for SHG and TCP of C LIP. Similar to the case of metal plasmas, the weak orthogonally-polarized second field did not improve the whole yield of harmonics and allowed the generation of only a few even harmonics, which were notably weaker than the odd ones. However, even in these unfavorable conditions, the HHG yield of the lower order harmonics was two to four times stronger compared with metal LIPs. The reasons for the appearance of enhanced harmonics in the case of carbon LIP could be attributed to the formation of the nanoparticles in plasma, which increased the recombination cross-section for the accelerated electrons and correspondingly amended the conditions of harmonic generation. The presence of nanoparticles during these

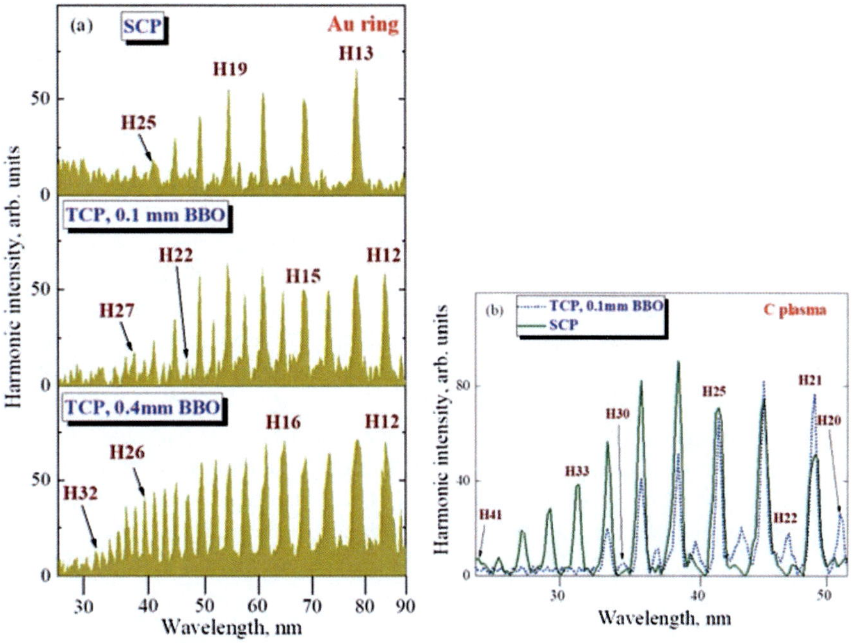

Fig. 3.13 SCP and TCP of different plasmas using 100 kHz laser. **a** HHG in Au LIP. Upper panel shows the harmonic spectrum obtained during SCP of the plasma produced by 250 fs pulses on the surface of gold ring. Two other panels were obtained using TCP of Au LIP when 0.1 and 0.4 mm thick crystals were used for SHG. **b** HHG in the carbon plasma produced on the graphite rod. The 0.1 mm thick crystal was used for SHG and TCP of C LIP. Reproduced from [46] with permission from The Optical Society of America

experiments was confirmed during the analysis of the debris from ablated graphite deposited on the nearby substrates.

The main goal of this work was to demonstrate the harmonic generation from the plasmas produced by a 100 kHz class laser. However, independently of which pulse repetition rate is used, the principles of optimal plasma formation are similar for the single shot, 10 Hz, 1 kHz, 100 kHz, etc., cases. The term "optimal" refers to the conditions when the driving beam propagates through the maximally dense plasma arriving on the pass of the driving pulse, with minimal factors decreasing harmonic conversion efficiency. The requirements to minimize the impeding factors, such as phase mismatch, strong incoherent plasma emission, and additional thermal loading on the target surface, remain the same for each of the abovementioned regimes of pulse repetition rate. One of the most important issues here is the determination of the optimal delay between heating and driving pulses. Once the same laser is used for heating of target by the part of radiation followed with the conversion of femtosecond laser pulses in plasma, the delay between these pulses becomes established by using the variable difference in the distances of propagation of the heating and driving beams towards the target. This is a very inconvenient procedure requiring the constant re-adjustment of the experimental scheme while adding some extension on the driving beam pass in the case of using materials with different atomic weights.

The most convenient option to define the optimal delay for different ablating materials is the application of the second source, like nanosecond Nd:YAG laser, digitally synchronized with the main laser. The application of such a scheme allowed precisely determining the optimal delay between heating and driving pulses to achieve the maximal harmonic yield from the materials possessing different atomic weights. The synchronization of two laser systems allows controlling the optimal conditions of HHG in laser-produced plasmas by changing the delay between heating and driving pulses using a digitally controlled facility rather than by applying the optical delay between pulses used in previous studies, which does not allow the dynamic variation of delay during experiments. The key feature of the used two-laser configuration is the possibility to control the delay between heating nanosecond pulses and driving femtosecond pulses over a long timescale (up to a few hundred microseconds, which is important once one uses complex molecules and large nanoparticles as targets) using the digital delay generator. The delay between the heating and driving pulses was monitored on an oscilloscope using a fast photodiode. The optimal delays (50, 75, 150, 150, and 220 ns) were determined in the case of C, Al, Ag, In, and Au plasmas respectively. The knowledge of the best delay between pulses for different targets obtained during these studies was then applied during HHG in different plasmas using a single laser when one adjusted the delay between heating (250 fs or 200 ps) and driving (35 fs) pulses using optical delay.

The additional task considered during these two-laser studies was the comparison of the "nanosecond" and "femtosecond" plasmas as the media for HHG. In the case of a 100 kHz laser, the nanosecond pulses from Nd: YAG laser (5 ns, 1064 nm, 10 Hz, 5 mJ) were used for ablation of targets by applying approximately the same fluence to compare the harmonic yield with regard to the ablation using femtosecond pulses. The discussed studies showed that the use of shorter pulses for ablation leads to better conditions of plasma formation, as well as lesser plasma emission, allowing the achievement of stronger harmonic yield at similar conditions of experiments. These studies confirmed the earlier reported results of harmonic generation in the plasmas produced by femtosecond and nanosecond pulses [45].

3.2.4 *Quasi-Phase Matching in the Plasmas Produced on the Rotating Targets*

This method is based on the formation of conditions when the coherent length of some harmonic and a few surrounding ones becomes equal to the sizes of a single plasma jet from a multi-jet LIP. The enhancement of a group of higher-order harmonics in plasmas due to the formation of QPM conditions has been demonstrated and analyzed using different approaches [52–54]. Most of the above QPM studies in LIPs were carried out using low pulse repetition rate lasers (10 Hz). It is obvious that the 1 kHz and higher class lasers provide an opportunity for enhancing the average power of the shorter wavelength components in the generated harmonics.

To create multi-jet plasma, the heating pulses from L2 (250 fs, 100 kHz) were focused using a cylindrical lens. The stainless steel mesh (sizes of open areas 1 × 1 mm^2) was installed in front of a cylindrical lens to modulate the spatial shape of the focused heating beam. The first set of experiments was carried out using the aluminum plane target.

The dynamics of plasma plumes ablated using femtosecond and nanosecond heating pulses is very important for achieving the efficient HHG in plasma plumes. The ICCD camera allowed analyzing the dynamics of plasma formation and spreading out from the target surface. The 20 ns frame gate was used for these purposes, which allowed measuring the position and emission of plasma at the moments of 0, 20, 40, 60 ns, etc. from the beginning of ablation, while also allowing varying the gate and acquisition time from target to target. Figure 3.14a presents the plasma images at 18, 50, 80, 100, and 400 ns from the beginning of ablation. Those images were obtained during ablation by a single pulse of 100 kHz laser. The left image shows the spreading plasma during ablation of the Al plane surface by cylindrically focused imperforated beam, while other images show the shapes of multi-jet plasmas at different time scales from the beginning of ablation. Based on these plasma maps, 50 and 80 ns delays between heating and driving beams were determined as the most suitable for the driving pulse to efficiently interact with the densest part of LIP.

In the case of extended imperforated plasma, HHG will be suppressed due to the phase mismatch between interacting waves since the length of the excited plasma containing a large number of free electrons is much longer than the coherence length of highest-order harmonics (i.e., the propagation length required for the phase mismatch to reach π). These conditions cause the notable decrease or even full vanishing of those harmonics. The division of such extended plasma on the group of jets equidistantly separated from each other (second to sixth images in Fig. 3.14a)

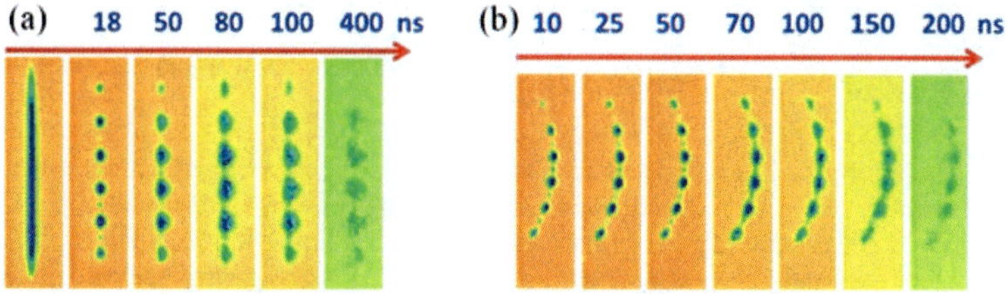

Fig. 3.14 Images of imperforated and perforated LIPs produced by 100 kHz, 250 fs pulses on the plane and curved surfaces of aluminum. **a** Images of spreading plasma during ablation of Al plane surface by cylindrically focused imperforated beam (left panel) and perforated beam at different delays from the beginning of ablation (other panels). **b** Images of spreading plasma during ablation of rotating Al rod by cylindrically focused perforated beam at different delays from the beginning of ablation. Reproduced from [46] with permission from The Optical Society of America

leads to the formation of the optimal phase-matching conditions between the driving harmonic waves for some orders while maintaining a high concentration of plasma components allowing larger conversion efficiency for the specific harmonic. Those images were taken using the digital generator to control the delay between the heating 250 fs pulses and the triggering time of the ICCD camera.

As has already been underlined, the application of the static plane target for plasma HHG is not suitable due to overheating of the target surface during the application of 100 kHz heating pulses. Because of this, the perforated plasma jets formed on the rotating targets were also analyzed to determine the conditions of the optimal ablation of targets to create suitable multi-jet plasmas. Figure 3.14b demonstrates the images of spreading plasma during ablation of the Al rod surface by the cylindrically focused perforated beam at different delays from the beginning of ablation. Such geometry of curved plasma jets was used for the HHG at the conditions of QPM in the case of a 1 kHz laser source. Notice that the formation of similar conditions for demonstration of QPM in the case of 100 kHz laser source is still under investigation and optimization due to the above-mentioned obstacles in the stabilization of plasma formation.

The following set of studies shows the enhanced harmonics in the shorter wavelength range of XUV from the plasmas suitable for QPM once one uses the multi-jet LIP produced on the rotating target instead of the extended imperforated plasma formations.

Figure 3.15 shows the setup and results of HHG experiments using multi-jet plasmas produced by 1 kHz, 200 ps pulses on the rounded surface. The experimental scheme shown in Fig. 3.15a explains how the driving pulse propagates through the imperforated plasma (left panel) and the set of plasma jets (right panel) produced on the rotating rod. In the latter case, different multi-slit masks (sizes of slits 0.5

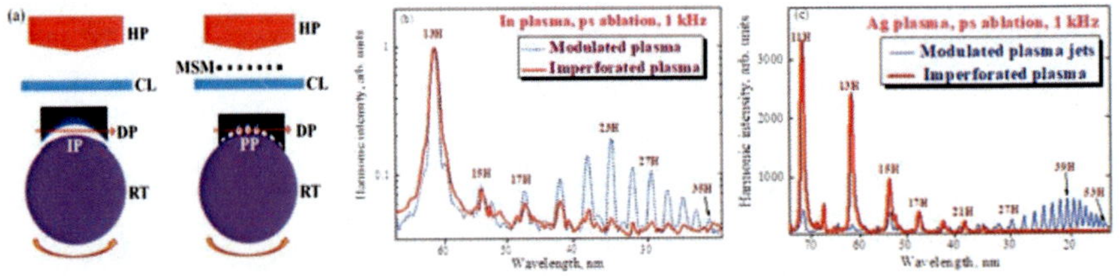

Fig. 3.15 HHG using multi-jet plasmas produced by 1 kHz laser. **a** Experimental scheme. Left pattern: HP, heating pulse; CL, cylindrical lens; DP, driving pulse; IP, imperforated plasma; RT, rotating target. Right pattern: HP, heating pulse; MSM, multi-slit mask; CL, cylindrical lens; DP, driving pulse; PP, perforated plasma; RT, rotating target. **b** HHG spectra from modulated (blue dotted curve) and imperforated (red solid curve) indium LIP produced on the rotating target. Resonantly enhanced 13th harmonic was normalized for two cases. The enhancement factor up to 18 × was achieved for the group of harmonics in the spectral range of 20–40 nm. **c** HHG spectra from modulated (blue thin curve) and imperforated (red thick curve) silver LIP produced by 1 kHz pulse repetition rate laser on the rotating target. The enhancement factor of up to 44 × was achieved for the group of harmonics centered at the wavelength of 20 nm. Reproduced from [46] with permission from The Optical Society of America

× 0.5 mm or 0.2 × 0.2 mm) were installed in front of the cylindrical lens. Two sets of HHG spectra from modulated (blue dotted curve) and imperforated (red solid curve) indium LIP produced by 1 kHz pulse repetition rate laser are shown in Fig. 3.15b. As already mentioned, indium plasma allows the resonance-induced enhancement of a single low-order (H13) harmonic of 800 nm radiation, while significantly suppressing the higher-order ones (Fig. 3.10a). In described QPM studies, the resonantly enhanced 13th harmonic was normalized for two cases of imperforated and perforated plasmas. One can see the following gradual decrease of harmonic orders, with harmonic cutoff at 31st order, in the case of imperforated plasma (Fig. 3.15b, red thick solid curve). The envelope of the harmonic spectrum drastically changed once the extended plasma was replaced by the multi-jet one using the 0.5 × 0.5 mmmulti-slit mask (Fig. 3.15b, blue dotted curve). The enhanced 13th harmonic remained to be the strongest one along the whole set of harmonics, while the notable growth of higher orders centered at around 23H was observed. The enhancement factor of up to 18 × was achieved for the group of harmonics in the spectral range of 20–40 nm.

Similar features of HHG spectra from modulated (blue thin curve) and imperforated (red thick curve) silver LIP are shown in Fig. 3.15c. In that case, the used 0.2 × 0.2 mm mask allowed the extension of the group of enhanced harmonics to the shorter wavelength region. One can see that, in the case of multi-jet plasma, the enhanced harmonics were observed in the spectral range where almost no harmonic emission was obtained in the case of extended imperforated plasma. Notice the decrease in the yield of low-order harmonics from multi-jet plasma. Meanwhile, the enhancement factor of up to 44 × was achieved for the group of harmonics centered at the wavelength of 20 nm.

Silver plasma is a very attractive medium for the generation of harmonics up to the sixties orders of 800 nm laser radiation. The increase of the fluence of heating pulses causes a two-sided effect on the harmonic distribution and harmonic intensity. Firstly, higher-order harmonics disappear from the spectra of collected emission, and secondly, lowest-order harmonics increase with regard to the above-described "optimal" plasma conditions. In that case, the harmonic distribution looks like the one shown in Fig. 3.15c (red thick curve). Notice that H11–H17 harmonics in this spectrum are notably (up to ten to twenty times) stronger than those obtained in the case of "optimal" plasma. The reason for the drop of the intensity of highest order harmonics is the appearance of a large number of free carriers (electrons) dramatically changing the phase-matching conditions for those harmonics while maintaining the affordable conditions for the lower-order harmonics. In other words, the coherence length for higher orders becomes notably smaller with regard to the coherence length for the lower orders. Thus at the fixed length of plasma, only lower orders become enhanced in the denser plasma.

Does the question arise on how to maintain the generation of strong higher-order harmonics at these conditions of the over-heated target? The answer is the use of extended dense plasma, which is spatially modulated in such a way that the modulation period coincides with the coherence length of high-order harmonics. In that case, the group velocities of fundamental and harmonic waves become equal to each other only in a relatively narrow range of the XUV spectrum. Correspondingly, these conditions cannot be considered as the phase-matched conditions along the whole studied spectral range but rather for a part of it. That is why these conditions of harmonic generation were classified as the nearly ("quasi") phase matching suitable only for the specific XUV range, rather than for the whole region of harmonics generation.

Thus QPM refers to the conditions when the phase matching between interacting waves occurs only in a relatively narrow spectral range of harmonics generation. In that case, the harmonics lying out of this region are notably suppressed compared with the case of QPM-free conditions, while the group of harmonics centered at the wavelength of some of them became strongly enhanced (thin blue curve in Fig. 3.15c). To demonstrate this effect, one has to use a strong ablation of a target, when in spite of a larger concentration of plasma the high concentration of free electrons causes a notable suppression of all harmonics, especially higher-order ones.

As it was mentioned, the difficulties encountered during experiments with 100 kHz laser, particularly the proper stabilization of plasma parameters, which is especially crucial in the case when collective effects play a decisive role in HHG, did not allow to achieve the stable QPM effect in the case of 100 kHz laser. The stable enhancement of a group of few harmonics was suppressed by target degradation.

The discussed studies (Figs. 3.15b and c) confirmed that the curved shape of multi-jet plasma did not prevent the demonstration of QPM, thus showing that variable density of different jets interacting with a driving pulse is not an unavoidable obstacle, which prevents the coherent addition of harmonics from different areas of rounded perforated ablation. The enhancement factors of QPM-enhanced harmonics in the case of curved jet structures were only slightly smaller compared with parallel multi-jet structures. This observation points out the limited amount of jets involved in the collective accumulating process of HHG allowing demonstration of the quadratic dependence of harmonic yield on the number of separated plasma jets. To overcome the problem of heterogeneous distribution of plasma density along the propagation of driving beam nearby to the rotating target, one can modify the position of the rotating motor in such a way that its axis becomes parallel to the direction of driving beam propagation like it was demonstrated in [41].

3.2.5 Discussion

The main goal of this study was the analysis of the high-order harmonics generation in different plasmas produced using 1 and 100 kHz laser sources. The advantages of plasma HHG over gas HHG have been demonstrated during the last decade in numerous laboratories. The distinction of the former approach over the latter one is not only the advantage in conversion efficiency demonstrated by different research teams [39–41] but rather the availability in studies of new optical processes, which were discovered during harmonics generation in plasmas and could not be realized, at least until now, using gases as the media for HHG. These include the resonance-induced enhancement of single harmonic in different plasmas and applications of nanoparticles and quantum dots for enhancement of harmonics. Meanwhile, the most important issue differing gas and plasma media as the environment for harmonics generation is not the availability of formation of the coherent XUV sources of emission, since similar sources are currently developed by different means, like X-ray lasers and free-electron laser-based technologies, but rather the availability in the analysis of the optical and physical properties of almost all non-radioactive solid elements of the periodic table through their transfer in the plasma state. Notice that, in the case of gas HHG, only a few noble and molecular gases could be used for these purposes, contrary to a few thousand complex solids, which notably increases the activity in the field of atomic physics, optics, and nonlinear spectroscopy of those species.

A significant advantage of gas HHG is the availability in demonstration of the attosecond timescale of harmonic pulses, though some early experiments (for example [55]) have demonstrated that the plasma-induced harmonics can also be considered as attosecond pulses. Notice that most previous attempts in measurements of pulse duration in the case of plasma HHG experiments were stalked due to the problem of delivering a suitable amount of harmonic shots for the analysis of their temporal characteristics. The larger stability of gas harmonics compared with plasma ones allows carrying out these measurements within a few ten minutes. Particularly, an attosecond streaking scan can be done in about 20 min at a 1 kHz pulse repetition rate. This time span is difficult to maintain for stable harmonic generation in the case of ablation using such lasers. Meanwhile, if one uses a 100 kHz laser then it should be possible to do these attosecond streaking scans within 0.2 min, assuming flux levels similar to gas harmonics. So 35 s of relatively stable harmonic yield demonstrated in analyzed studies, which diminishes only by two times during this time span, should be enough for attosecond streaking scans. Notice (a) the higher conversion efficiency of harmonics from plasmas compared with gases and (b) larger driving pulse energy of the used 100 kHz source with regard to earlier reported 100 kHzgas HHG experiments, when fewer photons per laser pulse compared to a higher pulse energy 1 kHz laser were employed for the attosecond streaking scans. These two advantages can also support the temporal characterization and applications of plasma harmonics.

It is obvious that the HHG efficiency depends on the plasma density, laser pulse parameters (intensity, pulse duration, contrast, focusability, etc.), background ionization, composition of the medium, etc. These parameters cannot be exactly reproduced in different experimental setups with different driving laser sources. However, previous experiments show that the ratio between the harmonic yields from different plasmas remains approximately the same even when being studied in different laboratories. The density of the plasma in the 100 kHz experiments was optimized to provide a maximum harmonic yield. The estimates of conversion efficiency (10^{-5}) approximately match with previous studies [56]. Correspondingly, in the case of the 25th harmonic of 1030 nm radiation ($\lambda = 41.2$ nm, $E = 30$ eV) generated during discussed experiments in carbon plasma, at 100 kHz and 40 W average power of driving pulses, 8×10^{13} XUV photons per second were emitted.

The measurements of conversion efficiency of high-order harmonics ($\sim 10^{-5}$) from silver LIP in the plateau region have also been reported in [57]. Thus one can assume that in discussed studies the 10^{-5} conversion efficiency for the harmonics in the plateau region was achieved. This value refers to a single harmonic in this XUV region. Another estimate of conversion efficiency is related to the comparison of the yields of harmonics from plasmas and gases in a single set of studies. In the discussed case, the intensity of harmonics from plasmas was at least one order of magnitude higher than the one obtained from argon gas at similar laser intensities and particle densities. By assuming the frequently reported efficiency of this process in argon ($\sim 10^{-6}$) one can estimate the 10^{-5} conversion efficiency for plasma harmonics. Thus, at the 40 W average power of driving pulses used in the discussed studies, the average power of a single harmonic was 10^5 times smaller, i.e., equal to 0.4 mW.

3.3 Application of 150 kHz Laser for High-Order Harmonic Generation in Different Plasmas

3.3.1 Scheme and Description of Targets

Using a motorized rotating target notably improves the stability of HHG from LIPs, particularly in the case of high pulse repetition rate lasers, and significantly minimizes the modification of the target surface that could cause degradation of harmonic yield. The periodic change of the ablation zone allows the cooling down of the heated area and maintains stable plasma formation. Correspondingly, the application of fiber chirped-pulse amplification systems, delivering higher pulse repetition rates (100 kHz and higher) compared with commonly used Ti: sapphire class lasers may open new opportunities for HHG from LIPs. Different processes, such as the local field enhancement in quantum dots, resonance enhancement of single harmonics in metal plasmas, the realization of the quasi-phase-matching conditions in periodically modulated plasma plumes, and nonlinear spectroscopy of ablated materials, could be more easily analyzed using the high pulse repetition coherent XUV sources.

The generation of high harmonics from gas jets allowed achieving μW harmonics between 45 and 61 nm (H23–H17), at a repetition rate of 50 kHz. Here, we discuss the application of the fiber laser system with a high repetition rate (150 kHz) for the studies of HHG in different LIPs [58]. The application of this laser allows achieving the harmonic flux of up to 8×10^{13} photons/s. The scheme (Fig. 3.16) was described in Sect. 3.2.

The harmonic output depended on the excitation state of laser-produced plasma. Highly excited plasma can destroy the phase-matching conditions due to the dispersion induced by the large concentration of free electrons in LIPs. The density of plasma at the focal point of driving femtosecond pulse was estimated to be $\sim 2 \times 10^{17}$ cm^{-3}, which corresponded to the pressure of ~10 mbar.

The focused driving pulses from the second port (37 fs, 1030 nm, 50–150 kHz, 1 mJ) propagated through LIP ($I_{\rm dp} = 2 \times 10^{14}$ W cm^{-2}), at a distance of 0.3 mm above the target surface. The delay between heating and driving pulses was installed

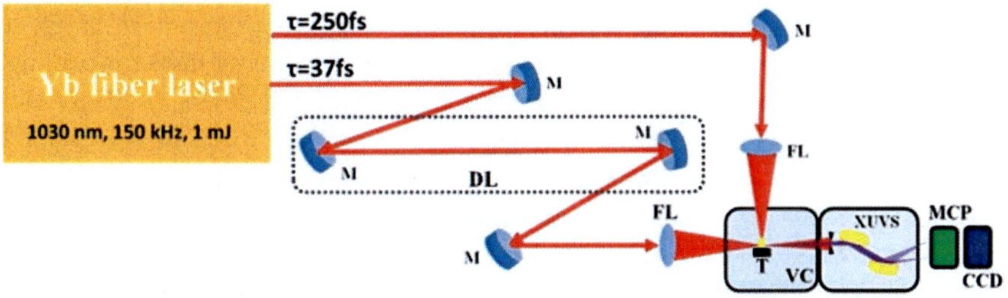

Fig. 3.16 Experimental scheme. Yb fiber laser, ytterbium-doped fiber chirped pulse amplification laser; M, mirrors; DL, delay line; FL, focusing lenses; VC, vacuum chamber; T, target; XUVS, extreme ultraviolet spectrometer; MCP, microchannel plate; CCD, CCD camera. Reproduced from [58] with permission from Springer

by using the optical delay line. The studies of HHG in different metal plasmas were carried out at different delays between the heating and driving pulses. At the used delays (25, 50, and 100 ns), the maximum density of ejected particles from three groups of studied materials possessing different atomic weights reaches the axis of propagation of the driving femtosecond beam, thus allowing the highest nonlinear optical response from the plasma plume. The optimal values of the delays were determined by the anticipated velocities of plasma clouds (~1.2×10^4, 6×10^3, and 3×10^3 m/s), for the particles with different atomic weights (B and C of similar weight (11 and 12), Mn and Zn of approximately similar weight (54 and 65), and Ti (204)), and the distance between the target surface and driving beam (0.3 mm). The harmonics and plasma emission were detected using a flat field grazing-incidence XUV spectrometer.

Carbon (C), titanium (Ti), boron (B), zinc (Zn), and manganese (Mn) rods were used as the targets for ablation. The cylindrical rods with a diameter of 15 mm and a length of 30 mm were installed on the rotating motor, with the driving pulse propagating at a distance of 200–300 μm above the target surface. The application of rotating targets allows for the improvement of harmonic stability in the case of 150 kHz ablation. Boron was also represented in the shape of the nanometer-sized powder (mean size of 100 nm), which was pressed together with silver microparticles (MPs, mean size of 70 μm) to form the tablets for better connectivity. The silver MPs were added to the boron nanoparticles (NPs) powder at the 1:3 weight ratio, to strengthen the tablets and to prevent them from fast disintegration during laser ablation. This powdered target was chosen for comparative studies with the ablated boron bulk target and the nanoparticles formed "in-situ" during the ablation of carbon, titanium, and manganese. The shape of this pressed target allowed for rotating during HHG experiments, to avoid the fast destruction of the optimal conditions of plasma formation. The discussed studies showed that, independently of the method of rotating target formation, the principle of rotation during ablation becomes crucial for maintaining the proper conditions of harmonics generation in the plasma plumes formed at a high pulse repetition rate.

The irradiation of the fresh surface of the target by each next pulse is not a necessary requirement for achieving harmonic stability. The heating shots can overlap on the target surface during its rotation until the overheating notably changes the conditions of plasma formation on the corrupted surface. Earlier, in the case of 1 kHz repetition rate lasers, the commonly used rotating speed was varied between 20 and 30 rotations per minute (rpm) for different targets [59]. This rotation speed allowed achieving the stable harmonic output. However, even at 5 rpm, the stability of harmonic yield was significantly better compared with the unmovable target. In the case of the 20 rpm speed of the targets ablated by 1 kHz pulses, approximately 30 pulses overlapped on the same spot.

In the case of ablation using 150 kHz pulses, the rotating speeds up to 1000 rpm were used to analyze optimal conditions for the generation of stable harmonics. Most of the discussed experiments were carried out at 150 rpm. However, the harmonic yield started to gradually decrease after some time from the beginning of ablation. In that case, approximately 400 laser shots were overlapped on the same spot. The increase of rotation speed up to 600 rpm did not allow the improvement of harmonic stability. Thus, in the case of 150 kHz ablation, the main factor of instability was the insufficient flow of heat out from the surface of the heated spot. However, the harmonic stability improved once the rotating target was dragged up and down. The optimization of this vertical movement of the target allows ensuring the reliable continuous generation of harmonics from LIP.

3.3.2 *Harmonic Spectra*

The excitation of plasma plumes was controlled by changing the fluence of heating femtosecond pulses. In these studies, a focusing lens was moved to change the heating beam spot size on the target surface, to control the fluence of the heating pulses. This method allowed the fine-tuning of the heating conditions on the target surface.

In the case of strong ablation, the plasma emission was too intense compared with harmonic emission. However, the most important obstacle was the appearance of the large concentration of the free electrons destroying the optimal phase-matching conditions between the driving and harmonic waves. The optimal fluence for plasma formation ($F = 0.5$ J cm^{-2}) was chosen by determining the strongest harmonic yield at different conditions of ablation. The optimal intensity of heating pulses at these conditions was equal to $I_{\text{hp}} = 2 \times 10^{12}$ W cm^{-2}. The corresponding intensities of heating pulses for two other used fluencies ($F = 1.0$ J cm^{-2} and $F = 2.0$ J cm^{-2}) were equal to $I_{\text{hp}} = 4 \times 10^{12}$ W cm^{-2} and $I_{\text{hp}} = 8 \times 10^{12}$ W cm^{-2}, respectively.

Figure 3.17a shows the harmonic and plasma emission spectra during the ablation of the carbon target at three different conditions of experiments. Upper panel demonstrates the pure and narrow harmonics extended up to the 41st order ($\lambda = 25.1$ nm, cutoff energy $E_c = 49.3$ eV) at 0.5 J cm^{-2} fluence of 250 fs heating pulses.

The middle panel spectrum was obtained at $F = 1$ J cm^{-2} fluence of heating pulses on the carbon target. This stronger ablation did not enhance the harmonic

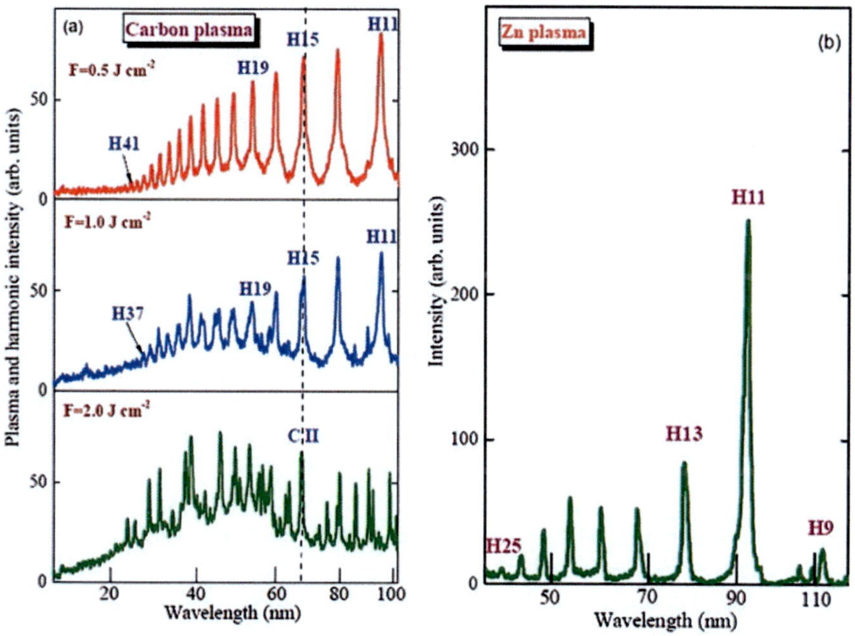

Fig. 3.17 **a** Harmonic emission from the carbon plasma produced at $F = 0.5$ J cm^{-2} (upper panel) and $F = 1$ J cm^{-2} (middle panel) fluencies of heating 250 fs pulses. Bottom panel shows the emission spectrum from carbon plasma during ablation using 2 J cm^{-2} fluence of the 250 fs pulses. The appearance of background points out the high excitation of laser-induced plasmas. The marks (H11, H15, etc.) on the panels of this and other figures refer to the harmonic orders. Dashed line corresponds to the strong emission (68.7 nm) of CII coinciding with H15. **b** Harmonic spectrum generated from zinc plasma. Reproduced from [58] with permission from Springer

yield due to the deterioration of the phase-matching conditions, despite the growth of the concentration of carbon plasma. One can see the appearance of some strong plasma emission lines. The emission from plasma still did not exceed the harmonic emission, since one had to avoid the formation of highly excited plasmas. While the harmonic cutoff expectably was not significantly changed, the larger amount of free electrons appearing during plasma formation, as well as the excitation of carbon atoms and ions, caused some deterioration of the phase-matching conditions between interacting waves achieved in the previous case (upper panel), which resulted in a decrease of harmonic yield.

The spectral resolution of the XUV spectrometer was equal to 0.5 nm. The natural width of harmonics in the 60 nm range was 0.7 nm, while the observed bandwidth was 2 nm and higher. Thus, the spectral resolution of the spectrometer allowed distinguishing the detuning of harmonic out from the resonance conditions. The appearance of the broadened harmonics (middle panel) was attributed to the propagation of the driving pulses through the denser plasma, containing a larger amount of free electrons.

The bottom panel of Fig. 3.17a shows the plasma emission spectrum in the case of twice higher fluence of 250 fs pulses (2 J cm^{-2}), without the propagation of 37 fs pulses through such plasma. At these conditions of plasma formation, only a few lowest-order harmonics appeared alongside the strong emission lines once the driving pulses propagated through such plasma. Note that, at these conditions of excitation, plasma emission is characterized by a broadband background.

One can see the coincidence of the strong emission lines with some of the harmonics. These conditions did not affect the envelope of harmonic distribution. Particularly, the presence of the strong ionic emission of CII in the XUV range at 68.7 nm (dashed line in Fig. 3.17a) did not affect the yield of nearby harmonic (H15), which points out the absence of the influence of ionic transitions on the harmonic efficiency. Notice that here we are talking about the single-particle response related to the micro-process of the resonance-induced enhancement of single harmonic, rather than multi-particle response, related with the macro-process of the phase mismatch, leading to a decrease of a whole set of harmonics.

In the case shown in the upper panel of Fig. 3.17a, the low fluence of heating pulses (0.5 J cm^{-2}) was used. At these conditions of ablation, no 68.7 nm plasma emission was observed, which has been proved by stopping the propagation of 37 fs driving pulses. Thus, here only harmonic emission was detected. The confirmation of the overlap of harmonic and emission lines and optimization of heating conditions was as follows. The driving pulses were blocked in front of the interaction zone of these pulses with LIP and collected the emission spectrum from plasma at different fluencies of heating pulses. Then, the blocker was removed and the fluence of heating pulses was reduced for optimization of the output of HHG emission from this plasma.

Actually, there is a very small difference between this emission (68.7 nm) and the wavelength of the 15th harmonic (68.66 nm), which cannot be distinguished by the spectrometer. Probably, the first signs of plasma emission could be seen in the second panel of this figure when one used twice stronger fluence of heating pulses (1 J cm^{-2}). However, the appearance of incoherent emission, which confirms the presence of such excited ions in the plasmas, did not result in the enhancement of either the 15th harmonic (compare this harmonic with the neighboring ones) or other harmonic orders.

There are a few previous studies of plasma HHG where the resonance enhancement of a single harmonic was attributed to the strong oscillator strength of the nearby transition (for example [60, 61]). The resonance enhancement of a single harmonic relies on the closeness or overlapping of the wavelength of this harmonic with the wavelength of the strong transition of an electron from the excited state to the ground state. The strength of transition, which is defined by its oscillator strength (*gf*), plays an important role in the enhancement of the nonlinear response of neutral or ionized atoms. The enhancement of a single harmonic can be observed in the vicinity of some ionic transition possessing high values of *gf*. The 68.7 nm C II line is actually made of three transitions with relatively large *gf*s (1.3, 0.71, and 0.14 [62]). However, the experiment using 1030 nm radiation showed neither the enhancement nor suppression of H15.

The resonance enhancement of the single harmonic of 800 nmfemtosecond driving pulses has initially been demonstrated at low pulse repetition rate conditions (10 Hz [63]). This mechanism shows the advantage in the enhancement of a single harmonic or a group of harmonics in the vicinity of some transitions possessing high *gf* s. To demonstrate this effect in the case of 150 kHz class laser sources, one has to resolve the problem of stable plasma formation. Otherwise, the impeding processes restricting HHG conversion efficiency using such lasers can strongly diminish the advantages in using the plasma media demonstrating the resonance effect. In the discussed studies, LIP was stabilized to analyze the conditions of the coincidence of the strong emission lines of CII and some harmonic orders (particularly, H15) of the 1030 nm laser. As it is seen in the middle panel of Fig. 3.17a, those studies showed the absence of resonance enhancement of harmonics, in spite of the closeness of some harmonic orders and emission lines.

Meanwhile, the studies confirmed that resonance-induced processes do not depend on the pulse repetition rate of driving radiation, and can be analyzed using 150 kHz class lasers. Here we discuss the enhancement of single (11th) harmonic (λ = 93.6 nm) of 1030 nm radiation generating in zinc plasma (Fig. 3.17b). The role of ionic resonances of Zn near the 9th and 10th harmonics of Ti: sapphire laser (800 nm) in amplifying these harmonics was discussed in [64]. Ablation of the zinc target allows generating harmonics in the presence of plasma emission lines. The observed HHG spectra under these conditions have shown the resonance-induced amplification at different wavelengths, corresponding to different transitions. In particular, the use of a near-infrared laser and it's second harmonic (1300 nm + 650 nm [65]) for pumping Zn plasma led to the generation of amplified H15 (86.7 nm), which was due to the influence of some strong Zn II transitions, particularly the $3d^{10}4s$–$3d^{9}4s4p$ transition [66]. The discussed studies suggest that this ionic transition can be responsible for the enhancement of H11 while using the 1030 nm laser.

Some experimental observations point out the collective nature of HHG from laser plumes. In resonance conditions, when the frequency of the harmonic is close to the frequency of an atomic or ion transition, the change in the wavenumber of a single harmonic can be significant, and the effect of free electron-induced phase mismatch can be compensated for by atomic dispersion for a certain harmonic order. There is no information in the literature about the wavenumber shift for a single harmonic in the resonance conditions. One can assume that the part of harmonic in resonance with the discussed ionic transition of zinc can be enhanced due to a resonance-induced increase of the nonlinear optical response of the laser-plasma. This means that propagation effects may play a decisive role in the formation of suitable phase-matching conditions between the driving and harmonic fields in the vicinity of the above-mentioned ionic transition. In that case, an improvement in the phase conditions for generating a single harmonic can be achieved.

The amplification of a particular harmonic can be associated with a modification of the refractive index of the plasma in the region of anomalous dispersion near the blue side of ionic transitions, which can create conditions for phase matching between the waves of the particular harmonic and the driving pulse. The influence of the single-particle-induced mechanism can be compensated for by the influence of the collective (i.e., multiparticle) processes related to the phase relations between the driving and harmonic waves. Notice that the absence of resonance-induced enhancement of 15th harmonic during the discussed studies using carbon LIP could be due to the insignificant excitation of the refereed C II transition, as well as the low concentration of carbon ions at the conditions corresponding to the maximal HHG conversion efficiency. Notice that no emission of the 68.7 nm line was observed alongside the emission of H15, as well as other plasma emission lines (see two upper panels of Fig. 3.17).

The process of HHG does not depend on the repetition rate of driving pulses. The reasons to carry out these studies are related to the demonstration of HHG in the plasmas produced at a high pulse repetition rate, which allowed one to achieve the highest flux of coherent XUV photons. The main restriction of plasma HHG during ablation of targets by high pulse repetition rate pulses is a difficulty in the formation of the stable laser-produced plasma plumes as the nonlinear media for the generation of stable harmonics. Two upper panels of Fig. 3.17 correspond to optimized HHG as a result of suitable excitation conditions of ablating target. At the high fluence of heating pulses, the concentration of free electrons is high enough to change the optimal phase conditions between the driving femtosecond pulses and generating high-order harmonics. Thus, the optimization of targets ablation at a high pulse repetition rate is important for the further application of such LIPs as the suitable media for HHG.

The other three LIPs containing B, Ti, and Mn species showed less efficient HHG compared to the carbon plasma while maintaining the same properties and dependences of the coherent XUV emission. Plasma emission from the ablated tablet containing the mixture of B NPs and Ag MPs, even at a three-fold increase of the fluence of heating pulse with regard to the optimal value of this parameter, was attributed to the transitions of the singly to triply ionized B, while no ionic lines from the heated Ag MPs were observed (Fig. 3.18a). As mentioned earlier, the weight ratio of these two components (B and Ag) was 3:1, so one would expect that the latter species should also emit in the XUV region. The absence of emission from Ag MPs can be attributed to insufficient excitation for evaporation compared with B NPs. One can assume that LIP at these conditions consisted of only B NPs and, correspondingly, those particles alongside the boron atoms and ions were responsible for HHG. Only low-order harmonics from this plasma were observed (H23, $\lambda = 44.7$ nm; Fig. 3.18b). One can note that silver being properly ablated serves as a medium for efficient harmonic generation, with the cutoffs in the range of 12 to 18 nm. The absence of higher-order harmonics points out the absence or insignificant amount of Ag particles in the plasma during the propagation of 37 fs pulses. The harmonic emission was analyzed at similar experimental conditions while ablating the bulk boron, which did not contain silver MPs (Fig. 3.18c). One can see the approximately similar spectral distribution of lower-order harmonics, like in the case of ablated B NP + Ag MPs target. The harmonic emission was compared from the ablated B NPs + Ag MPs and bulk boron target. Though the harmonic efficiency in the former case was larger, the cutoffs in both cases were approximately similar to each other (Fig. 3.18b, c).

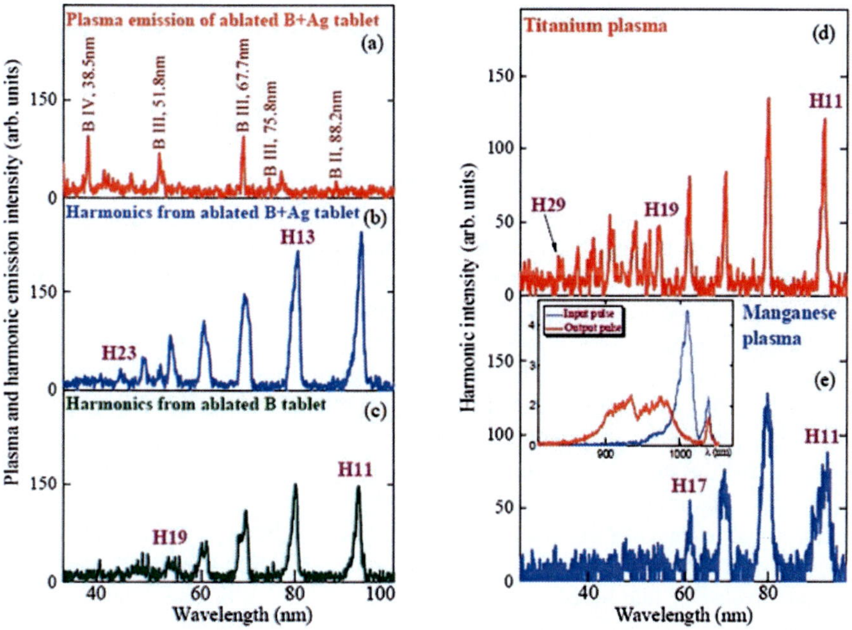

Fig. 3.18 **a** Plasma emission from the ablated B + Ag tablet at strong intensity of 250 fs heating pulses ($I_{hp} = 4.4 \times 10^{12}$ W cm^{-2}), **b** harmonic emission from the ablated B + Ag tablet at optimal intensity of heating pulses ($I_{hp} = 1.5 \times 10^{12}$ W cm^{-2}), **c** harmonic emission from the ablated bulk B at similar conditions, **d** harmonics generated from titanium LIP, **e** harmonics generated from manganese LPP. Reproduced from [58] with permission from Springer

3.3.3 Discussion

These studies showed the peculiarities of harmonics generation in the complex plasmas produced on the mixed powdered targets. Similar to the cases of carbon and zinc plasmas, the repetition rate of driving pulses at proper conditions of plasma formation allows observing the harmonic emission from the powdered NPs, while at the same time, increasing the average power of generated high-order harmonics from the nanostructured LIPs.

Harmonic spectra from titanium and manganese plasmas are shown in Fig. 3.18d, e. The broadening of the harmonic spectra was observed in the case of Mn and nanoparticle plasma plumes. The appearance of the broadened harmonics was attributed to the propagation of the driving pulses through the denser plasma, containing a larger amount of free electrons. The main reason for this process is the phase modulation of driving radiation, which led to the rearrangement of the spectrum and intensity of generated harmonics. The self-phase modulation (SPM) of laser pulses in LIP is widely used for the explanation of the frequently reported generation of additional frequencies in the spectrum of the harmonics of laser radiation [67]. The conditions of harmonic spectra broadening were analyzed at a comparatively "weak" influence of the SPM on the spectral properties of harmonics. This term refers to the conditions under which the spectral and spatial parameters of radiation to be converted change insignificantly upon passage through the nonlinear medium. The SPM of a laser pulse increases with the increase of laser intensity or the particle density in the plasma plume.

In the case of Ti plasma, no impeding processes prevented the phase-matched HHG during propagation through the LIP, in spite of the appearance of the ionic lines alongside the shorter wavelength harmonics (Fig. 3.18d; see the spectral range between H19 and H29). Contrary to that, the used Mn LIP caused a strong broadening of the harmonic spectra (Fig. 3.18e). SPM leading to the broadening of the harmonic spectra from Mn plasma can also be attributed to the strong excitation of LIP on the surface of the Mn target at a high fluence of heating pulses. In that case, the broadened spectra of harmonics compared with those generated in Ti plasma were observed. SPM depends on the condition and parameters of LIP. The excitation of plasma on the surface of the Mn target requires less fluence compared with other metal targets. In this case, the intensity of heating pulses was high enough for excitation of the double charged ions of Mn. The presence of additional free electrons in LIP led to the SPM of driving pulses, which caused the broadening of the harmonic spectra.

The broadening and blue-shifting of the HHG spectra were also observed at the high intensity of driving pulses. It can be explained by the appearance of the blue-shifted components in the spectrum of the driving pulse (see inset in Fig. 3.18). One can see the appearance of the blue components in the spectra of the driving pulses propagated through LIP during HHG in plasma. This figure shows modulation of the spectrum of driving pulses during propagation through the plasma plumes excited on the surface of metal targets. The spectrum of driving pulses was examined before (input pulses; blue thin curve) and after (output pulses; red thick curve) propagation through the plasmas produced by strong heating pulses. One can see the spectral broadening and shift of femtosecond driving pulses after propagation of plasma plumes, due to the group velocity dispersion of different spectral components of the 37 fs pulses. Notice that, at optimal conditions of laser ablation and plasma formation, the spectral shape of the driving pulse after propagation through the plasma plume remains unchanged. The term "optimal conditions" refers to the case when one was able to achieve the maximal yield of harmonics.

There is no accumulation effect on the broadening of femtosecond driving pulses at the high repetition rate. The interval between pulses at 150 kHz ablation is ~6 μs. During this period, the plasma entirely disappears, and a new ablation will be produced on the fresh surface. Thus, from the context of pulse repetition rate, the broadening of high-order harmonics does not depend on this parameter, provided that the interval between pulses becomes larger than the lifetime of plasma (~100–300 ns). Notice that this time interval corresponds to the MHz class lasers.

The availability in the formation of the stable B, Ti, and Mn plasmas at the used pulse repetition rate allowed making the conclusion that the latter parameter does not play important role in the relative intensities of harmonics from different plasmas. Particularly, carbon plasma has been demonstrated to be the efficient medium using other wavelengths (1300 nm, 800 nm, and 400 nm) and different pulse repetition rates of laser sources. The stabilization of plasma formation at 150 kHz ablation allowed concluding about the smaller HHG conversion efficiency in above metal plasmas with regard to the carbon plasma. Thus, the high repetition rate of driving pulses at proper conditions of plasma formation leads to increasing the average power and flux of generating high-order harmonics, while also allowing the analysis of different processes demonstrated using the low pulse repetition rate lasers.

Notice that the application of the 1030 nm wavelength in these studies did not allow the observation of the resonance enhancement of harmonics in Mn plasma. The difference in harmonic spectra using various laser sources points out the role of proper matching of the wavelengths of some harmonics and ionic transitions possessing high oscillator strengths.

The measurements of pulse energy in the XUV range are difficult to carry out in each HHG study. Previous studies of harmonics from different plasmas included the quantitative measurements of coherent XUV yield. Taking into account the similarity of relative intensities of harmonics from different plasmas, one can estimate the flux of harmonic emission using previous measurements of HHG conversion efficiency ($\sim 10^{-5}$, [68]). The flux of single (H25, $E = 30$ eV) harmonics in analyzed studies using carbon LIP was estimated to be 8×10^{13} photons/s. The corresponding fluxes obtained from other LIPs were 4×10^{13} (B NP + Ag MP), 2×10^{13} (Mn, B, and Ti), and 1.5×10^{13} (Zn) photons/s.

The average flux of harmonics achieved in this study is high enough to compare with the maximal harmonic fluxes reported in the case of gas HHG. Recently, high flux from argon gas was reported using a similar high pulse repetition source [69]. They produced $\sim 7 \times 10^{14}$ photons/s of high-order harmonic in argon, which corresponds to ~ 2 mW average power in the case of their 166 kHz laser source. However, this flux was obtained using the third harmonic (344 nm) of the 1030 laser and can be compared with the discussed case taking into account the λ^{-5} rule, which gives ~ 250 × -fold growth of harmonic yield in the case of 344 nm pump compared with 1030 nm pump. Meanwhile, the maximal flux reported in the case of the 1030 nm pump was less than 10^{11} photons/s, which is significantly smaller than the flux achieved during the studies. It is not surprising, since the HHG conversion efficiency of 1030 nm pulses in the plateau range in their case was measured to be about 1×10^{-8} (500 nW from 45 W) in the 30–40 eV spectral range. Notice that, at the 1030 nm pump, a higher cutoff than that in the refereed paper at the same wavelength (>62 eV and < 50 eV, respectively) was achieved. Furthermore, notice that the concentration of argon gas in those studies was 15 times larger than the concentration of particles in the plasmas. This comparison also points out the higher conversion efficiency in the case of HHG, in plasmas compared with gas media.

3.4 Summary

In this Chapter, the application of the two-laser system approach allows analyzing the slowly spreading plasma, which contains molecular species, in a broad time scale by temporal matching with the propagation of the driving femtosecond pulses through the plasma plume. The advantage of HHG in the plasma created by the nanosecond laser system digitally synchronized with the femtosecond laser system is the delay-dependent analysis of the harmonic yield generated from different components of plasma plumes. The discussed experiments demonstrated the dependence of harmonic yield from different materials on the delay between the heating and driving pulses. The optimal delays for B_4C, ZnO, GaP, GaAs, and Ag_2S plasmas were determined to be approximately 200, 300, 350, 500, and 700 ns, respectively. These variations of delays correspond to the $t \propto (m)^{0.5}$ dependence for both molecular and atomic species. This diagnostic technique can be applied to the analysis of the presence of nanostructures in LIPs through HHG with high spatiotemporal resolution. We discussed the applicability of the proposed method for analysis of the resonance-enhanced harmonics in atomic and molecular plasmas (Mo and MoS_2) and for studies of the large perovskite aggregates as potential emitters of harmonics. This technique can also be useful for the study of the contribution of different evaporated molecular species for HHG using extremely short pulses. Additionally, the joint application of the harmonic generation and mass-spectroscopy techniques can be considered as a highly versatile tool for in situ diagnostics of the laser plasmas containing a large population of clusters, molecules, and nanoparticles.

We analyzed the high-order harmonics generation using 1 and 100 kHz lasers by ablating different rotating targets, as well as discussed the high average flux of coherent short-wavelength radiation using the latter laser while comparing the plasma formation conditions at different pulse repetition rates. The analysis of harmonic stability in the case of 100 kHz experiments has shown the two-fold decay of the 27th harmonic generating in silver plasma after 3.5×10^6 shots, while no decay in harmonic yield was observed in the case of 1 kHz ablation. The two-color pump of plasma, resonance enhancement of single harmonic, and quasi-phase-matching scheme were presented in the case of 1 and 100 kHz pulse repetition rates of target ablation. The formation of modulated multi-jet plasma on the plane and curved surfaces during ablation by 100 kHz pulses has been demonstrated. In the case of the 25th harmonic of 1030 nm radiation ($E = 30$ eV) generated during experiments in carbon plasma, at 100 kHz and 40 W average power of driving pulses, 0.4 mW of average power in the 40 nm spectral range was achieved.

Notice that these studies are the first step towards the formation of the stable plasma plume using 100 kHz class lasers, which needs additional efforts for maintaining stable harmonic yield from such laser-produced plasma. We would like to mention the incompleteness in resolving the stability issue for generating harmonics in the case of 100 kHz plasma. The possible scenarios to overpass this issue by dragging the target up and down, alongside target rotation at optimal velocity, are suggested. The preliminary experiments with the manually movable target (e.g., by pressing and releasing the target holder from the top of the vacuum chamber) have shown the perspective in the stability improvement for harmonic generation from high-pulse repetition rate plasma. The following plans include the programmable movement of rotating targets up and down, which can further diminish the instability issue.

Finally, we discussed the study of HHG using a 150 kHz laser and the details of the application of the rotating target for improvement of stability of the low and harmonics from plasma in the case of low and high pulse repetition rates of laser-matter interactions. In discussed studies, the harmonics were generated in carbon, boron, manganese, zinc, and titanium plasmas using a 150 kHz laser. The flux of single (H25, $E = 30$ eV) harmonics using carbon LIP was estimated to be 8×10^{13} photons/s. The corresponding fluxes obtained from other LPPs were 4×10^{13} (B NP + Ag MP), 2×10^{13} (Mn, B, and Ti), and 1.5×10^{13} (Zn) photons/s. The discussed studies also demonstrated the resonance enhancement of single harmonic, HHG in nanoparticles, and other harmonic generation effects, using 1030 nm, 150 kHz, 37 fs laser source.

References

1. A. L'Huillier, P. Balcou, Phys. Rev. Lett. **70**, 774 (1993)
2. D. von der Linde, K. Rzàzewski, Appl. Phys. B **63**, 499 (1996)
3. F. Krausz, M. Ivanov, Rev. Mod. Phys. **81**, 163 (2009)
4. C. Vozzi, R. Torres, M. Negro, L. Brugnera, T. Siegel, C. Altucci, R. Velotta, F. Frassetto, L. Poletto, P. Villoresi, S. De Silvestri, S. Stagira, J.P. Marangos, Appl. Phys. Lett. **97**, 241103 (2010)
5. M. Oujja, R. de Nalda, M. López-Arias, R. Torres, J.P. Marangos, M. Castillejo, Phys. Rev. A **81**, 043841 (2010)
6. H. Singhal, V. Arora, B.S. Rao, P.A. Naik, U. Chakravarty, R.A. Khan, P.D. Gupta, Phys. Rev. A **79**, 023807 (2009)
7. R.A. Ganeev, M. Suzuki, M. Baba, M. Ichihara, H. Kuroda, J. Phys. B **41**, 045603 (2008)
8. R.A. Ganeev, M. Suzuki, M. Baba, M. Ichihara, H. Kuroda, J. Appl. Phys. 103, 063102 (2008)
9. Y. Pertot, S. Chen, S.D. Khan, L.B. Elouga Bom, T. Ozaki, Z. Chang, J. Phys. B 45, 074017 (2012)
10. S. Haessler, V. Strelkov, L.B. Elouga Bom, M. Khokhlova, O. Gobert, J.-F. Hergott, F. Lepetit, M. Perdrix, T. Ozaki, P. Salières, New J. Phys. 15, 013051 (2013)
11. N. Rosenthal, G. Marcus, Phys. Rev. Lett. **115**, 133901 (2015)
12. M.A. Fareed, V.V. Strelkov, N. Thiré, S. Mondal, B.E. Schmidt, F. Légaré, T. Ozaki, Nat. Commun. **8**, 16061 (2017)
13. R.A. Ganeev, M. Suzuki, M. Baba, M. Ichihara, H. Kuroda, J. Opt. Soc. Am. B **25**, 325 (2008)
14. R.A. Ganeev, L.B. Elouga Bom, T. Ozaki, J. Phys. B 42, 055402 (2009)
15. M.A. Fareed, V.V. Strelkov, M. Singh, N. Thiré, S. Mondal, B.E. Schmidt, F. Légaré, T. Ozaki, Phys. Rev. Lett. **121**, 023201 (2018)
16. R.A. Ganeev, *High-Order Harmonic Generation in Laser Plasma Plumes* (Imperial College Press, London, 2012)
17. R. de Nalda, M. Lopez-Arias, M. Sanz, M. Oujja, M. Castillejo, Phys. Chem. Chem. Phys. **13**, 10755 (2011)
18. I. Lopez-Quintas, M. Oujja, M. Sanz, M. Martın, R.A. Ganeev, M. Castillejo, Appl. Surf. Sci. **278**, 33 (2013)
19. J.J. Camacho, M. Oujja, M. Sanz, A. Martìnez-Hernandez, I. Lopez-Quintas, R. de Nalda, M. Castillejo, J. Anal. At. Spectrom. **34**, 489 (2019)
20. D.W. Hahn, N. Omenetto, Appl. Spectrosc. **64**, 335A (2010)
21. M. Oujja, M. Sanz, F. Agua, J.F. Conde, M. García-Heras, A. Dávila, P. Oñate, J. Sanguino, J.R. Vázquez de Aldana, P. Moreno, M.A. Villegas, M. Castillejo, J. Anal. At. Spectrom. **30**, 1590 (2015)

22. G.S. Boltaev, R.A. Ganeev, V.V. Kim, K. Zhang, M. Venkatesh, C. Guo, Phys. Plasmas **26**, 100703 (2019)
23. M. Oujja, I. Lopez-Quintas, A. Benítez-Cañete, R. de Nalda, M. Castillejo, Appl. Surf. Sci. **392**, 572 (2017)
24. B. Verhoff, S.S. Harilal, J.R. Freeman, P.K. Diwakar, and A. Hassanein J. Appl. Phys. **112**, 093303 (2012)
25. M. Masnavi, M. Nakajima, K. Horioka, H.P. Araghy, A. Endo, J. Appl. Phys. **109**, 123306 (2011)
26. W. Marine, J.M. Scotto d'Aniello, M. Gerri, Mat. Sci. Eng. B 13, 57 (1992)
27. J.P. Zheng, Z.Q. Huang, D.T. Shaw, H.S. Kwok, Appl. Phys. Lett. **54**, 280 (1989)
28. S.M. Park, J.Y. Moon, J. Appl. Phys. **86**, 7139 (1999)
29. P.B. Corkum, Phys. Rev. Lett. **71**, 1994 (1993)
30. M. Lewenstein, P. Balcou, M.Y. Ivanov, A. L'Huillier, P. Corkum, Phys. Rev. A **49**, 2117 (1994)
31. D.B. Milošević, J. Phys. B **40**, 3367 (2007)
32. V. Strelkov, Phys. Rev. Lett. **104**, 123901 (2010)
33. M. Tudorovskaya, M. Lein, Phys. Rev. A **84**, 013430 (2011)
34. R.A. Ganeev, H. Singhal, P.A. Naik, V. Arora, U. Chakravarty, J.A. Chakera, R.A. Khan, P.V. Redkin, M. Raghuramaiah, P.D. Gupta, J. Opt. Soc. Am. B **23**, 2535 (2006)
35. R.A. Ganeev, K.S. Rao, Z. Yu, W. Yu, C. Yao, Y. Fu, K. Zhang, C. Guo, Opt. Mater. Express **8**, 1472 (2018)
36. G.S. Boltaev, R.A. Ganeev, P.S. Krishnendu, K. Zhang, C. Guo, Sci. Rep. **9**, 11414 (2019)
37. G.S. Boltaev, R.A. Ganeev, N.A. Abbasi, M. Iqbal, V.V. Kim, H. Al-Harmi, X.M. Tong, A.S. Alnaser, New J. Phys. **22**, 083031 (2020)
38. Y. Akiyama, K. Midorikawa, Y. Matsunawa, Y. Nagata, M. Obara, H. Tashiro, K. Toyoda, Phys. Rev. Lett. **69**, 2176 (1992)
39. Z. Abdelrahman, M.A. Khohlova, D.J. Walke, T. Witting, A. Zair, V.V. Strelkov, J.P. Marangos, J.W.G. Tisch, Opt. Express **26**, 15745 (2018)
40. R.A. Ganeev, G.S. Boltaev, V.V. Kim, M. Venkatesh, A.I. Zvyagin, M.S. Smirnov, O.V. Ovchinnikov, M. Wöstmann, H. Zacharias, C. Guo, J. Appl. Phys. **126**, 193103 (2019)
41. M. Wöstmann, L. Splitthoff, H. Zacharias, Opt. Express **26**, 14524 (2018)
42. C. Hutchison, R.A. Ganeev, T. Witting, F. Frank, W.A. Okell, J.W.G. Tisch, J.P. Marangos, Opt. Lett. **37**, 2064 (2012)
43. R.A. Ganeev, G.S. Boltaev, V.V. Kim, M. Venkatesh, C. Guo, OSA Continuum **2**, 2381 (2019)
44. Y. Pertot, L.B. Elouga Bom, V.R. Bhardwaj, & T. Ozaki, Appl. Phys. Lett. 98, 101104 (2011)
45. R.A. Ganeev, M. Suzuki, M. Baba, H. Kuroda, Phys. Rev. A **76**, 023805 (2007)
46. G.S. Boltaev, R.A. Ganeev, V.V. Kim, N.A. Abbasi, M. Iqbal, A.S. Alnaser, Opt. Express **28**, 18859 (2020)
47. F. Brygo, C. Dutouquet, F. Le Guern, R. Oltra, A. Semerok, J.M. Weulersse, Appl. Surf. Sci. **252**, 2131 (2006)
48. R.A. Ganeev, L.B. Elouga Bom, J.-C. Kieffer, T. Ozaki, Phys. Rev. A 76, 023831 (2007)
49. I.J. Kim, G.H. Lee, S.B. Park, Y.S. Lee, T.K. Kim, C.H. Nam, T. Mocek, K. Jakubczak, Appl. Phys. Lett. **92**, 021125 (2008)
50. R.A. Ganeev, V.V. Strelkov, C. Hutchison, A. Zaïr, D. Kilbane, M.A. Khokhlova, J.P. Marangos, Phys. Rev. A **85**, 023832 (2012)
51. J. Bechtel, J. Appl. Phys. **46**, 1585 (1975)
52. R.A. Ganeev, V. Tosa, K. Kovács, M. Suzuki, S. Yoneya, H. Kuroda, Phys. Rev. A **91**, 043823 (2015)
53. V.V. Strelkov, R.A. Ganeev, Opt. Express **25**, 21068 (2017)
54. P.V. Redkin, R.A. Ganeev, C. Guo, J. Phys. B **52**, 075601 (2019)
55. L.B. Elouga Bom, S. Haessler, O. Gobert, M. Perdrix, F. Lepetit, J.-F. Hergott, B. Carré, T. Ozaki, P. Salières, Opt. Express 19, 3677 (2011)
56. L.B. Elouga Bom, Y. Pertot, V.R. Bhardwaj, T. Ozaki, Opt. Express 19, 3077 (2011)
57. R.A. Ganeev, M. Baba, M. Suzuki, H. Kuroda, Phys. Lett. A **339**, 103 (2005)

58. G.S. Boltaev, V.V. Kim, M. Iqbal, N.A. Abbasi, V.S. Yalishev, R.A. Ganeev, A.S. Alnaser, Photonics **7**, 66 (2020)
59. M. Wöstmann, P.V. Redkin, J. Zheng, H. Witte, R.A. Ganeev, H. Zacharias, Appl. Phys. B **120**, 17 (2015)
60. R.A. Ganeev, P.A. Naik, H. Singhal, J.A. Chakera, P.D. Gupta, Opt. Lett. **32**, 65 (2007)
61. M. Suzuki, M. Baba, H. Kuroda, R.A. Ganeev, T. Ozaki, Opt. Express **15**, 1161 (2007)
62. An Atomic Database for Spectroscopic Diagnostics of Astrophysical Plasmas. CHIANTI Database Version 9.0 released. https://chiantidatabase.org. Accessed Mar 2019
63. R.A. Ganeev, H. Singhal, P.A. Naik, V. Arora, U. Chakravarty, J.A. Chakera, R.A. Khan, I.A. Kulagin, P.V. Redkin, M. Raghuramaiah, P.D. Gupta, Phys. Rev. A **74**, 063824 (2006)
64. R.A. Ganeev, M. Baba, M. Suzuki, S. Yoneya, H. Kuroda, J. Appl. Phys. **116**, 243102 (2014)
65. R.A. Ganeev, M. Suzuki, S. Yoneya, V.V. Strelkov, H. Kuroda, J. Phys. B **49**, 055402 (2016)
66. A.M. Crooker, K.A. Dick, Canad. J. Phys. **46**, 1241 (1968)
67. R.A. Ganeev, H. Singhal, P.A. Naik, J.A. Chakera, M. Tayyab, M. Baba, H. Kuroda, P.D. Gupta, J. Opt. Soc. Am. B **26**, 2143 (2009)
68. R.A. Ganeev, C. Hutchison, T. Witting, F. Frank, W.A. Okell, A. Zaïr, S. Weber, P.V. Redkin, D.Y. Lei, T. Roschuk, S.A. Maier, I. López-Quintás, M. Martín, M. Castillejo, J.W.G. Tisch, J.P. Marangos, J. Phys. B **45**, 165402 (2012)
69. A. Comby, D. Descamps, S. Beauvarlet, A. Gonsalez, F. Guichard, S. Petit, Y. Zaouter, Y. Mairesse, Opt. Express **27**, 20383 (2009)

Chapter 4
Applications of Clusters and Quantum Dots for HHG in LIPs

4.1 Low- and High-Order Nonlinear Optical Properties of Ag_2S Quantum Dot Thin Films

Recently, significant efforts have been made in the syntheses of metal chalcogenide nanocrystals (NCs) due to their attractive size-tunable optical properties and possible applications in various technological fields including light-emitting diodes [1, 2], photovoltaic devices [3, 4], and fluorescent biological labels [5, 6]. Several other sulfide precursors were subsequently reported, but elemental sulfur has been the most widely used as a less air-sensitive precursor in the syntheses of metal sulfide NCs. Decreasing the size of such materials allows to tune their physical properties and observe new phenomena. Taking this into account, newly synthesized semiconductor nanoparticles should be studied under different conditions to understand the nonlinear optical mechanisms and use their attractive properties for practical applications [7–9].

Thin films of different materials have also attracted much attention due to their various promising applications [10–15]. Analyses of linear and nonlinear optical features of the thin films comprising metal sulfide NCs are important for their possible use in various fields of optoelectronics and laser physics. To verify the generalized effective-medium theories for linear and nonlinear optical properties of metal sulfide nanoparticles and quantum dots (QDs) requires the knowledge of the linear and nonlinear susceptibilities of the constituent materials Ag_2S QDs are non-cadmium and non-lead fluorescent nanomaterials developed in recent years, and Ag_2S QDs are known for displaying large optical nonlinearities [16–18]. Ag_2S QD thin films can have even higher nonlinearities.

The QDs can also be efficiently used as the sources of high harmonics generation (HHG). HHG in gases, plasmas, and solids has been extensively studied due to its possible applications in the generation of attosecond laser pulses and the formation of short-wavelength coherent sources [19–27]. Note that the size variation of the polyatomic systems like QDs can efficiently manipulate the HHG spectra.

In this section, we discuss the variations of the optical nonlinearities of Ag_2S QD thin films of different thicknesses [28]. We also analyze the ablation of Ag_2S QD films

R. A. Ganeev, *High-Order Harmonics Generation in Plasmas*, Springer Series on Atomic, Optical, and Plasma Physics 122, https://doi.org/10.1007/978-3-031-09040-0_4

for efficient emission of high-order harmonics of femtosecond pulses. Particularly, we discuss HHG using the electronically driven delay between the heating and the driving lasers for the generation of harmonics from Ag_2S QD-contained plasmas in the range of 30–80 nm.

4.1.1 Experimental Arrangements of Ag_2S QD Thin Films

In the typical process of synthesis 17 g $AgNO_3$ was dissolved in 100 mL tri-n-octyl-phosphine (TOP) to obtain the 1 M Ag-TOP. 6.4 mL of oleic acid, 5.4 g of 1-octadecylamine (ODA), and 12.8 mL of 1-octadecene were mixed in a 100 mL round-bottom flask under N_2 flushing three times to remove the water and oxygen and heated up to 70 °C. With the continuous N_2 supply and stirring, 72 mg of sulfur powder was added into the flask immediately and heated to the temperature of 190 °C. Then 4 mL of Ag-TOP was quickly injected in the flask, and the reaction was allowed to proceed for 5 min at 160 °C. 10 ml of N-butyl alcohol was injected to provide solidification from ODA. The synthesized QDs were isolated by precipitation with methyl alcohol and redispersed in methylbenzene. This process was repeated three times to ensure a clean product. The obtained QDs dispersed in either hexane or toluene were rinsed by the ultrasonic process to achieve better dispersion of QDs. The Ag_2S QD solution was then spin-coated on the glass slides to produce films of different thicknesses (80 and 500 nm).

In order to study the low-order nonlinear optical properties of these two films, the single-beam z-scan technique was employed [29]. Ti:sapphire laser (800 nm wavelength, 30 fs pulse duration, 1 kHz repetition rate) was used in these studies. The schematic diagram of the experimental setup is shown in Fig. 4.1a. The nonlinear optical properties of films were also analyzed using the second harmonic of this radiation ($\lambda = 400$ nm).

Before the z-scan measurements, the spatial characteristics of the probe beam were analyzed using a CCD camera and confirmed that the beam profile in the focal area was close to Gaussian, which is a prerequisite for the analysis of z-scan traces using the relations developed for this technique. The full width of focused femtosecond beams at $1/e^2$ level of intensity distribution was measured to be 72 μm. Before measuring the signal using a large area photodetector with an aperture in front of it [closed-aperture (CA) scheme with 1 mm aperture allowing propagation of ~10% of output radiation and open-aperture (OA) scheme with fully opened aperture] a detector was used to measure the incident pulse energy. The measurements of propagated radiation allowed the determination of the nonlinear refractive indices (γ) and the nonlinear absorption coefficients (β) of samples. The CA scheme allowed calculation of the sign and magnitude of γ and β of films. The OA z-scan was used for the measurements of β, which were compared with the measurements of this parameter using the CA scheme.

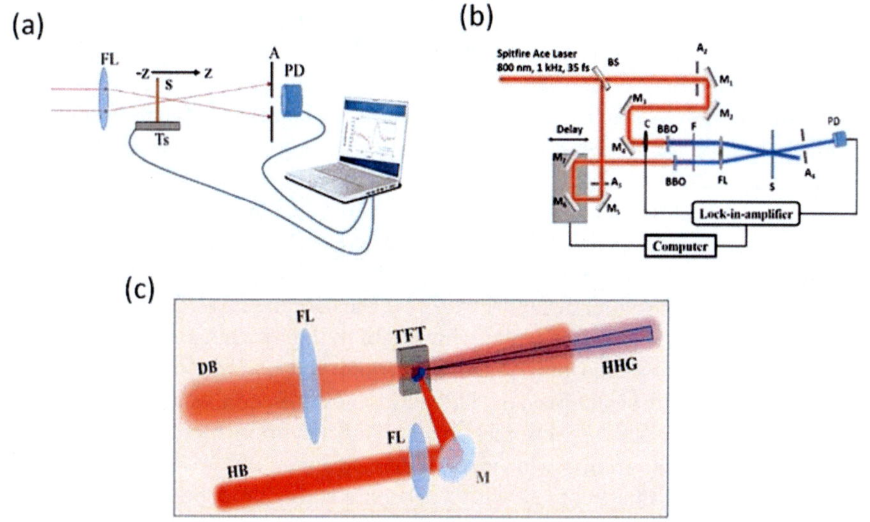

Fig. 4.1 Experimental setup. **a** Experimental setup for the measurement of the low-order nonlinear optical characteristics of the films. **b** Transient absorption experimental setup. FL, focusing lens; S, sample; Ts, translation stage; A, aperture; PD, photo detector; F, filter; M, mirrors; BS, beam splitter; BBO, β-barium borate crystal. **c** HHG setup. DB driving beam; HB, heating beam; FL, focusing lenses; M, mirror; TFT, thin film target; HHG, HHG beam. Reproduced from [28] with permission from Walter de Gruyter GmbH & Science Wise Publishing

The transient absorption (TA) measurements were carried out using the noncollinear pump–probe technique at $\lambda = 400$ nm. The schematic of the experimental setup is shown in Fig. 4.1b. Femtosecond pulse ($\lambda = 800$ nm, $t = 30$ fs) was split at the ratio of 30:70 for the pump and probe pulses. The BBO crystals were used to generate the 400 nm pump and probe radiation. These two pulses were focused on the Ag_2S QD thin film using a 300 mm focal length lens. The energies and beam waists of the pump and the probe pulses at the focus were adjusted to be 10:1 and 1:2 respectively to match the condition required for observation of TA. In order to introduce the delay in arrival time of probe pulses with respect to pump pulses, the former radiation was reflected by a pair of mirrors mounted on the motorized translational stage. The arrival time of probe pulses with respect to pump pulses was controlled by moving the motorized translational stage connected to the motion controller. The fast photodiode and lock-in-amplifier were used to measure the transmittance of the probe pulses with respect to the position of the motorized stage. Lock-in-amplifier was externally triggered by an optical chopper running at 300 Hz.

The Ti:sapphire femtosecond laser and Nd:YAG nanosecond laser were used in HHG studies. The nanosecond Nd:YAG laser operating at a 10 Hz pulse repetition rate is advantageous for plasma formation compared with the earlier used picosecond pulses. The nanosecond pulses allow the formation of less ionized and less excited plasma during a longer period of laser-matter interaction compared with shorter pulses. Additionally, the delay between the heating and the driving pulses could be electronically adjusted in a broad range (0–10^6 ns), contrary to the earlier used optically driven delay.

The driving femtosecond pulses (800 nm, 30 fs, 1 kHz) propagated through the plasma produced on the Ag_2S NP film at different delays from the beginning of target irradiation by nanosecond heating pulses (1064 nm, 5 ns, 10 Hz). The variable delay between heating and driving pulses was established to generate harmonics in plasma at the used geometry of experiments, when the 800 nmfemtosecond pulses were focused onto the area of plasma orthogonally to ablating radiation, at a distance of ~200 μm above the target surface (Fig. 4.1c). The harmonics yield was maximized by adjusting the position of the target with regard to the optical axis of propagation of the driving femtosecond pulses and by varying the position of focus for this radiation with regard to the plasma. The harmonics emission was directed to the extreme ultraviolet (XUV) spectrometer containing a cylindrical mirror and a 1200 grooves/mm flat field grating with variable line spacing. The XUV spectrum was recorded by a micro-channel plate with a phosphor screen, and the harmonics were imaged by a CCD camera.

4.1.2 Z-Scans and Transient Absorption Measurements of Ag_2S QD Films

Figure 4.2 shows the absorption spectra of two QD films of different thicknesses (80 and 500 nm). The inset shows the TEM image of Ag_2S QDs. The average size of QDs was ~4 nm. Figure 4.3 shows the CA z-scans of two Ag_2S QD films using 30 fs pulses at $\lambda = 800$ nm. The pulse energy was 0.36 μJ. The maximum fluence at the beam waist during these z-scans was 7×10^{-3} J cm^{-2} and the peak intensity was $I_0 = 2 \times 10^{11}$ W cm^{-2}.

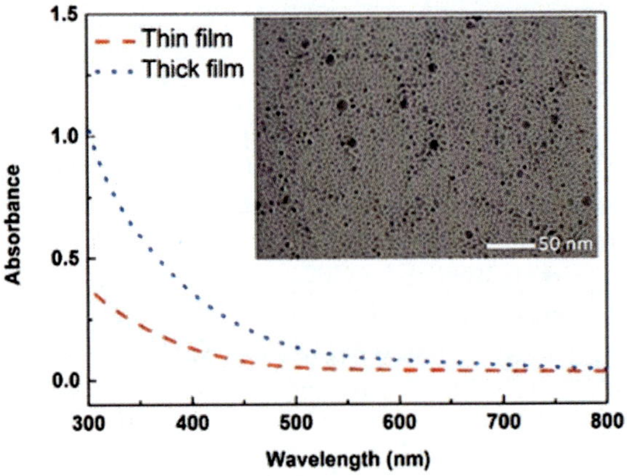

Fig. 4.2 Absorption spectra of thin (80 nm) and thick (500 nm) Ag_2S films. Inset: TEM image of synthesized silver sulfide QDs. Reproduced from [28] with permission from Walter de Gruyter GmbH & Science Wise Publishing

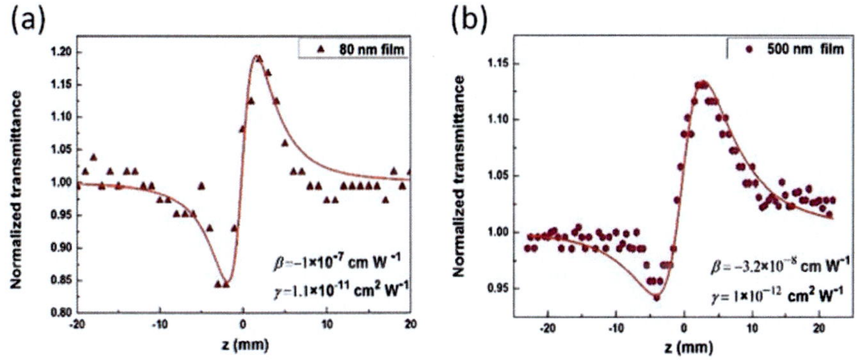

Fig. 4.3 Z-scan measurements at 800 nm. CA z-scans of **a** 80 nm and **b** 500 nm thick films. Solid curves show the fits using Eq. (4.2). Reproduced from [28] with permission from Walter de Gruyter GmbH & Science Wise Publishing

In the general case of joint contribution of nonlinear refraction and absorption the normalized transmittance of samples along z-axis, T(z), can be presented as follows [30]:

$$T(z) = 1 + \frac{4x}{(x^2+9)(x^2+1)}\Delta\Phi_0 - \frac{2(x^2 3)}{(x^2+9)(x^2+1)}\Delta\psi_0 \qquad (4.1)$$

where $\Delta\Phi_0 = k\gamma I_0 L_{\rm eff}$ and $\Delta\psi_0 = \beta I_0 L_{\rm eff}/2$ are the phase variations due to nonlinear refraction and nonlinear absorption, respectively, $x = z/z_0$, $z_0 = k\,(w_0)^2/2$ is the Rayleigh length, $k = 2\pi/\lambda$ is the wave number, w_0 is the beam waist radius at the 1/e^2 level of intensity distribution, I_0 is the intensity in focal plane, $L_{\rm eff}$ = [1-exp(-$\alpha_0 L$)]/α_0 is the effective length of the medium, α_0 is the linear absorption coefficient and L is the thickness of sample.

By making the substitution $\rho = \beta/2k\gamma$ one can get the relation between $\Delta\Phi_0$ and $\Delta\psi_0$ ($\Delta\psi_0 = \rho\Delta\Phi_0$). In that case Eq. (4.1) can be re-written as follows:

$$T = 1 + \frac{2(-\rho x^2 + 2x - 3\rho)}{(x^2+9)(x^2+1)}\Delta\Phi_0. \qquad (4.2)$$

This formula was fitted with the experimental CA data.

The thinner films had higher values of γ and β. The β of the 80 and 500 nm films was measured to be -1×10^{-7} and -3.2×10^{-8} cm W^{-1}. The negative nonlinear absorption was attributed to saturable absorption (SA). The nonlinear refractive index of the thicker film ($\gamma = 1 \times 10^{-12}$ cm^2 W^{-1}) was 11 times smaller than that of the thin film ($\gamma = 1.1 \times 10^{-11}$ cm^2 W^{-1}). Notice that z-scan measurements of the Ag_2S

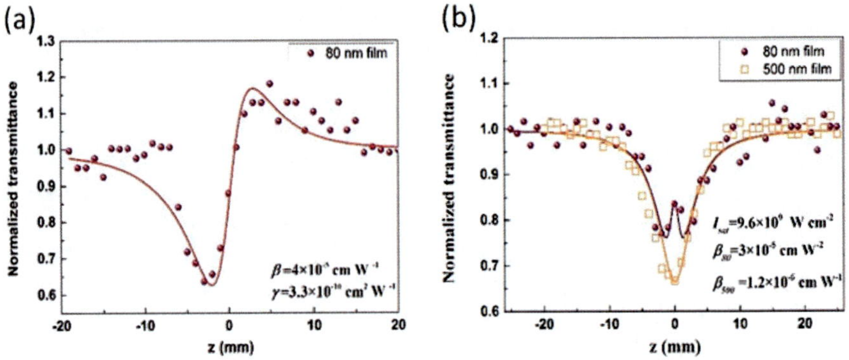

Fig. 4.4 Z-scan measurement using 400 nm radiation. Normalized transmittances in the case of **a** CA z-scan of 80 nm film and **b** OA z-scans of 80 nm and 500 nm films measured using the 400 nm, 35 fs pulses. Theoretical curves are the fits deduced from Eqs. (4.2), (4.4) and (4.5). Reproduced from [28] with permission from Walter de Gruyter GmbH & Science Wise Publishing

QDs dissolved in the hexane showed that the nonlinear absorption coefficient of this suspension at $\lambda = 800$ nm was 8.8×10^{-12} cm W^{-1}, and the nonlinear refractive index was 3×10^{-16} cm^2 W^{-1} [31] that was correspondingly 10^4 and 10^5 times smaller than the same parameters of 80 nm film.

In the case of 400 nm, 35 fs probe pulses the nonlinear refraction was larger than in the case of 800 nm pulses. The CA z-scan of 80 nm film showed the positive nonlinear refraction and positive nonlinear absorption when the valley was deeper than the peak (Fig. 4.4a). The nonlinear refraction index of this thin film was calculated to be $\gamma = 3.3 \times 10^{-10}$ cm^2 W^{-1}, while in the case of the thicker film (500 nm) it was 25 times smaller ($\gamma = 1.3 \times 10^{-11}$ cm^2 W^{-1}).

For the OA z-scans the normalized transmittance is given by [29]

$$T(x) = \sum_{m=0}^{\infty} \frac{\left[\frac{-\beta(I) I_0 L_S^{eff}}{1+x^2}\right]^m}{(m+1)^{3/2}}. \tag{4.3}$$

This formula can be presented in a simplified form

$$T(Z) \approx 1 - \frac{q}{2\sqrt{2}}, \tag{4.4}$$

where $q = \beta I_0 L_{\mathrm{eff}}/(1 + z^2/z_0^2)$.

In the case of SA, the nonlinear absorption coefficient and absorption coefficient are determined by the relations

$$\beta(I) = \frac{\alpha(I) - \alpha_0}{I}, \tag{4.5}$$

$$\alpha(I) = \alpha_0 \frac{1}{1 + (\frac{I}{I_{sat}})} \quad (4.6)$$

Here I_{sat} is the saturation intensity for the SA, which is related to the concentration of the active centers in the medium, the effective cross-sections of involved transitions, and the lifetimes of excited states. In the case of low intensity, the SA at $\lambda = 800$ nm dominated over other nonlinear optical processes in the used samples.

Contrary to the 800 nm probe pulses, the use of 400 nm radiation led to a change of nonlinear absorption from SA to reverse saturable absorption (RSA). The CA measurements revealed the high value of nonlinear absorption coefficient in the case of 80 nm film ($\beta = 4 \times 10^{-5}$ cm W^{-1}). An almost similar value was obtained during the OA z-scan of this film ($\beta = 3 \times 10^{-5}$ cm W^{-1}, Fig. 4.4b). Note that simultaneously with this process the growth of normalized transmittance at the highest intensities was observed that can be attributed to the higher-order SA effect (see the variation of normalized absorption in the vicinity of z = 0, Fig. 4.4b). The nonlinear absorption in thicker (500 nm) film ($\beta = 1.2 \times 10^{-6}$ cm W^{-1}, in the case of OA z-scan, and $\beta = 1.6 \times 10^{-6}$ cm W^{-1}, in the case of CA z-scan) was 25 times smaller compared with the β of the thin film.

There are various reports on the TA phenomenon in the Ag_2S QDs and nanocrystals measured in the near IR region [32, 33], which was attributed to the Auger recombination process. Meantime, the decay time constant for Ag_2S QDs was reported neither at 800 nm nor at 400 nm wavelength regions. TA in the 80 nm and 500 nm thin films containing the Ag_2S QDs deposited on the glass substrates was measured using the 400 nm, 30 fs pulses. The energy of the pump pulse was used to be 35 nJ to excite the films. Prior to the measurements of the TA of films, similar measurements of the glass substrate were performed, to separate its contribution from the TA data obtained in the case of Ag_2S QD thin films, which showed the absence of the change in transmittance with time delay in the case of substrates.

Figure 4.5a shows the temporal evolution of transmittance in the case of the 80 nm Ag_2S QD thin film. The TA profile demonstrated the transient bleaching due to SA as compared to the reverse process in the 500 nm thick Ag_2S QD film (Fig. 4.5b). In the case of 500 nm Ag_2S QD film, transient transmittance decreases with delay time, which is due to RSA during the temporal overlap of pump and probe pulses. In order to confirm the RSA at 400 nm wavelength, the OA Z-scans of these two films (Fig. 4.5b), which depicted the positive nonlinear absorption at employed pulse energy and at the area out from the maximal intensity region (i.e. out from the focal area), were analyzed. However, as already mentioned, in the vicinity of the focal area, the transmittance of laser pulses increased for 80 nm Ag_2S QD thin film, which is probably due to the dominance of SA phenomenon over RSA, contrary to the case of the 500 nm film showing the RSA at these conditions. Thus TA measurements were found to be consistent with the OA z-scan studies of two films. The decay time constants were determined to be 150 and 245 fs in the case of 80 and 500 nm films respectively. The difference of the decay times for these two films was attributed to different lifetimes of excited states and absorption probabilities in the case of thin and thick Ag_2S QD films.

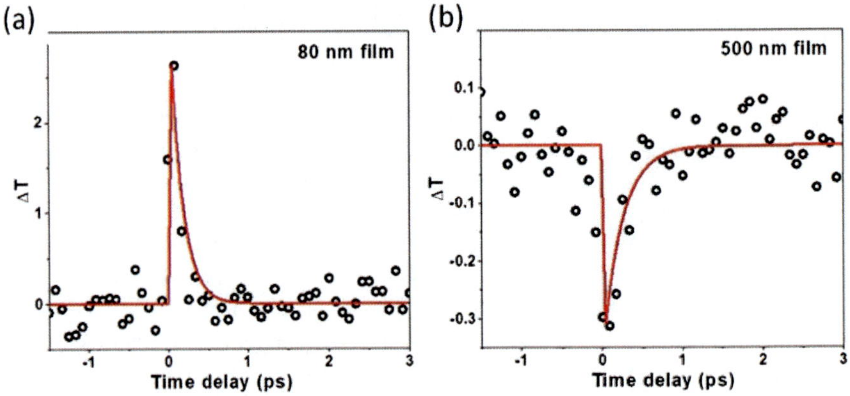

Fig. 4.5 Temporal evolution of probe pulse transmittance through the Ag_2S films in the case of 400 nm, 30 fs pulses. **a** 80 nm thick film and **b** 500 nm thick film. Reproduced from [28] with permission from Walter de Gruyter GmbH & Science Wise Publishing

4.1.3 HHG in the Plasmas Produced on the Ag_2S QD Films

A search for new applications of QDs is an important task for the optical community. One of the ideas is to find the optimal conditions in the application of such QDs as effective emitters of the high-order harmonics of femtosecond pulses for the development of efficient sources of coherent XUV radiation. To study the high-order nonlinear optical properties through HHG the Ag_2S QD suspension was dried on the quartz substrates and placed in the vacuum chamber for the creation of the plasma plume by laser ablation of this film. Below we discuss the influence of the delay between the heating and the driving pulses and the application of different energies of heating pulses on the harmonics yield.

Figure 4.6 shows the spectra of the high-order harmonics in the spectral range of 30–80 nm generated from the ablated QD thin film. The optimal delay between the heating and the driving pulses was found to be 400–500 ns. This figure also shows the plasma emission spectrum of ablated Ag_2S QD film. Note that the conditions of ablation at a relatively high pulse energy of the heating radiation ($E_{ns} = 9$ mJ), when the plasma emission shown in Fig. 4.6c dominates in the XUV spectra, are unsuitable for efficient HHG due to the presence of a large number of free electrons, which prevent the maintenance of optimal phase relation between the generated harmonics.

Lesser excitation of the QD-containing films (by 3 and 5 mJ pulses) allowed formation of the plasmas suitable for harmonics generation (Fig. 4.6a, b). Odd harmonics of 800 nm radiation up to the 23rd order (H23) were achieved using the Ag_2S QD plasmas. The appearance of strong harmonics emission did not coincide with the plasma emission from highly charged particles in the studied spectral range (30–80 nm) that confirmed that these experiments were carried out in a soft regime of

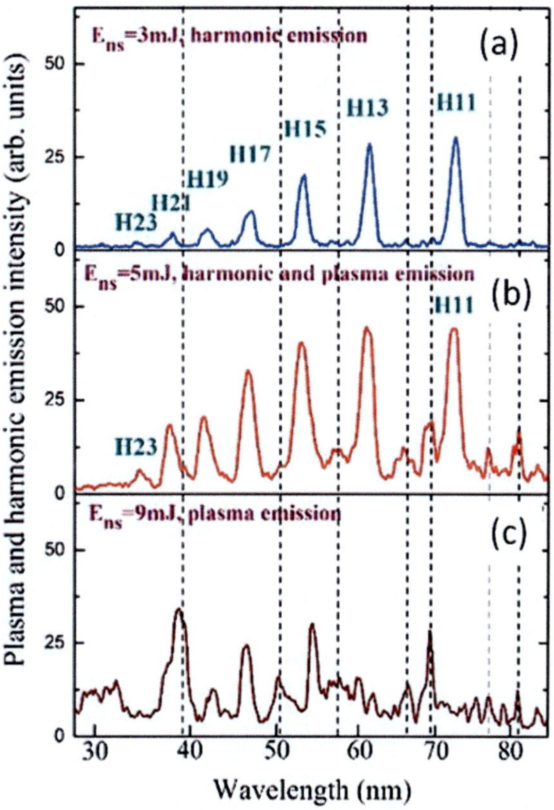

Fig. 4.6 Harmonics and plasma emission spectra. **a** Harmonics emission spectrum in the case of Ag_2S QD ablation of thin films using 3 mJ, 5 ns pulses. **b** Harmonics emission spectrum in the case of Ag_2S QD ablation of thin films using 5 mJ, 5 ns pulses. **c** Plasma emission spectrum in the case of Ag_2S QD ablation of thin films using 9 mJ, 5 ns pulses. Dashed lines show the similarity in plasma emission lines in these three conditions of plasma formation. Reproduced from [28] with permission from Walter de Gruyter GmbH & Science Wise Publishing

ablation of QD-containing targets. The driving pulse [$I \approx (2–5) \times 10^{14}$ W cm^{-2}] did not cause the appearance of strong ionic emission from higher charged molecules and QDs as well.

The estimate of HHG conversion efficiency was carried out using the comparison with known results from other plasmas (silver and carbon [34, 35]). In the case of silver plasma at similar conditions of experiments, the almost three times smaller conversion efficiency with regard to Ag_2S QD plasma was observed. By knowing the conversion efficiency from previous measurements of harmonic generation in the plasmas produced on the surface of bulk Ag at similar experimental conditions (8×10^{-6} [34]) the HHG conversion efficiency in Ag_2S QD plasmas was determined to be ~2×10^{-5}.

The application of two electronically separated pulses from different lasers synchronized by a digital delay generator allows analyzing the involvement of various ablated species in the HHG process. Application of this approach for HHG in multi-particle plasmas, alongside other methods of harmonic enhancement, requires the analysis of the ablated species, to temporally match them with the propagation of driving femtosecond pulses through the plasma.

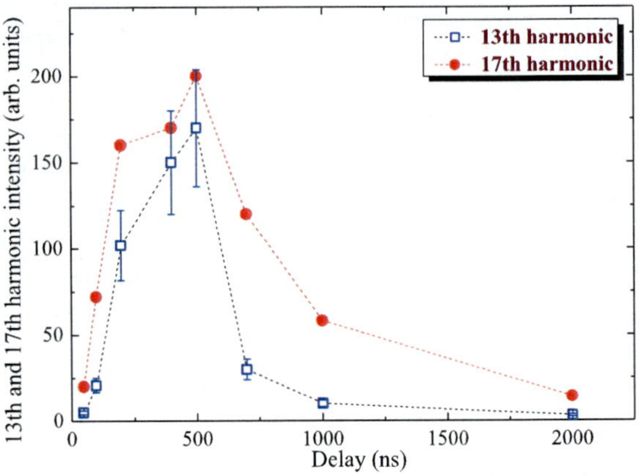

Fig. 4.7 The 13th and the 17th harmonics emission variations at different delays between heating nanosecond pulses and driving femtosecond pulses. Reproduced from [28] with permission from Walter de Gruyter GmbH & Science Wise Publishing

The highest harmonics yield was observed in the plasmas of ablated Ag_2S QD thin film (atomic weight 247.8 amu) after 400 to 500 ns delay from the beginning of ablation. Note that the use of carbon plasma (atomic weight 12 amu) allowed maximal harmonic intensity after a significantly smaller delay (about 30 to 50 ns between the heating and the driving pulses using similar pulse energies). Further increase of the delay led to a gradual decrease of the HHG efficiency from Ag_2S QDs plasmas (see variations of 13th and 17th harmonic yields at different delays, Fig. 4.7). The harmonics almost disappeared once the delay exceeded 2 μs. No harmonics were observed after the maximum delays explored in these studies (25 μs).

Experiments using ablated Ag_2S bulk target led to similar dependence of the harmonics yield on the delay between the heating and the driving pulses. These studies showed that independently of whether molecules or QDs spread out from the target surface of the same material, they appear in the area of femtosecond beam propagation at approximately the same time. The harmonics cut-off was higher for the ablated bulk silver sulfide (H33) with regard to the QD species. Plasma from the ablated Ag_2S QD thin film allowed the generation of up to H19 at similar experimental conditions of ablation (Fig. 4.8). In the meantime, the conversion efficiency for the lowest orders, in that case, was three to four times larger compared with the ablation of bulk silver sulfide.

4.2 Effective High-Order Harmonic Generation from Metal Sulfide Quantum Dots

A search for new applications of QDs is an important task for the optical community. The interesting idea is to find the optimal conditions in the application of such QDs as effective emitters of the high-order harmonics of femtosecond pulses for the development of efficient sources of coherent XUV radiation. In this section, we analyze HHG in such QDs using electronically driven delays between the heating

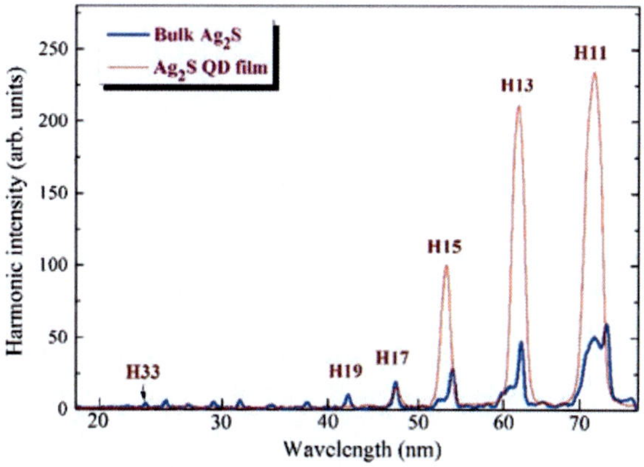

Fig. 4.8 Comparison of harmonics spectra in the case of ablation of the Ag_2S bulk target (thick curve) and Ag_2S QD thin film (thin curve). Reproduced from [28] with permission from Walter de Gruyter GmbH & Science Wise Publishing

and driving lasers [36]. Ag_2S, CdS, and ZnS QDs were ablated using a nanosecond laser and demonstrated efficient HHG in plasmas using a femtosecond laser. With this approach, effective harmonics generation was achieved in the spectral range from 20 to 115 nm using such QDs.

4.2.1 *Preparation of QD-Containing Targets*

Improving the efficiency of HHG in a plasma containing QDs requires solving two problems: (1) developing the methods for the synthesis of metal sulfides QDs, followed by the formation of solid-state materials containing large concentrations of these species, and ensuring the stability of the morphology of vaporized materials under laser ablation, and (2) optimizing the process of converting the wavelength of laser pulses during the HHG in a laser-plasma containing such multi-particles.

The synthesis of QDs was carried out taking into account a number of criteria for their use to generate harmonics of ultrashort pulses. The first criterion is the formation (as a result of ablation of samples by either picosecond or nanosecond heating pulses) of a plasma cloud consisting of QDs, as well as atoms, ions, and clusters at the time of the passage of the converting femtosecond pulses above the target surface. The second criterion is the presence of the maximal concentration of QDs at the time of transmission of the femtosecond pulses through the plasma. The third criterion is the size characteristics of quantum dots. Previous studies of harmonic generation in multi-particle containing plasmas, as well as qualitative assessments, show that the most optimal sizes of such species should be in the range of 1–4 nm (see also the discussion section).

Synthesis of colloidal QDs of ZnS and CdS was carried out by mixing the $CdBr_2 \times 4H_2O$, $ZnBr_2 \times 4H_2O$, and thioglycolic acid (TGA) in the required proportions. The corresponding salts of cadmium and zinc bromide were dissolved in water (200 ml) and then the TGA was added. Then the 50 ml of Na_2S aqueous solution (1.3 mM) was injected into the prepared solution. The synthesized colloidal solution of QDs was dissolved by adding ethanol to a 50% solution, centrifuged, and re-dissolved in water. The cleaning procedure was repeated several times. A similar procedure was used for the synthesis of Ag_2S QDs.

The synthesized samples of colloidal solutions of zinc, cadmium, and silver sulfides were further used to produce the solid targets for laser ablation and generation of harmonics. A new method of increasing the filling factor Ag_2S, ZnS, and CdS QDs in the stabilizing polymer was developed to form solid samples of multi-particle species for laser ablation in vacuum and formation of the plasma torches containing a large amount of QDs required for efficient generation of harmonics. The method is based on the compatibility of synthesized colloidal QDs and gelatin, as well as the possibility of increasing the concentration of QDs during centrifugation in the presence of acetone. Gelatin's properties of high filling ability as a polymer and the possibility of drying such samples to a solid-state are among the advantages of this method. The preparation of solid tablets was accomplished in the following sequence: the prepared concentrated solution of QDs was introduced into the gelatin melted at a temperature of 40–50 °C using intensive stirring and then the suspensions were dried at a temperature of 80 °C for 24 h to a solid-state and then placed for further vacuum drying where they were kept for 10 days. Control of the obtained samples was carried out by analysis of their luminescence, which was compared with the luminescence of the same QDs in TGA. A small difference in the spectra of these two cases was observed.

4.2.2 Plasma HHG Setup and Characterization of QDs

The metal sulfide QD plasmas were used for HHG. The driving femtosecond pulses (800 nm, 1 kHz) propagated through the plasma at different delays from the beginning of target irradiation by nanosecond heating pulses (1064 nm, 10 Hz). The variable delay ($0-10^6$ ns) between 5 ns heating pulses and 30 fs driving pulses was established to generate harmonics in plasma at the used geometry of experiments when the 800 nmfemtosecond pulses or 800 nm + 400 nm pulses were focused onto the plasma area from the orthogonal direction with regard to ablating radiation, at a distance of ~200 μm above the target surface. Additionally, the harmonic yield was maximized by adjusting the position of the target with regard to the optical axis of propagation of the driving femtosecond pulses and by varying the focusing position of this radiation with regard to the plasma. The harmonic emission was directed to the XUV spectrometer.

The originality of the general approach to increase the HHG efficiency is based on the choice of quantum confined multi-atomic particles to facilitate an increase of the nonlinear optical response of the plasma medium. Qualitatively, an increase of the cross-section for the recombination of accelerated electrons with the multi-particle species with increasing size of the parent particle can be expected. However, this assumption is valid only for certain sizes of multi-particle species. A further increase of the number of atoms or molecules in the particle leads to conditions where the atoms or molecules located inside the particle become completely shielded and thus eliminated from the three-step process of HHG, i.e., ionization, acceleration, and recombination. In other words, the internal atoms and molecules do not supply electrons for acceleration and recombination. Thus, the optimal size of the particles is one of the most important characteristics in order to achieve an increase in the efficiency of harmonic generation. Moreover, its importance rises with the increasing order of the generated harmonics. Early studies carried out in gases and plasmas have shown an increase of HHG intensity using clusters containing up to 1000 noble gas atoms, as well as using nanoparticles containing tens of thousands to a few million atoms. It turned out that the optimal number of atoms in the particles seems to lie in an intermediate size region. If one divides the aggregates into three categories like clusters (0.1–1 nm), QDs (1–6 nm), and nanoparticles (10–100 nm), then the second group seems to be the most promising for HHG. The QDs synthesized in this work are satisfying this criterion.

Samples of colloidal QDs were characterized by x-ray diffraction, transmission electron microscopy (TEM), optical absorption, and luminescence spectroscopy. The composition of the particles was determined by x-ray and electron diffraction. The dispersion of the particle sizes was determined by statistical data analysis of the TEM images of the QDs. TEM also provided information on the morphology of the QDs. The analysis of the structure, phase composition, and morphology of the samples allowed improvements to the synthesis techniques.

The TEM images of synthesized metal sulfide QDs are shown in Fig. 4.9a. The size distribution is shown in the inset as histograms. From the histograms, one can conclude that Ag_2S QDs show mean sizes in the range of 1.7 to 2.0 nm, CdS QDs about 3.0 to 3.2 nm, and ZnS QDs about 2 to 2.2 nm diameter. The morphological studies reveal that Ag_2S QDs condensate in a monoclinic lattice, while CdS and ZnS QDs condensate in a cubic lattice and Ag_2S QDs in a monoclinic lattice. Figure 4.9b presents images of the QDs ablated by 10 mJ and 5 ns pulses and deposited on nearby substrates. The collected particles confirm the presence of metal sulfide QDs within the plasma plume. The sizes of deposited particles were larger than those of the original ones. Optical absorption and photoluminescence spectra are shown in Fig. 4.9c. The spectra were significantly blue-shifted with regard to the bulk materials of similar materials. The CdS, ZnS and Ag_2S QDs were embedded in gelatin with about 20% weight and prepared in the form of $5 \times 5 \times 3$ mm plates.

4.2.3 Harmonic Generation

Figure 4.10 shows high-order harmonics generated from these QDs in the spectral range of 20–120 nm. HHG at different delays between the heating and driving pulses from 100 to 2000 ns is shown for pulse energies of 4 and 0.4 mJ, respectively. At initial stages of plasma formation and spreading out from the target, i.e., at delays less than 50 ns, the concentration of particles (neutrals and singly charged ions) was

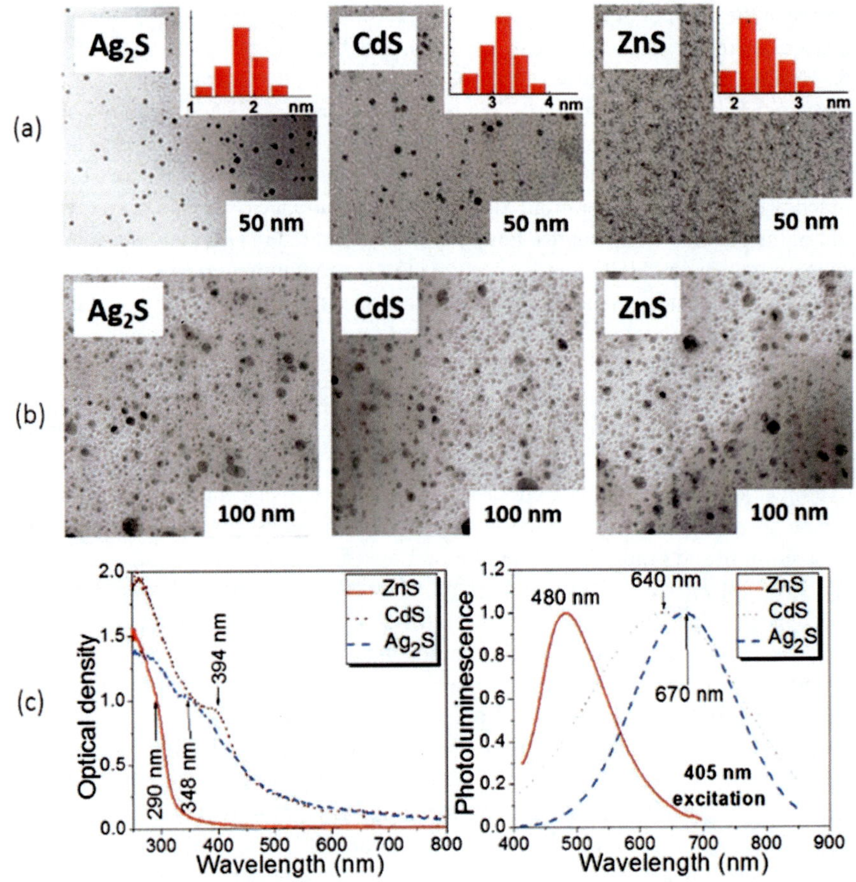

Fig. 4.9 **a** TEM images of synthesized metal sulfide QDs. **b** TEM images of the Ag_2S, CdS and ZnS QDs ablated at the 10 mJ energy of heating nanosecond pulse and deposited on nearby glass and silicon substrates. **c** Spectra of (left graph) optical absorption and (right graph) photoluminescence of synthesized QD. Reproduced from [36] with permission from Optical Society of America

insufficient for HHG since the whole ensemble of particles possessing velocities of ~2×10^4 m s^{-1} cannot reach the spatial region of the driving beam. At larger delays (above 100 ns) sufficient amounts of QDs appeared in the path of the femtosecond beam which allowed the generation of harmonics. The optimal delay was different for three QD plasmas studied. At about 400, 300, and 250 ns the highest harmonic yield was observed for Ag_2S QD (atomic weight 247.8 amu), CdS QD (144.5 amu), and ZnS QD (97.5 amu), respectively. Notice that the use of a carbon plasma (atomic weight 12 amu) yielded maximal harmonic intensity at a significantly smaller delay of 30 ns between the heating and driving pulses using similar pulse energies. It should also be mentioned that harmonics generated from pure gelatin were almost

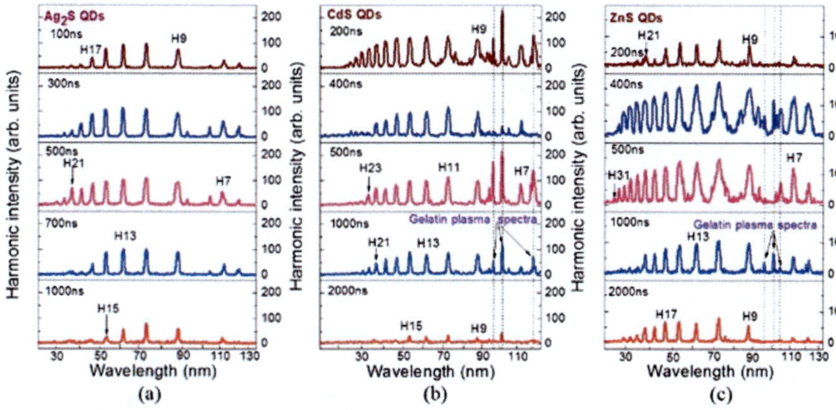

Fig. 4.10 Harmonic spectra from QD-containing plasmas [**a** silver sulfide QDs, **b** cadmium sulfide QDs, **c** zinc sulfide QDs] at different delays between heating and driving pulses. Dash-dotted lines show the plasma elision attributed to overexcited gelatin. Reproduced from [36] with permission from Optical Society of America

one order of magnitude lower than those observed from the QDs. The harmonic cut-off is higher for lighter QD species, e.g., at the 31st harmonic for ZnS while Ag_2S showed the 25th as the highest harmonic. A further increase of the delay for each of these plasmas led to a gradual decrease in the HHG efficiency. The harmonics almost disappeared once the delay exceeded 2 to 3 μs. No harmonics were observed at delays up to 50 μs.

Experiments using bulk targets of the same materials led to similar dependencies of the harmonic yield on the delay between the heating and driving pulses. These experiments showed that independently of whether molecules or QDs spread out from the target surface of the same material they appear in the area of the femtosecond beam approximately at the same time. Figure 4.11a shows the harmonic generation in the laser-produced plasma after ablation of bulk ZnS by 4 mJ pulses. The harmonic yield maximizes at about 200–500 ns delay.

Figure 4.11b shows a comparison of the harmonic yield in the case of ablation of bulk and QD-containing ZnS targets. It is obvious that the QD plasma yield significantly stronger harmonics which also extend to higher orders. In the case of QD-containing plasma, the harmonics extended to ~50 eV (25 nm) or 31st order, while for the plasma produced from bulk ZnS the highest generated harmonic order was the 16th corresponding to a photon energy of ~26 eV (47 nm). The weight concentration of ZnS QDs in gelatin (0.2) was taken into account to carry out the comparison of the HHG spectra produced from monoparticles and multi-particle plasma. The overall ratio of intensities of the lower-order harmonics generated in these two plasmas was ~7.

Figure 4.12 shows the plasma emission spectra of Ag_2S, CdS, and ZnS QDs in gelatin (three upper panels) and pure gelatin (bottom panel) using larger energy of heating pulses (15 mJ). The majority of lines belong to gelatin emission, except the

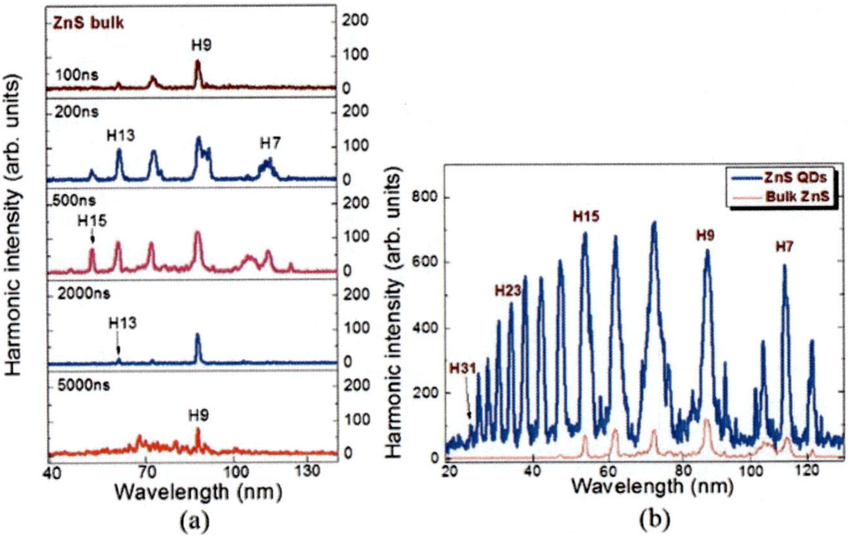

Fig. 4.11 **a** Harmonic spectra from the plasmas produced on the bulk ZnS surface at different delays between heating and driving pulses. The bottom panel was obtained at stronger excitation of the target (10 mJ). One can see the plasma emission lines from bulk zinc sulfide. **b** Comparison of harmonic distribution and intensities in the case of ablation of the QD (thick curve) and bulk (thin curve) ZnS targets at 500 ns delay between the heating and driving pulses. The weight part of QDs in gelatin (20%) is taken in to account for comparison of HHG at similar concentrations of plasmas. Reproduced from [36] with permission from Optical Society of America

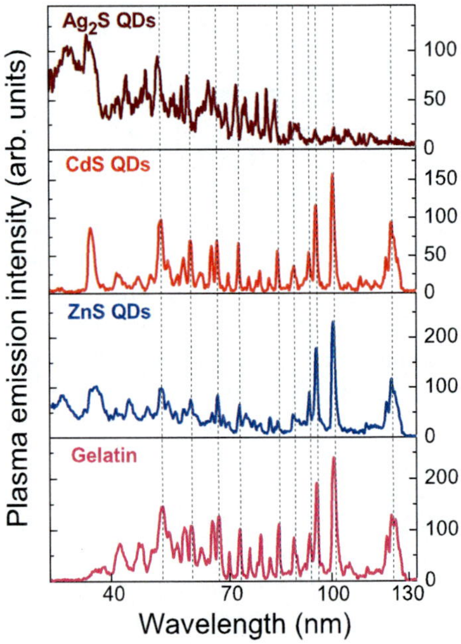

Fig. 4.12 Plasma emission spectra of Ag_2S, CdS and ZnS QDs in gelatin (three upper panels) and pure gelatin (bottom panel). Dash-dotted lines show similar emission in all four plasmas. Reproduced from [36] with permission from Optical Society of America

shorter-wavelength part of this spectrum. Most of the lines can be assigned to the emission from ionized carbon, as expected for gelatin. Those in the range of 45–90 nm seem mostly related to the C II emission. Probably, the existence of singly charged carbon caused the harmonic generation in pure gelatin plasma. These spectra underline that the metal sulfides were weakly excited and ionized during ablation. This condition of "mild" ablation, when the existing density of free electrons in the plasma does not prevent the formation of phase-matching conditions between the waves of the driving and harmonic fields, is a prerequisite for efficient HHG. Notice that the conditions of ablation, when the plasma emission shown in Fig. 4.12 dominates in the XUV spectra, are unsuitable for efficient HHG due to the presence of a large number of free electrons which prevent an optimal phase relation between the driving and harmonic waves.

Less excitation of the QD-containing targets (4 mJ) allowed the formation of plasmas suitable for a harmonic generation. Odd harmonics of 800 nm radiation up to H25, H29, and H31 were achieved using Ag_2S, CdS, and ZnS QD plasmas, respectively (see Fig. 4.10). The appearance of strong harmonic emission did not coincide with the plasma emission from highly charged particles in the studied spectral range (20–120 nm), which confirms that these experiments were carried out at a soft ablation regime of QD containing targets. The driving pulse [$I \approx (2\text{–}5) \times 10^{14}$ W cm^{-2}] also did not cause the appearance of strong ionic emission from higher charged molecules and QDs.

A final set of HHG studies was performed using two orthogonally polarized driving pulses. A 0.4 mm thick BBO crystal inserted inside the vacuum chamber on the path of the 800 nm fundamental radiation allowed the generation of 400 nm pulses while maintaining sufficient temporal and spatial overlap of 800 and 400 nm fields in the preformed plasmas. Notice that even only a small conversion efficiency (~5%) into the 400 nm radiation allowed the generation of the lower-order even harmonics alongside the odd ones (Fig. 4.13, dash-dotted curves). Even harmonics generation up to H14 were observed. The seemingly relatively small yield of low-order harmonics (H6 to H9) was caused by a weaker sensitivity of the MCP detector for wavelengths above 100 nm.

The conversion efficiency of even harmonics in the case of TCP was higher than of odd harmonics in the case of SCP, despite very low pulse energy of the 400 nm radiation with respect to the 800 nm one (~1:20). Notice the relatively weaker H12 compared to H10 and H14. This heterogeneity in the even harmonic yields was caused by group velocity dispersion in the 0.4 mm thick BBO crystal, which causes a notable delay between the 400 and 800 nm pulses at the output of the crystal in the case of 30 fs pulses. Each 0.1 mm of BBO causes ~19 fs delay between the driving (800 nm) and second harmonic (400 nm) waves. Thus, only a small part of the second harmonic remains within the pulse duration of the driving fundamental 800 nm pulse. This causes a further decrease in the influence of the second harmonic field on the whole pattern of harmonic distribution. On the other hand, this second harmonic pulse generates relatively strong low-order odd harmonics: $H5_{400nm}$ which corresponds to $H10_{800nm}$, and $H7_{400nm}$ which corresponds to $H14_{800nm}$, see Fig. 4.13. An introduction of either positive or negative chirp in the femtosecond pulses by

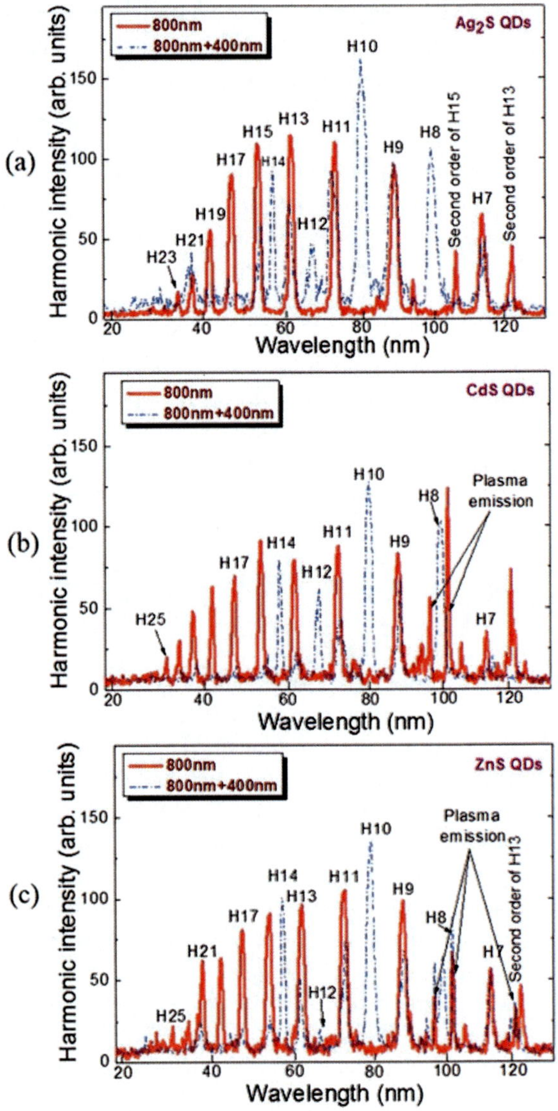

Fig. 4.13 Two-color (800 nm + 400 nm, dash-dotted blue line) and single-color (800 nm, full red line) pump induced harmonic spectra from Ag_2S (upper panel), CdS (middle panel), and ZnS (bottom panel) QDs at optimal delays between the heating and driving pulses for each species. Reproduced from [36] with permission from Optical Society of America

varying the distance between gratings in the compressor led to an increase of the pulse duration and a better overlap of the two orthogonally polarized pulses in the plasma. However, in that case, a decrease in the intensity of the driving pulse led to a general decrease of the whole harmonic yield.

The peculiarity of the gelatin plasma emission and TCP-induced harmonic spectra was the proximity of some ionic transitions of gelatin with the wavelengths of even harmonics, see H8 and the nearby emission lines of gelatin lines in the case of CdS QD (Fig. 4.13, middle panel) and ZnS QD (Fig. 4.13, bottom panel) plasmas.

Neither enhancement nor suppression of this harmonic generated in both plasmas was observed. These studies showed that the proximity of strong emission lines and harmonics does not necessarily lead to a variation of the harmonic yield.

The estimates of conversion efficiency were carried out using the comparison with known results from other plasmas). In the case of silver plasma at similar conditions, almost equal conversion efficiencies with regard to ZnS in the range of 40–110 nm were achieved. By knowing the conversion efficiency from previous measurements of harmonic generation in the plasmas produced on the surface of bulk Ag (4×10^{-6}), one can deduce that conversion efficiency in ZnS QD plasmas (which was 7 times higher than in bulk ZnS plasma) was ~3×10^{-5}.

4.2.4 Discussion

The discussed experiments have two key features: ablation by nanosecond pulses and QDs-containing targets. So, the observed enhanced HHG conversion efficiency can have three origins: HHG from QDs, HHG from larger concentrations of neutral atoms, and HHG enhancement due to the interaction of QDs with atoms. The analyzed studies show that particles with sizes of the order of a few nanometers can effectively generate high-order harmonics. HHG directly from QDs has not yet been studied theoretically, but a simulation of HHG from small clusters showed the increased HHG yield under the assumption of possible recombination of an electron on different atoms in the cluster. The cutoff harmonics were also enhanced much more strongly than the plateau harmonics, similarly to the discussed results in the case of ZnS QDs (Fig. 4.11). The local field enhancement in the vicinity of larger nanoparticles also increases the intensity of cutoff harmonics generated from the surrounding atoms [37].

Early experiments with larger nanostructures [carbon nanotubes, nanoparticles of metals and semiconductors, and other agglomerates of relatively large sizes (20–80 nm)] have shown both merits and drawbacks of this method for increasing the HHG efficiency. An increased recombination cross-section of the parent particle with the accelerated electron which, in the case of relatively large particle sizes, may result in an increased number of harmonic emitters is accompanied by a reduced number of atoms located on their surface. In traditional gas HHG, therefore, fewer atoms are involved in the generation of the harmonics because the atoms located inside the nanoparticles are unlikely to participate in HHG. As a consequence, the amount of harmonic generation processes per single atom falls short for large nanoparticles.

Laser ablation-induced HHG spectroscopy of semiconductors can reveal the resonance-induced enhancement of some harmonic orders in the XUV range as well as the cluster-induced growth of harmonic yield. The harmonic generation in such plasmas can prove or disapprove the role of the complex composition of ablated species in this process. A signature that would reveal the nature of the emitters is the growth of harmonics emission with the time delay with respect to the ablation laser pulse. This delay is converted to the time of propagation of QDs or clusters to the interaction region with the ultrashort driving pulse.

One can expect from this assumption the arrival of ZnS ions, atoms, and QDs in the area of interaction with the driving beam at $(M_{Ag2S}:M_{ZnS})^{0.5}$ ~1.59 times earlier with regard to Ag_2S-containing particles taking into account the ratio between the atomic weights of these two molecules ($M_{Ag2S}:M_{ZnS} \approx 2.54$). Once one compares the "optimal" delays for these two species, their ratio (400 ns: 250 ns = 1.6) becomes close to the above estimates. Thus, one can assume that the above rule properly explains the dynamics of material spreading out from the ablated target, once one analyzes the particles' movement during laser ablation at relatively moderate fluencies (5–20 J cm^{-2}) of heating nanosecond pulses.

As already mentioned, an increase of the recombination cross-section for QDs with respect to atoms or single molecules can enhance the HHG efficiency in multi-particle plasmas. An additional cause for strong harmonics generation from QDs compared to single atoms or ions could be the higher concentration of neutral metal sulfide molecules inevitably accompanying the presence of multi-particle species. The QDs present the extreme case of solid-state density in a very small volume. Compared to solids, they still do not absorb all the HHG radiation produced inside them due to their very small thickness. QDs can further improve phase-matching conditions for harmonics generated from atoms and ions [38] using compensation of free-electron dispersion of the driving pulse by dispersion of clusters.

4.3 Summary

In this chapter, we analyzed different methods of the applications of clusters and quantum dots for HHG in LIPs.

The nonlinear optical studies of Ag_2S QD thin films are discussed. Using the open-aperture and closed-aperture z-scan technique and theoretical fitting, the nonlinear absorption coefficient was calculated to be 4×10^{-5} cm W^{-1} for the 80 nm thick film in the case of 400 nm probe pulses. The nonlinear absorption coefficients and nonlinear refractive indices of 80 and 500 nm thick films were significantly larger than for the Ag_2S QDs suspension. In the case of transient absorption measurements, the two distinct characteristic time constants of 150 fs and 245 fs were obtained for the films of different thicknesses. We also discussed the studies of high harmonics emission from the ablation of 80 nm films using two laser sources. The conversion efficiency in the plasmas produced by ablation of the Ag_2S QD thin film (2×10^{-5}) was several times higher compared to the case of bulk Ag_2S ablation. The analyzed approach establishes a novel and efficient way of generating high-order harmonic yield.

We analyzed the possibility and conditions of using quantum dots as efficient emitters for high-order harmonics generation by 30 fs pulses using optimally delayed heating and driving laser pulses. We presented the method for the aqueous synthesis of Ag_2S, ZnS, and CdS QDs that enable, by their size and dispersion, as well as their structural properties, the effective generation of harmonics in plasma plumes. The study of the morphological properties of the samples demonstrated the formation of ZnS and CdS QDs with mean sizes of 2–3.2 nm in a cubic lattice, and of Ag_2S QDs with mean sizes of 1.7–2.0 nm in a monoclinic lattice, all with a weight fraction of about 20–25% in gelatin.

A comparison of HHG using QDs and ablation of solid-state targets for the same materials (Ag_2S, ZnS, and CdS) was carried out. In the case of QDs, the conversion efficiency into harmonics is higher despite the lower concentration of these species in the plasma compared to the concentration of molecules and ions for bulk ablation. The observed increase of HHG conversion efficiency could be caused by a larger concentration of harmonic emitters and the specific properties of QDs, particularly their plasmonic properties. The maximum HHG conversion efficiency is achieved much earlier than one would expect from the kinetic model of QDs spreading out from the target surface. Therefore similar velocity distributions of molecules and QDs of these metal sulfides are assumed in the ablation process.

References

1. M. Yarema, S. Pichler, M. Sytnyk, R. Seyrkammer, R.T. Lechner, G. Fritz-Popovski, D. Jarzab, K. Szendrei, R. Resel, O. Korovyanko, M.A. Loi, O. Paris, G. Hesser, W. Heiss, ACS Nano **5**, 3758 (2011)
2. P. Jiang, C.N. Zhu, Z.L. Zhang, Z.Q. Tian, D.W. Pang, Biomater. **33**, 5130 (2012)
3. I. Hwang, M. Seol, H. Kim, K. Yong, Appl. Phys. Lett. **103**, 023902 (2013)
4. I. Hwang, K. Yong, ChemPhysChem **14**, 364 (2013)
5. X. Gao, Y. Cui, R.M. Levenson, L.W.K. Chung, S. Nie, Nature Biotechnol. **22**, 969 (2004)
6. B. Dubertret, P. Skourides, D.J. Norris, V. Noireaux, A.H. Brivanlou, A. Libchaber, Science **298**, 1759 (2002)
7. H. Linnenbank, Y. Grynko, J. Förstner, S. Linden, Light: Sci. Appl. **5**, e16013 (2016)
8. L. Gao, C. Chen, K. Zeng, C. Ge, D. Yang, H. Song, J. Tang, Light: Sci. Appl. **5**, e16126 (2016)
9. S. Abe, J.J. Joos, L.I.D.J. Martin, Z. Hens, P.F. Smet, Light: Sci. Appl. **6**, e16271 (2017)
10. C.E.R. Disney, S. Pillai, C.M. Johnson, M.A. Green, ACS Phot. **2**, 1108 (2015)
11. P. Kuang, S. Eyderman, M.-L. Hsieh, A. Post, S. John, S.-Y. Lin, ACS Nano **10**, 6116 (2016)
12. A. Oskooi, M. De Zoysa, K. Ishizaki, S. Noda, ACS Phot. **1**, 304 (2014)
13. R.A. Ganeev, M. Baba, M. Morita, D. Rau, H. Fujii, A.I. Ryasnyansky, N. Ishizawa, M. Suzuki, H. Kuroda, J. Opt A **6**, 447 (2004)
14. T. Karali, N. Can, L. Valberg, A.L. Stepanov, P.D. Townsend, C. Buchal, R.A. Ganeev, A.I. Ryasnyansky, H.G. Belik, M.L. Jessett, C. Ong, Physica B **363**, 88 (2005)
15. R.A. Ganeev, U. Chakravarty, P.A. Naik, H. Srivastava, C. Mukherjee, M.K. Tiwari, R.V. Nandedkar, P.D. Gupta, Appl. Opt. **46**, 1205 (2007)
16. O.V. Ovchinnikov, M.S. Smirnov, A.S. Perepelitsa, T.S. Shatskikh, B.I. Shapiro, Quantum Electron. **45**, 1143 (2015)
17. H. Aleali, N. Mansour, Optik **127**, 2485 (2016)
18. R. Karimzadeh, H. Aleali, N. Mansour, Opt. Commun. **284**, 2370 (2011)
19. T. Brabec, F. Krausz, Rev. Modern Phys. **72**, 545 (2000)
20. K.J. Schafer, B. Yang, L.F. DiMauro, K.C. Kulander, Phys. Rev. Lett. **70**, 1599 (1993)
21. M. Lenzner, J. Krüger, S. Sartania, Z. Cheng, C. Spielmann, G. Mourou, W. Kautek, F. Krausz, Phys. Rev. Lett. **80**, 4076 (1998)
22. T. Apostolova, B. Obreshkov, Diam. Relat. Mater. **82**, 165 (2018)

23. T.T. Luu, M. Garg, S.Y. Kruchinin, A. Moulet, M.T. Hassan, E. Goulielmakis, Nature **521**, 498 (2015)
24. S. Ghimire, A.D. DiChiara, E. Sistrunk, P. Agostini, L.F. DiMauro, D.A. Reis, Nature Phys. **7**, 138 (2011)
25. R.A. Ganeev, M. Suzuki, M. Baba, M. Ichihara, H. Kuroda, J. Phys. B **41**, 045603 (2008)
26. R.A. Ganeev, M. Suzuki, M. Baba, M. Ichihara, H. Kuroda, J. Opt. Soc. Am. B **25**, 325 (2008)
27. R.A. Ganeev, P.A. Naik, H. Singhal, J.A. Chakera, M. Kumar, M.P. Joshi, A.K. Srivastava, P.D. Gupta, Phys. Rev. A **83**, 013820 (2011)
28. Y. Fu, R.A. Ganeev, G.S. Boltaev, S.K. Maurya, V.V. Kim, C. Zhao, A. Rout, C. Guo, Nanophot. **8**, 849 (2019)
29. M. Sheik-Bahae, A.A. Said, T.H. Wei, D.J. Hagan, E.W. Stryland, IEEE J Quantum Electron. **26**, 760 (1990)
30. X. Liu, S. Guo, H. Wang, L. Hou, Opt. Commun. **197**, 431 (2001)
31. Y. Fu, R.A. Ganeev, C. Zhao, K.S. Rao, S.K. Maurya, W. Yu, K. Zhang, C. Guo, Appl. Phys. B **125**, 1 (2019)
32. J. Sun, W. Yu, A. Usman, T.T. Isimjan, E. Alarousu, K. Takanabe, O.F. Mohammed, Phys. Chem. Lett. **5**, 659 (2014)
33. S. Chand, E. Sharma, P. Sharma, J Alloys Compd. **770**, 1173 (2019)
34. R.A. Ganeev, M. Baba, M. Suzuki, H. Kuroda, Phys. Lett. A **339**, 103 (2005)
35. R.A. Ganeev, C. Hutchison, T. Witting, F. Frank, W.A. Okell, A. Zaïr, S. Weber, P.V. Redkin, D.Y. Lei, T. Roschuk, S.A. Maier, I. López-Quintás, M. Martín, M. Castillejo, J.W.G. Tisch, J.P. Marangos, J. Phys. B **45**, 165402 (2012)
36. R.A. Ganeev, G.S. Boltaev, V.V. Kim, K. Zhang, A.I. Zvyagin, M.S. Smirnov, O.V. Ovchinnikov, P.V. Redkin, M. Wöstmann, H. Zacharias, C. Guo, Opt. Express **26**, 35013–35025 (2018)
37. M.F. Ciappina, S.S. Aćimović, T. Shaaran, J. Biegert, R. Quidant, M. Lewenstein, Opt. Express **20**, 26262 (2012)
38. J.W.G. Tisch, Phys. Rev. A **62**, 041802 (2000)

Chapter 5
HHG Using Carbon-Contained Nanoparticles

5.1 Role of Carbon Clusters in High-Order Harmonic Generation in Graphite Plasmas Using 800 nm Femtosecond Radiation

High-order harmonic generation in isotropic media can provide table-top sources in generating coherent extreme ultraviolet radiation [1]. Therefore, it is important to analyze high-order nonlinear optical properties of materials through the search for enhancement of harmonic yield from different isotropic media (gases [2], and plasmas [3]). Particularly, the conditions for efficient emission of coherent XUV radiation through HHG in different clusters using ultrashort laser pulses should be studied, which have further application in a range of fields from physics, to chemistry and to biology.

HHG conversion efficiency in the XUV range is limited due to a range of factors that prevent the generation of sufficient fluencies of coherent short-wavelength radiation. The best results (10^{-5} in the case of gas and plasma HHG and 10^{-4} for resonantly enhanced single harmonic of 800 nm radiation) show some limitations of harmonic generation efficiency. The phase mismatch, low density of media, and limitation in laser intensity further restrict harmonic generation efficiency in the isotropic medium [4]. Currently, significant research efforts are shifted towards attosecond pulse generation in the case of gas media [5], nonlinear spectroscopy of plasmas through single harmonic generation in the case of plasmas [6], quasi-phase-matching (in the cases of gases [7] and plasmas [8]) and HHG studies using clusters, quantum dots, and nanoparticles in the cases of plasmas [9] and gases [10].

Another emerging field is HHG in solid bulk media, for example in ZnO or ZnSe [11]. In that case, different channels including intraband and interband processes were considered for the explanation of the harmonic generation using mid-infrared driving pulses. HHG in solids could also become relevant for nanoparticles. From this perspective, clusters, quantum dots, and large nanoparticles may act as an intermediate material between atoms and solids.

R. A. Ganeev, *High-Order Harmonics Generation in Plasmas*, Springer Series on Atomic, Optical, and Plasma Physics 122, https://doi.org/10.1007/978-3-031-09040-0_5

The studies of HHG using clusters, quantum dots, and nanoparticles require analysis of the plasmas, which contain multi-particle species, to temporally match them with the propagation of driving femtosecond pulses through the plasmas. In the past, the plasma formation for HHG was achieved with a picosecond laser pulse heating [6, 8, 9], and the optical delays between the heating picosecond pulses and driving femtosecond pulses originated from the same laser were usually less than 100 ns. Meanwhile, to match the propagation of the driving pulses and the highest concentration of the studied groups of multi-particle species and to confirm or disapprove the assumption of the variation of the velocities of different components, one has to find other methods to form the optimal delay between the heating and the driving pulses to maximize HHG efficiency.

The earlier reported studies of HHG from gases and during ablation of cluster and nanoparticle targets have revealed the advantages of using such species for frequency conversion in the XUV range. Larger cross-section of recombination, the possibility of recombination of an accelerated electron with parent multi-particle through either recombination with the same atom or neighboring atom, or the multi-particle as a whole were considered as the most probable reasons in enhancing HHG yields from such clustered plasmas. However, there has been no study in optimizing the delay between heating and driving pulses in the microsecond timescale assuming the hypothesis of the slower velocities for larger-sized species. Notice that, even without this optimization, the harmonic yields from ablated multi-particle targets are higher than from ablation of bulk targets of the same material leading to the formation of single atomic/ionic plasmas. Here the meaning of multi-particle target refers to the one containing the clusters, quantum dots, or nanoparticles glued on the surface of different substrates.

The most probable reason for harmonic enhancement was attributed to the appearance of small clusters in the area of femtosecond pulse propagation at relatively small optical delays. Their appearance in that area was assumed once estimating their velocities and delays between the beginning of ablation and propagation of femtosecond pulse at the fixed distance above the target surface. Particularly, for most HHG experiments, these distances were maintained at 0.2 mm, while the delay for each target was chosen to be 20–80 ns depending on the atomic weight of the elements comprising different targets. Particularly, the maximal harmonic yield from ablated graphite has earlier been achieved at ~30 ns, which allows calculating the velocity of atoms and ions spreading out from the target (~6×10^3 m s^{-1}). At the same time, ablation of the fullerenes comprising 60 atoms of carbon led to a similar "optimal delay" at which maximal yield of harmonics was achieved. Correspondingly, one could assume the similarity in the velocities of carbon single atoms and C_{60} at the conditions of ablation using picosecond heating pulses. A similar feature has also been observed in the case of carbon nanotubes.

Another important concept allowing the enhancement of harmonic yield is the application of the two-color pump (TCP) of generating medium when two fields simultaneously participate in the HHG process. This concept has earlier been demonstrated in the case of gases and ablated bulk materials. Earlier studies of HHG in gases have anticipated that stronger harmonics generation in the case of TCP was possible due to the formation of a quasi-linear field, selection of the short quantum path component, which has a denser electron wave packet, and higher ionization rate compared with single-color pump (SCP) [12]. The orthogonally polarized second field can also be responsible for the modification of the trajectory of the accelerated electrons from being two-dimensional to three-dimensional which may lead to the removal of the medium symmetry. With suitable control of the relative phase between the fundamental and second harmonic radiation, the latter field enhances the short path contribution, resulting in a clean spectrum of harmonics. One can expect that TCP of clusters can further enhance the harmonic yield even at small ratios of second and first waves' intensities provided the sufficient overlap of these two fields in the medium under consideration.

The first attempts in frequency conversion at such conditions were reported in the case of the studies of the lowest-order harmonic generation from the ablated carbon-containing materials [13]. The third harmonic (TH) in the laser plasmas produced on the surfaces of different materials was analyzed using the delays ranging between a few tens nanosecond to a few hundred microseconds between a 7-ns heating pulse and a 17-ns driving pulse, both of 1064 nm in wavelength. The observation of TH emission, though extremely weak, at the microsecond timescale of delays allowed concluding that the process of third harmonic generation (THG) occurs not only from atoms and ions of carbon materials at small delays but also from their aggregates at large delays.

This approach is extremely useful for the analysis of multi-atomic systems as the media for HHG. In this section, we analyze HHG in graphite plasma using a broad range of delays between a 5-ns heating pulse and a 40-fs driving pulse [14]. We discuss the advantages of using nanosecond pulses for carbon nanoparticles (NP) plasma formation to achieve efficient HHG compared with picosecond pulses induced ablation. Those studies have shown that small carbon clusters arrive at the area of interaction with femtosecond beam notably earlier than one can expect assuming the kinetic consideration when the components are assumed to reach thermodynamic equilibrium during the expansion. These observations allow concluding that all carbon clusters acquire from the very beginning similar kinetic energy and spread out of the surface with the velocity approximately similar to the one that possesses a single carbon atom ablating from the bulk material.

5.1.1 Experimental Arrangements

The graphite plasma was used for HHG (Fig. 5.1). The femtosecond pulse propagated through the plasma 30 ns from the beginning of target irradiation by 200 ps or 5 ns heating pulses. This delay between heating and pump pulses was proven to be optimal for HHG in graphite plasma at the used geometry of experiments when the 800 nm driving femtosecond pulses were focused onto the plasma from the orthogonal direction, at a distance of 0.2 mm above the target surface.

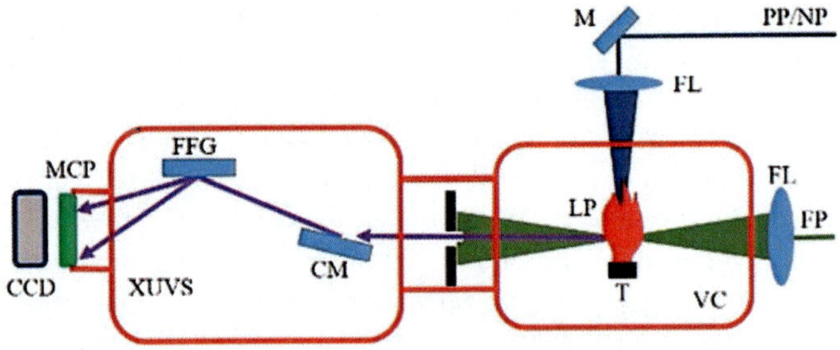

Fig. 5.1 Setup for high-order harmonic generation. FP, converting femtosecond pulses, PP/NP, picosecond or nanosecond heating pulses; FL, focusing lenses; VC, vacuum chamber; T, target; LP, laser plasma; XUVS, extreme ultraviolet spectrometer; CM, cylindrical gold-coated mirror; FFG, flat field grating; MCP, microchannel plate; CCD, CCD camera. Reproduced from [14] with permission from the Optical Society of America

The harmonic yield was maximized by adjusting the position of the target with regard to the optical axis of propagation of the driving femtosecond pulses and by varying the focusing position relative to the plasma. The harmonic emission was directed to the XUV spectrometer, which contained a cylindrical mirror and a 1200 grooves/mm flat field grating with variable line spacing. The XUV spectrum was recorded by a micro-channel plate (MCP) with a phosphor screen, which was imaged by a CCD camera.

Some of the experiments were carried out using a two-color pump (TCP) of graphite plasma. The reason for using TCP instead of the single-color pump (SCP) was related to earlier demonstrated advantages of this approach in generating the odd and even harmonics, as well as larger efficiency of harmonic yield. TCP using the 800 nm radiation as the first field and 400 nm radiation as the second field was applied to carry out the comparative analysis of TCP and SCP schemes. The 0.4-mm-thick barium borate crystal (BBO, type I) was installed inside the vacuum chamber on the path of focused the 800 nm pulse. The conversion efficiency of H2 pulses (λ = 400 nm) was ~5%. The overlap of these two pulses in the plasma area was sufficient for observation of the role of the second orthogonally polarized field influencing the whole process of HHG due to small group velocity dispersion in the thin BBO crystal.

5.1.2 *High-Order Harmonic Generation*

XUV emission spectra of carbon plasma at different regimes of graphite ablation in the range of 30–130 nm are shown in Fig. 5.2a, b. Either picosecond (upper panel) or nanosecond (middle panel) pulses were used for laser ablation (Fig. 5.2). Various emission lines attributed to CII–CIII ions were identified and used for calibration of the XUV spectrometer. These conditions of ablation were unsuitable for efficient

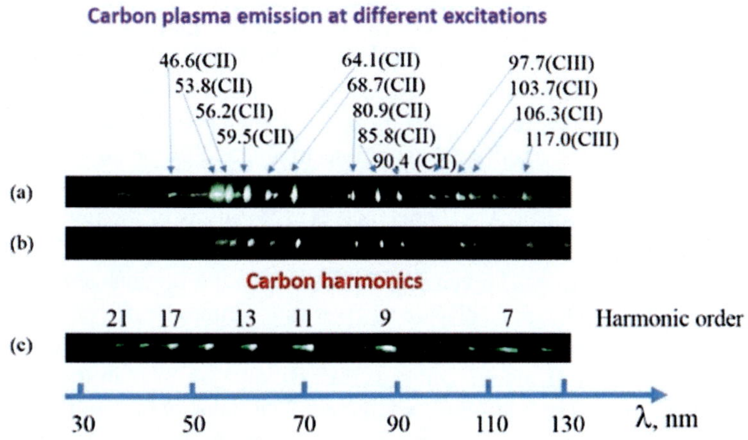

Fig. 5.2 Raw images of plasma and harmonic spectra. **a** Graphite plasma emission using picosecond pulses induced ablation. **b** Graphite plasma emission using nanosecond pulses induced ablation. **c** Harmonic spectra using single-color pump (800 nm) of C plasma at 30 ns delay between heating and driving pulses. The fluence of heating pulses on the target surface was maintained at 5 J cm^{-2}. The distance between the target and the axis of propagation of the driving beam was 0.2 mm. Reproduced from [14] with permission from the Optical Society of America

HHG due to the presence of a large number of free electrons preventing optimal phase relations between the driving and harmonic waves. Nanosecond pulses induced ablation was chosen for further analysis of HHG from different species of graphite plasma since in that case, contrary to ablation by picosecond pulses, one can able to analyze the delay-dependent harmonic emission a the delays between heating and converting pulses ranging between 0 and 5×10^4 ns.

Lesser excitation of graphite target by nanosecond pulses allowed the formation of plasma suitable for a harmonic generation (Fig. 5.2c). Odd harmonics of 800 nm radiation up to 23rd order (H23) were observed. It is worth noting that those observations of harmonic spectra did not show a simultaneous presence of plasma emission from high-charged particles in the studied spectral range (30–130 nm) thus confirming that the experiments were carried out at a soft ablation regime of the carbon-containing target. The absence of the plasma emission associated with the higher-charged ions at these conditions was accomplished by the moderate excitation of graphite. The driving pulse at $I \approx 2 \times 10^{14}$ W cm^{-2} also did not cause the appearance of strong ionic emission from the higher-charged carbon ions.

The important peculiarity of the plasma emission and harmonic studies was the proximity of some ionic transitions of carbon with the wavelengths of harmonics, particularly strong CII transitions (53.8 and 59.5 nm) and 15th and 13th harmonics (53.3 and 61.5 nm) respectively. Neither enhancement nor suppression of these harmonics generated in graphite plasma was observed. These studies showed that the proximity of strong emission lines and harmonics did not necessarily lead to the variation of the harmonic yield.

The reason for presenting the experimental data as the images appearing on the screen of the computer rather than the line-outs of HHG spectra is justified by better viewing of obtained data and a clear definition of the difference between the plasma and harmonic spectra. The goal of this group of studies was the qualitative determination of the influence of different conditions of ablation on the plasma emission and the harmonics conversion efficiency rather than the quantitative measurements of harmonics in the plasmas. Also notice that visual analysis of the raw images of harmonics and plasma emission allows distinguishing these sources of XUV radiation by their divergences. The vertical dimensions of the raw images of plasma emission, which characterize its divergence, were a few times larger compared with those of harmonics. An additional advantage of this method of data presentation is a direct demonstration of the insignificant role of emission lines in the harmonic conversion efficiency. The approximately plateau-like shape of harmonics distribution remained unchanged in spite of the coincidence of plasma emission limes and some harmonics orders.

The λ-axis in Fig. 5.2 is shown on the basis of the calibration of the XUV spectrometer for distinguishing the distribution of harmonics along the short-wavelength region. The saturated images were intentionally chosen to present the spectra for better viewing. Note that the unsaturated images were used for the line-outs of the HHG spectra.

The next set of HHG studies was performed using two orthogonally polarized driving beams. The 0.4 mm thick BBO crystal was inserted on the path of 800 nm radiation inside the vacuum chamber that allowed the generation of 400 nm pulses (H2) while maintaining sufficient temporal and spatial overlap of 800 and 400 nm pulses in C plasma. Even small conversion efficiency (~2%) of 400 nm radiation allowed the generation of the lower-order even harmonics alongside the odd ones (Fig. 5.3). The even harmonics generation up to H14 were observed (see middle panel). A relatively small yield of lower-order harmonics in these spectra (H6 to H9) was caused by the weaker sensitivity of MCP in the range of the wavelengths above 100 nm. The use of different fluencies of heating nanosecond pulses ranging between 5 and 9 J cm^{-2} led to variations of generated spectra.

The conversion efficiency of odd harmonics in the case of a two-color pump was approximately three times higher than in the case of a single-color pump. One can see that even at a very small ratio of 400 and 800 nm pulse energies (1:20) the influence of weak field was sufficient to strongly modify the whole harmonic spectrum.

The application of two-color sources for HHG has also another advantage. It allows the two-fold increase of the number of generated harmonics, which in turn may lead to the observation of the resonance-induced enhancement of single components of those multi-harmonic sources. The latter process is mostly related to the ionic transitions of various materials possessing high oscillator strengths. In the case of graphite plasma, no enhancement of even harmonics was observed, while in the case of indium plasma H12 was notably enhanced alongside H13 due to proximity with the ionic transition possessing large oscillator strength [15–17].

In the case of stronger excitation of carbon plasma, the small redshift of odd harmonics was observed, while even harmonics were not shifted due to notably

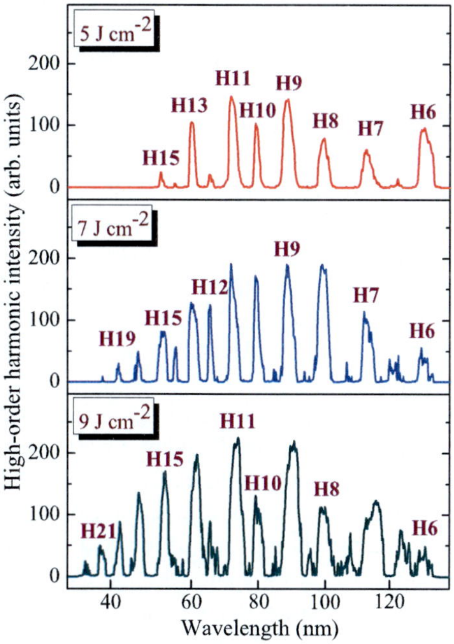

Fig. 5.3 Harmonic spectra using the two-color pump (800 nm + 400 nm) at different fluencies of nanosecond pulses on the target surface. One can see some red shift of odd harmonics at highest fluence while even harmonics remained at fixed wavelengths. Reproduced from [14] with permission from the Optical Society of America

weaker energy of 400 nm radiation compared with 800 nm pulses (bottom panel of Fig. 5.3). Previous reports on gas HHG experiments have shown a distinct difference between the HHG spectra driven by negatively and positively chirped pulses, which was related with both the influence of the dynamically induced negative chirp and the positive chirp induced by strong laser radiation and the ionized medium due to the self-phase modulation of the laser pulse [18]. The observation of red-shifted harmonics has been analyzed in [19] during gas HHG studies, while the blue-shifted harmonics is a common feature observed in multiple studies of HHG from the gases and plasmas. Notice that the red-shift of chirp-free pulses in the case of dense plasmas has already been reported in the case of plasma harmonics studies [20].

Different scenarios can be involved in the case of clustered media in regards to the observed redshift of odd harmonics with higher intensities of the nanosecond ablating pulse. It has been shown in [19] that the chirp-induced shifts of harmonics generated in a Xe gas jet exhibit a trend toward either the blue or red, depending on the particle density. The authors point out that redshift was observed at low gas densities. They suggest that this process might be related to the Kerr effect, electron–ion recombination, and ion expansion, and cluster formation. The latter mechanism seemed consistent with the experimental findings. The chirp-induced redshift and blueshift were explained by the model including both atoms and ions. The neutrals were assumed to be responsible for the blue shift, while the ions were the main reason for the redshift. The redshift can also be attributed to the self-phase modulation (SPM) of the laser pulse during the propagation of the leading part of the pulse through the medium consisting of neutrals and singly charged ions. As one can see, the broader spectra of odd harmonics (Fig. 5.3, bottom panel) from C clusters can also be attributed to the influence of SPM, leading to the redshift of those harmonics. Thus the observations of the redshifted odd harmonics were consistent with the model described in [19].

The plasma density was not measured in these experiments. The heating radiation intensity on the target surface was maintained at 2×10^9 W cm^{-2}. The particles density in plasma was estimated using the hydrodynamic code HYADES [21] to be a few units of 10^{17} cm^{-3}. The ionization potentials of carbon atoms and singly charged ions are equal to 11.25 and 24.38 eV, respectively. Thus such plasma mainly consists of excited neutral atoms and singly ionized carbon. Notice that the presence of excited atoms and ions can increase a nonlinear optical response of laser-produced plasma.

A typical dependence of the 9th and the 17th harmonics intensities on the delay between pulses is presented in Fig. 5.4. At the initial stages of plasma formation and spreading out of the target surface, i.e. at the delays less than 10 ns, the concentration of neutrals and singly charged ions is insufficient to produce a measurable amount of harmonic photons since the whole ensemble of particles possessing the velocities of $\leq 1 \times 10^4$ m s^{-1} cannot reach the optical axis of the propagation of driving beam, which was ~0.2 mm above the graphite surface in the discussed experiments. The increase of delay up to 10–50 ns caused the appearance of a large number of carbon atoms and ions, as well as light clusters, along the path of the femtosecond beam that allowed maximizing the yield of harmonics. The increase of delay above 50 ns led to a decrease in HHG efficiency. A similarity in the delay dependence for H9 and H17 was observed. The harmonics almost disappeared from XUV spectra once the delay exceeded 300 ns. The attempts to observe the harmonic emission in the microsecond time scale of delays were unsuccessful.

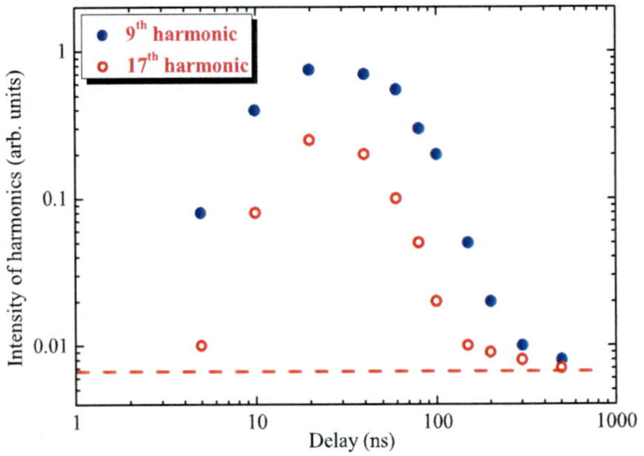

Fig. 5.4 Dependences of H9 and H17 yields on the delay between nanosecond and femtosecond pulses. Dashed line shows the threshold of registration of harmonic emission. Reproduced from [14] with permission from the Optical Society of America

5.1.3 Discussion

Clusters play important role in the modification of the harmonic spectra generated in different media. The harmonics' enhancement in the gas jets containing clusters has been reported in [10]. In the case of clusters, the cross-section of recombination of the accelerated electron with the parent particle is higher compared with the atoms [22]. In the meantime, the experiments with gas clusters have revealed some difficulties in disentangling the harmonics produced by different species (monomers and clusters of different sizes). The uncertainty in the exact mechanism of the HHG from clusters has previously been underlined in [22, 23].

Laser ablation of carbon has been proven to be a versatile method for the synthesis of clusters, which allowed characterization and dynamics of the multi-particle species formation [24]. The authors of [25] have concluded that the appearance of clusters during ablation is a result of clusters' ejection from the target and collision-related aggregation in the expansion of the plume. The growth of clusters is based on the addition of many small neutral species to the ions in a stepwise fashion. The processes that can cause the formation and disintegration of multiatomic species include photomechanical ejection, liquid phase ejection, and fragmentation, homogeneous nucleation and decomposition, heterogeneous decomposition, etc. [26].

A detailed understanding of the basic physical processes governing the composition of carbon ablation and the reliable methods for controlling the plume species became a necessity for the selection of the optimal plasma conditions for HHG. The following arguments supporting the explanation of higher yield of harmonics from such plasmas were considered: (i) carbon species allow the formation of multi-particle clusters during laser ablation, which can enhance HHG yield, (ii) the first ionization potential of carbon is high enough to prevent the appearance of high concentration of free electrons, (iii) the graphite target allows easier generation of a relatively dense carbon plasma and the production of phase-matching conditions for a lower-order harmonic generation. The carbon-containing materials like graphite, fullerenes, and carbon nanotubes have proven that, at optimal ablation conditions, they can be considered as attractive plasma media for efficient harmonic generation in the 40–80 nm spectral range using the femtosecond laser pulses [27].

To define the role of clusters in a variation of the nonlinear optical response of medium during frequency conversion of ultrashort pulses one has to analyze both lowest- and higher-order harmonic generation conditions. As already mentioned, the necessity in lowest-order harmonic generation studies is as follows. It provides a guide for the selection of the targets suitable for the efficient generation of short-wavelength coherent radiation using different delays between heating and driving pulses [28]. This method also allows the application of low-order harmonic generation for plasma diagnostics, which is less complex compared with the HHG-based method. The additional assumption for strong harmonics generation from clusters and nanoparticles compared with single atoms or ions could be the higher concentration of neutral atoms inevitably accompanying the presence of aggregated multi-atomic species. It was suggested in [29] that small carbon clusters play an important role in the generation of high-order harmonics of plasmas. A similar conclusion was made in [30] with regard to TH emission. It was underlined that harmonics generation in plasmas, when performed with full temporal delay control and under conditions of spatial segregation typical of laser ablation plasmas, provides a method to qualitatively assess cluster size distributions for different materials and irradiation conditions.

In graphite, the ablation plasma plume may contain various species of carbon, i.e. neutrals and ions, small molecules, clusters, aggregates, etc., which can contribute to harmonic generation to various extents. It is important to determine their presence in the region, where the driving laser pulse interacts with the expanding plasma. In particular, the production of clusters in the laser-plasma during laser ablation of various targets has a high probability, while their presence in the plasma area where the frequency conversion occurs was yet to be confirmed directly.

Earlier, the analysis of carbon cluster composition of the ablation plume produced by nanosecond laser pulses for HHG was accomplished by time-of-flight mass spectroscopy (ToFMS) [31], which confirmed the presence of carbon clusters ranging from C_5 to C_{25}. The attempts to find higher mass clusters failed, though they were searched for over a long range of delays up to a few μs between the onset of laser ablation and the switching on the triggering pulse in ToFMS. Earlier studies of carbon plasma have also shown that the $4n + 3$ clusters in C_5–C_{25} range had a higher abundance than neighboring ones [32]. The conclusion was made that those clusters (C_7, C_{11}, C_{15}, and C_{19}) could be considered as the most effective emitters of harmonics.

The harmonic generation in such plasma can prove or disapprove the role of the complex composition of ablated species in the former process. A signature revealing the nature of the emitters could be the growth of harmonics emission arising at some specific delay from the beginning of ablation. This delay may characterize the time of propagation of the clusters from the target's surface to the optical axis of ultrashort pulse propagation. It was suggested in [33] that if the components could be assumed to reach thermodynamic equilibrium during the expansion so that a similar average kinetic energy $E = mv^2/2$ could characterize all plasma components, the average arrival times could be assigned to different cluster sizes. This formula rather determines the condition of the similar velocities for the atoms and multiatomic systems of similar elements, like carbon and fullerene. A similar assumption can be applied to atoms and ions of different materials, which was confirmed during the comparative THG studies using indium and graphite plasmas. The ejection of lighter atoms and ions from the target surface does not allow them to reach the axis of the driving beam earlier as compared with heavier species.

Following the logic of Ref. [33], the large clusters like C_n should appear in the laser-plasma interaction zone $n^{0.5}$ times later compared with single atoms and ions of carbon. In other words, the appearance of the group of C_7–C_{19} clusters in the interaction zone could be expected at delays of 2.7–4.4 times later compared to the C_1 particles. However, this scenario does not happen, since no enhancement of harmonics occurred at these delays (90–140 ns). The discussed studies showed that carbon clusters arrive at the area of interaction with femtosecond beam notably earlier than one can expect assuming the above kinetic equation. In other words, from the very beginning, all atoms in clusters acquire similar kinetic energy and spread out of the surface with a velocity approximately equal to the one of a single atom of carbon.

Below, we address the difference in the expected and actual optimal delays between heating and driving pulses in the case of graphite-ablated plasma. In the case of a thermalized ablation plume, i.e. at very slow heating conditions, the average arrival times can be assigned to different cluster sizes. Another situation occurs in the case of the fast, either nanosecond or picosecond, ejection of energy in the tiny area of the target. In that case, the delay between heating and driving pulses at which the harmonic yield reaches its maximum should scale as a square root of the atomic or molecular weight of the constituents. The additional studies revealed that, for bulk gold ablation, the maximum high-order harmonic yield from single gold atoms and ions occurred at a delay of about 180–300 ns. Meantime, the ablated 20-nm Au nanoparticles allowed efficient generation at about 200–350 ns delay, which is approximately equal to the delay in the case of Au atoms and ions. Furthermore, attempts to observe HHG at the delays of up to 50 μs, i.e., at the expected delay for thermalized larger nanoparticles, did not show any harmonic emission. Thus, the discussed studies demonstrate that clusters, quantum dots, and relatively small nanoparticles arrive in the area of interaction with the femtosecond laser beam notably earlier than one would expect assuming the thermalized regime of evaporation. In other words, all multi-atomic species acquire, from the very beginning, similar kinetic energy and spread out from the surface with velocities approximately similar to those of the single atoms and ions ablating from bulk material. This conclusion reconciles the similarity in the optimal delays for HHG from the plasmas produced on the bulk target leading to evaporation of atoms and ions and on the targets containing nanoparticles of the same material.

The velocities of the multi-atomic carbon clusters, which assume to be the main and most efficient emitters of third and higher-order harmonics, are estimated by knowing the "optimal" delays for highest harmonic yields in both cases (~100 and ~25 ns) and the distances from targets in these two groups of experiments (0.6 and 0.2 mm). These particles possessed the 6×10^3 and 8×10^3 m s^{-1}. One can see a similarity of the velocities determined from the two groups of experiments using different facilities.

A more in-depth physical explanation of why clusters of various masses seem to spread with a velocity similar to that of a single atom is required in spite of the experimental confirmation of this fact. To resolve this issue, the ToFMS studies can give a convincing answer on the presence of both ion and cluster clouds at the same moment in the area of propagation of the converting pulses. In the meantime, one can assume the difficulties in establishing such sort of experiments requiring the combined HHG and mass spectrometry facilities.

5.2 Efficient High-Order Harmonics Generation Using 1030 nm Laser in Carbon Nanostructures

LIPs may have a complex composition containing neutrals, ions, clusters, quantum dots, NPs, and large fragments of source materials. Each of these components can play a specific role in the process of HHG. Studies of HHG from cluster-containing gases and from the ablation of quantum dots and NPs have revealed the advantages of using such species for efficient frequency conversion in the XUV range [34, 35]. Only a few HHG studies in gases using clusters comprising 10^3–10^4 atoms (with the particles sizes between 0.8 and 4 nm) have been reported so far (for example [34, 36, 37]), while some theoretical simulations predict a significant increase in the harmonic intensity as compared to monoatomic media [38–41].

The pulse duration is one of the most important parameters when considering the *in-situ* synthesis of NPs in LIP during the ablation of metals. The same can be said when one considers materials like graphite and carbon-containing nanomaterials, such as fullerenes and nanotubes. LIP produced during graphite ablation has proven to be a medium containing different carbon clusters [42].

Here we discuss the application of a two-laser scheme to investigate HHG in carbon-based plasmas, where the nanosecond laser is used for LIP formation and the high pulse repetition femtosecond 1030 nm laser is used for harmonics generation in plasma [43]. Three materials were used as targets: bulk graphite, fullerene powder (C_{60}, size of fullerenes 0.7 nm), and multi-walled carbon nanotube (CNT) powder (15 nm diameter; the length of tubes was in the range of 500 nm–20 μm). The powders were mixed with epoxy resin (ER) and glued on the glass substrates (5 mm × 5 mm × 1 mm). After drying they were placed along with bulk graphite, on the translating stage inside the vacuum chamber for laser ablation. The laser ablation fluencies varied between 1 and 3 J cm^{-2} depending on the properties of targets used for LIP formation. In each case, the conditions of ablation were determined to achieve the maximal harmonic yield. The targets were continuously moved during HHG experiments to maintain the stability of harmonic emission.

The experimental setup for the HHG experiments was similar to those described in previous chapters. The first chamber contained targets, which were ablated by nanosecond pulses, and the second chamber comprised an XUV spectrometer allowing the analysis of the plasma and harmonic emission. Femtosecond DP (40 fs, 1030 nm, 0.3 mJ, 100 kHz) from Yb-doped fiber laser were focused inside the plasma plume to generate harmonics. For laser plasma formation, the 6 ns HP ($\lambda = 1064$ nm, 5 mJ, 10 Hz) from Nd:YAG laser were used. HP were focused by a 300 mm focal length spherical lens on the target surface at the spot with a diameter of 0.3 mm providing intensity up to 10^9 W cm^{-2} and fluence up to ~6 J cm^{-2}.

The femtosecond pulses were delayed with respect to those from Nd:YAG laser to propagate through the formed plasma at a maximal density of the ejected particles (Fig. 5.5). The synchronization of femtosecond and nanosecond lasers was accomplished using a digital delay generator, and the delay between HP and DP was varied in the range from 0 to 1000 ns. DPs were focused by a 400 mm focal length spherical

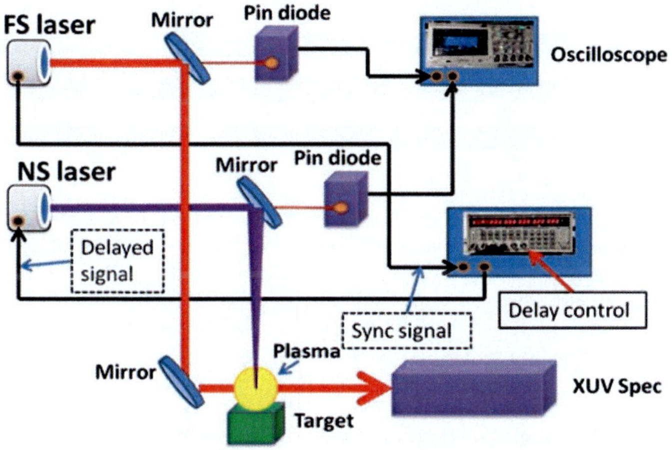

Fig. 5.5 Experimental setup for the formation of LIP by nanosecond pulses followed by generation of high-order harmonics in carbon atoms and nanostructures at different delays between HP and DP. Reproduced from [43] with permission from the American Institute of Physics

lens inside LIP ~150 μm above the target surface. The intensity of the driving pulses in the plasma area was maintained to be ~2×10^{14} W cm^{-2}. The harmonics and plasma emission were detected using a flat field grazing-incidence XUV spectrometer.

The time map of Yb-doped fiber laser and Nd:YAG laser synchronization was as follows. The synchronization signal was taken from the femtosecond laser operating at a 100 kHz pulse repetition rate. The electric signal from this laser triggered the digital delay generator, which provided a synchronization pulse to start up the nanosecond laser operating at a significantly smaller pulse repetition rate (10 Hz). The latter synchronization pulse was adjusted in such a manner that nanosecond pulse arrived on the target surface after the initial femtosecond pulse, but ~60 ns prior to the next femtosecond pulse of 100 kHz laser. The same operation occurred each 100 ms. Thus only ten pulses out of 10^6 pulses of femtosecond laser emitting during one second participated in a harmonic generation, while other pulses did not participate in this process. This adjustable delay between heating nanosecond pulse and driving femtosecond pulse allowed carrying out all delay-related experiments.

5.2.1 Calibration of HHG Using Carbon Plasma

Here we analyze the harmonic emission obtained during ablation of the "simplest" carbon-containing target (graphite, Fig. 5.6). The delayed 40 fs driving pulses propagated through laser-induced plasma 70 ns from the beginning of target ablation. Both single-color pump (SCP, 1030 nm) and two-color pump (TCP, 1030 nm + 515 nm) schemes were used in those studies. The 0.1 mm and 0.4 mm thick barium borate (BBO) crystals inserted inside the vacuum chamber on the pass of focused DP were

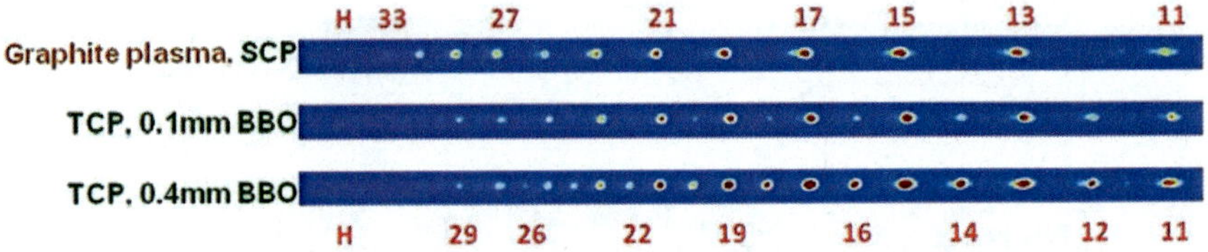

Fig. 5.6 Raw images of harmonic spectra from the ablation of graphite by 6 ns heating pulses. The 40 fs pulses propagated through laser-induced plasma 70 ns from the beginning of target ablation. Upper panel: Single-color pump (SCP, 1030 nm) of LIP. Middle panel: Two-color pump (TCP, 1030 nm + 515 nm) of LIP using 0.1 mm thick barium borate (BBO) crystal. Bottom panel: TCP of LIP using 0.4 mm thick BBO. The numbers above and below raw images show the harmonic orders. Reproduced from [43] with permission from the American Institute of Physics

used for second harmonic generation (515 nm) and TCP (1030 nm + 515 nm) of carbon-containing LIP. Second harmonic generation efficiencies in these comparative experiments were 1.5% (in the case of 0.1 mm thick BBO) and 4.7% (in the case of 0.4 mm thick BBO).

The upper panel Fig. 5.6 shows the harmonic spectrum after propagation of the 1030 nm femtosecond pulses through the graphite plasma produced using the optimal ablation fluence ($F = 1.7$ J cm^{-2}). The term "optimal ablation fluence" refers to the conditions of HP-target interaction allowing the formation of LIP where the maximal harmonic yield can be achieved. Obviously, the fluence satisfying the above requirement can be distinguished from target to target. The value of this parameter depends on various factors related to the properties of ablated material and defines empirically. Particularly, in the case of graphite ablation, the optimal ablation fluence was equal to 1.7 J cm^{-2}, while for CNT and C_{60} cases the optimal fluence was ~1.2 J cm^{-2}, i.e. 1.5 times smaller.

TCP of this plasma using 0.1 mm thick BBO crystal led to the generation of relatively weak lower-order even harmonics (H10–H20; Fig. 5.6, middle panel), alongside the odd ones. Significantly stronger even harmonics were generated while applying the 0.4 mm thick BBO crystal for the formation of the second field (Fig. 5.2, bottom panel). Though the ratio of second harmonic and fundamental pulses in the case of using 0.4 mm BBO was still low (~1:20), the sufficient overlap of two pulses in plasma area allowed significant amendment of TCP at these conditions. Moreover, in that case, the odd harmonics were also enhanced. Earlier, it has been suggested that the stronger harmonic generation in the case of TCP is possible due to the selection of a short quantum path component, which has a denser electron wave packet, and higher ionization rate compared with the SCP [44].

The thickness of the nonlinear crystal influences the conversion efficiency of the second field used during the two-color pump of plasma. For example, in the case of 0.1 mm thick BBO crystal, the second harmonic generated with the efficiency not exceeding 1.5%, which gives the ratio 1.5:98.5 $\approx$ 1:66 of the energies of second harmonic (515 nm) and fundamental (1030 nm) waves. This ratio also indicates a significant difference in the intensities of two waves in the nonlinear medium, which causes a lesser response of plasma on the joint influence of these waves during generation of H = 4(n + 1) harmonics (H12, H16, H20, etc.), as well as smaller conversion efficiency of other even harmonics (H = 2(2n + 1); H10, H14, H18, etc.) corresponding to the odd harmonics of 515 nm pulses (see middle panel of Fig. 5.6). Another pattern was observed in the case of application of 0.4 mm thick BBO allowing a notable increase of above ratio (4.7:95.3 ≈ 1:20) leading to better interaction of interacting pumps in LIP and higher conversion efficiency towards the odd and even harmonics (bottom panel of Fig. 5.6).

The important notice during consideration of two-color interaction in LIP is the overlap of fundamental and second harmonic pulses in the studied medium. This overlap depends on the thickness of BBO, the wavelength of fundamental radiation, and pulse duration. Thus in the case of TCP containing 806 nm and 403 nm waves, the IR pulse is delayed by 38 or 76 fs relative to the second harmonic pulse when using a 0.2 or 0.4 mm crystal due to the group velocity dispersion in BBO crystal. The mismatch between the driving ω and 2ω pulses in the laser-plasma region did not prevent them from being converted to odd and even harmonics in the case of a 0.2 mm thick crystal. Meanwhile, in the case of a 0.4 mm crystal, the HHG spectrum using 800 nm class lasers consisted of the odd harmonics of ω and 2ω waves, while the even harmonics determined by the ratio $H = 4(n + 1)$ were considerably weaker due to insufficient overlap of 800 and 400 nm waves in the plasma region. At the same time, the appearance of all even harmonics in the case of the thinner (0.2 mm) BBO crystal revealed sufficient overlap of 800 nm and 400 nm pulses in LIP [45].

In the case of longer wavelength fundamental pulses (like in discussed case), the influence of group velocity dispersion in BBO on the overlap of pulses in LIP decreases compared with the case of 800 nm pump. The delay between 1030 and 515 nm pulses after propagation of 0.4 mm crystal becomes equal to 48 fs, which does not prevent the overlap of the trailing part of 2ω pulses and leading part of ω pulses thus resulting in the efficient generation of all even harmonics (bottom panel of Fig. 5.6).

The cutoff orders in the three cases shown in Fig. 5.6 were approximately close to each other (H33 in the case of SCP (upper panel), H31 in the case of TCP using 0.1 mm BBO (middle panel), and H29 in the case of TCP using 0.4 mm BBO (bottom panel)). No resonance-induced enhancement of odd and even harmonics was observed since all generated orders of harmonics showed a featureless gradual decrease of intensity towards the shorter-wavelength range.

The wavelength of the harmonics for SCP was slightly blue-shifted compared with those for TCP (compare upper and middle panels of Fig. 5.6). This phenomenon manifests itself on a case-by-case basis. The probable reason is the self-phase modulation of the strong fundamental wave (ω) during propagation through the plasma resulting in, particularly, a blue shift of its wavelength and the wavelengths of odd harmonics. In the meantime, in the case of TCP, the intensity of the fundamental pump decreases due to different factors, such as a decrease of pulse energy due to second harmonic generation and an increase of pulse duration. The propagation of 40 fs, 1030 nm wave through relatively thick (0.4 mm) BBO causes some growth of pulse duration due to the group velocity dispersion in this crystal. The longer-wavelength component of laser pulse spreads through the crystal at a smaller velocity compared with the shorter-wavelength component. This process becomes crucial in the case of ultrashort laser pulses; however, even in the case of relatively long (40 fs) pulses, such an effect can cause the increase of pulse duration leading to the decrease of the intensity of laser pulse. These factors lead to lesser self-phase modulation of the driving pulses and correspondingly lesser blue shift of the harmonics originated from this pump. The blue shift of the odd harmonics of notably weaker second harmonic wave (2ω) was not observed since this component of the two-color pump does not undergo the self-phase modulation in LIP.

The transmission electron microscopy (TEM) studies of the debris deposited on nearby glass slides during ablation of graphite showed the presence of small NPs ranging from 4 to 14 nm. These NPs appeared in TEM images at the optimal fluence of ablation ($F = 1.7$ J cm^{-2}) when the maximal harmonic yield was achieved.

HHG requires a coincidence of the phase velocities of the driving wave and harmonic waves along the whole length of the nonlinear medium. Without fulfilling these conditions the harmonic efficiency becomes too small (10^{-9} or even less). The maintenance of the phase-matching conditions when the wavefronts of driving and harmonic pulses spread with equal velocity allows drastically improving HHG efficiency (up to 10^{-5} and higher). This requirement is the prerequisite for the studied process of HHG, similar to the one required for the efficient second harmonic generation in the nonlinear crystal.

Intuitively, as higher ionization rate, which can be achieved at a relatively high fluence of HP, and as the larger concentration of harmonic emitters appears in the plasma then correspondingly as larger will be the nonlinear optical response of the medium. The incorrectness of this assumption has been demonstrated during early HHG experiments in plasma when the appearance of a large number of electrons at high fluencies of heating pulses entirely canceled the advanced features of denser LIP [46–48]. The excess of free electrons leading to significant growth of group velocity dispersion of plasma medium leads to the sharp disruption of the phase-matching conditions between the waves of DP and harmonics. That is why one has to optimally ablate the targets to create conditions when the ionization degree becomes as small as possible while keeping a sufficient amount of harmonic emitters. This condition allows finding the compromise in creating the "optimal" LIP. The ionization degree at those conditions does not exceed 1 and actually becomes notably less. In other words, to create the optimal conditions for HHG in LIP one has to maintain ablation of targets when only 5 to 10 percentages of atoms become ionized, which allows keeping the concentration of free electrons below 10%.

HHG in LIP was studied at the conditions when plasma emission lines coincided with some harmonic orders to analyze the resonance enhancement of those harmonics. Previous studies of resonance-induced enhancement of single harmonic in metal plasmas have revealed that the amplification of single harmonic strongly depends on its detuning with respect to the ion transition possessing strong oscillator strength *gf* [49]. One can expect that the intensity ratio between the "resonant" and neighboring harmonics will be obviously changed as soon as the wavelength of the amplified harmonic is tuned towards or out of this transition.

Meanwhile, the presence of ion transitions with large *gf*s cannot be considered as a sufficient prerequisite for the observation of resonance-induced amplification of harmonics. An example that supports this statement is graphite plasma. The 68.7 nm CII line, which coincides with H15 ($\lambda = 68.66$ nm) of 1030 nm radiation, actually consists of three transitions whose *gf*s (1.3, 0.71, and 0.14) are not small. However, the experiments using 1030 nm radiation showed neither amplification nor suppression of H15. Thus the strong oscillator strength of ion transitions does not necessarily cause harmonic amplification, since other processes, like phase mismatch, can reduce the role of the resonance-induced processes.

5.2.2 HHG in Fullerene and Carbon Nanotube Plasmas

For graphite, CNT, and C_{60} cases, one can notice a similarity in the harmonic cutoffs during SCP of LIP (H33, Fig. 5.7), in spite of the difference in the ionization potentials. The main expected difference is the prevalence of harmonic yield in the cases of CNT and C_{60} plasmas compared with graphite one. These expectations are referred to as the single-color pump scheme. Meanwhile, the application of TCP of these plasmas can reveal the enhancement of odd harmonics, the appearance of even harmonics, and an insignificant decrease of harmonic cutoff due to a decrease of the intensity of the driving pulses (like in the case shown in the bottom panel of Fig. 5.6).

The presence of highly charged ions (i.e. those above C III) does not provide sufficient contribution to the extension of HHG cutoff, since the harmonic conversion efficiency quickly reduces with increasing ionization potential. HHG spectra from ablated graphite demonstrated the repeatable uniform distribution of harmonics with the same cutoff for the delays between HP and DP varying from 50 up to 200 ns, as well as the gradual decrease of harmonic intensity in the vicinity of the cutoff region. Meanwhile, C_{60} showed higher conversion efficiency even for the long delays. For example, at 300 ns delay, 17th harmonic generating in fullerene plasma was more than 20 times stronger than the same harmonic from the ablated graphite. A similar feature was observed in the case of CNT plasma compared with the graphite plasma. One can assume that enhanced lower-order harmonics were attributed to the presence of small fragments of CNTs in LIP. One can also suppose that, in the case of CNT, the fragmentation process also induces the appearance of small-sized carbon molecules (C_2) and clusters (C_5–C_{25}) leading to the larger yield of coherent XUV radiation, since at similar delay between HP and DP the integrated harmonic yield from CNT plasma exceeded that from graphite LIP (compare middle and bottom panels of Fig. 5.7).

Overall, among the two studied nanostructures, fullerenes (upper panel of Fig. 5.7) can be considered as the most efficient harmonic emitters compared with carbon atoms and small-sized carbon molecules. Their presence in LIP leads to stronger enhancement of the lower-order part of the HHG spectrum comprising the harmonics between 13 and 25th orders. Both fullerene and nanotube LIPs allow for the enhancement of harmonics compared with the plasma produced on the graphite surface while

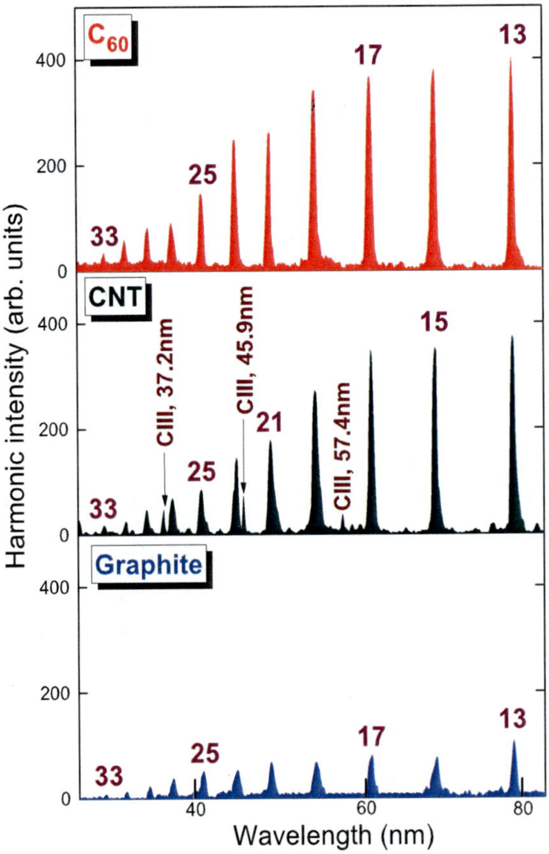

Fig. 5.7 Comparative spectra of harmonics generated in fullerene (upper panel), carbon nanotube (middle panel) and graphite (bottom panel) LIPs using SCP. Reproduced from [43] with permission from the American Institute of Physics

using optimal fluencies of the heating nanosecond pulses for ablation of each target. The harmonics at these conditions were enhanced with a factor ranging between 3 × (in the case of higher orders) and 7 × (in the case of lower orders).

The delay-dependent measurements of harmonic yield from CNT and C_{60} LIPs (Fig. 5.8) were almost similar to the one measured for the graphite plasma. The maximal yield of harmonics was observed at 60–70 ns delay from the beginning of ablation. The only difference compared with similar dependence observed in the case of graphite plasma was the extended range of delays at which the lower-order harmonics were observed (up to ~1000 ns, Fig. 5.8). The probable reason for the appearance of the extended delay curve is the involvement of the molecules of epoxy resin in HHG, which could extend the range of delays at which the harmonics could be observed. The attempts to observe harmonics from C_{60} and CNT LIPs at larger delays (up to a few tens microseconds) did not reveal any coherent emission in this timescale.

In separate experiments, HHG from ablated pure epoxy resin (i.e. resin free from carbon-contained nanostructures) was analyzed. However, those harmonics

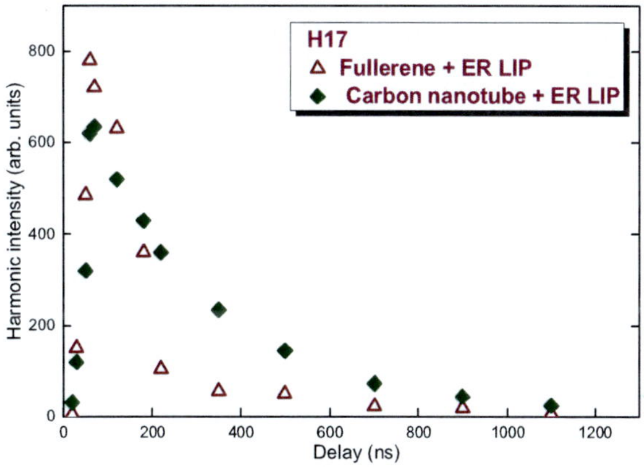

Fig. 5.8 Delay dependences of H17 generating in C_{60} (empty triangles) and CNT (filled rhombs) LIPs. The fluence of heating pulses on the ablated surfaces was 1.2 J cm^{-2}. Reproduced from [43] with permission from the American Institute of Physics

were significantly weaker compared with the harmonics obtained from the plasmas produced during ablation of the mixtures of epoxy resin and carbon nanostructures. Nevertheless, even weak harmonic yield from the former component can cause the extension of the delay-dependent curve towards the longer delays (Fig. 5.8). Being ablated, heavier molecules of epoxy resin (a few hundred g/mol) move slower compared with carbon particles thus arriving at the DP axis of propagation 500 to 1000 ns later with regard to the latter species. Some extension of this dependence in the long delay range can be attributed to the presence of different disintegrated components of the epoxy resin molecules.

5.2.3 Analysis of Harmonics Generation in Carbon Nanostructures

The odd harmonics of the second harmonic wave (515 nm) corresponded to the even harmonics of the 1030 nm wave. However, (a) the efficiency of those harmonics ((H = 2(2n + 1), like H10, H14, H18, H22, etc.) was smaller compared with the case of the two-color pump (1030 nm + 515 nm) and (b) the H = 4(n + 1) harmonics (H12, H16, H20, etc.) were missed in those XUV spectra. Thus some even-order harmonics can be attributed to the generation from 515 nm pump (H = 2(2n + 1)), while H = 4(n + 1) even harmonics can be generated only due to the interaction of two pumping waves in the plasma region. The most interesting peculiarity is the similarity of the energies of the nearby H = 2(2n + 1) and H = 4(n + 1) even harmonics in the case of stronger conversion efficiency of the second harmonic in the thicker (0.4 mm) crystal. Obviously, a stronger second harmonic wave should cause better interplay between 1030 and 515 nm waves in the laser-plasma. However, this assumption is correct only in the case when the group velocity dispersion in BBO does not result in a large delay between these waves in the plasma medium. The latter is correct in the case of the longer wavelength sources of coherent radiation, like 1030 nm laser.

Contrary to that, in the case of 800 nm wave and its second harmonic, the application of thicker crystals results in reverse effect, i.e. weaker interplay between two waves in plasma.

The motivation of these delay-dependent studies was to understand the nanostructured plasma spreading dynamics by analyzing harmonics generation from different ablated carbon-containing species. Though previous HHG studies of different allotropes of carbon, as well as commercially available nanostructures based on the carbon atoms, have already been reported [50], none of them was aimed at distinguishing the difference in plasma spreading in the case of clusters, nanoparticles, nanotubes, and fullerenes containing different amount of carbon atoms. Intuitively, each of the above species should arrive in the area of driving beam propagation at a significantly different time from the beginning of ablation. One can assume from the $E = mv^2/2$ relation that the velocities of each of these multi-particle materials will be proportional to the square root of their masses. Meanwhile, the "optimal" delays corresponding to maximal harmonic yield for each studied sample were, to some extent, approximately the same and were restricted below one hundred nanoseconds. Thus the energy transfer from the heating pulse to the ablating compound represents the separate transfer of energy to each atom, which causes the elevation of multi-particle species with velocities similar to the velocity of a single carbon atom. The availability of different carbon nanostructures allowed analyzing this process using particles of different masses. As it was mentioned, each of those particles showed, to some extent, similar optimal delay, which allowed concluding on the approximate similarity in the velocities of evaporated species.

If larger NPs would tend to arrive later at the area of DP propagation due to slower expansion then the harmonics could be registered at longer delays from the beginning of ablation. In the timescale language, CNT should arrive after 600 μs from the beginning of ablation. Correspondingly, the fullerenes should arrive at ~500 ns timescale from the beginning of ablation. However, neither of these scenarios was realized in those experiments. No harmonics were observed at large delays up to a few tens μs. Thus the same average arrival time toward the area of DP propagation is expected for carbon clusters and NPs of different sizes since no harmonic emission was detected in the microsecond timescale of delays.

The difference in assumptions of NPs velocities (for example in [51], which used the 100 fs pulses and the fluencies up to 5 J cm^{-2}) with regard to the estimates based on the nonlinear optical studies of the plasma spreading can be attributed to the longer ablation pulses (10 ns) and smaller fluencies (<1.7 J cm^{-2}) in analyzed studies. These two regimes of ablation can be distinguished as blast-like expansion and mild heating and evaporation. Correspondingly, the dynamics of plasma spreading can differ from each other.

The integrity of the used nanostructures ablated off the surface should be intact until the driving pulse arrives. So, the heating pulse intensity is a very sensitive parameter, which was kept in the experiments between 2×10^8 and 4×10^8 W cm^{-2}. These intensities correspond to the fluencies of nanosecond pulses in the range of 1–2 J cm^{-2}. At smaller intensities, the concentration of studied species in the ablation plume becomes too low, while at higher intensities one can expect the fragmentation of studied nanostructures and the appearance of a large number of free electrons.

Below we discuss what has happened in the three targets after the ablation by HP, describe the plasma status of those targets, and clarify the mechanism of the HHG.

The morphology of ablated surface was changed by different means and was depended on the target properties. In the case of graphite, the formation of craters after a few thousand shots of nanosecond pulses operated at a 10 Hz pulse repetition rate was observed. Dragging the target up and down, as well as moving it back and forth, allowing keeping stable harmonic emission during ablation of graphite. Meantime, the ablation of soft targets like fullerenes and CNT embedded in epoxy resin caused fasted degradation of ablating surfaces, thus prompting the faster movement of those targets.

The plasma status can be characterized by the plasma and electron density, the velocity of spreading out from the target, temperature, spectrum of emission, etc. The most important characteristics here, from the point of view of the studied HHG process, are the first two parameters. It is difficult to provide the quantitative parameters of plasma density, especially in the case of ablation of the nanoparticle-contained targets. One can estimate them by different means, i.e. using code HYADES, or other approaches. Particularly, in the case of solids like metals, the average parameters of the concentrations of "optimal" plasma and electrons calculated by code HYADES were ~10^{17} and 10^{16} cm^{-3}, respectively [52].

The particle concentration in the interaction zone can also be determined by the thermodynamic parameters of the plasma [53]. In the refereed paper, the relation between the surface and the vapor thermal parameters was defined on the basis of the solution of the kinetic equation. On the basis of those relations, the plasma density in the interaction zone was determined to be varied from 5×10^{16} to 2×10^{17} cm^{-3}, while electron density was estimated to be ten times smaller depending on the ablating sample, experimental conditions, and distance between the target and the laser beam. The calculated vapor flow speed (2.1×10^3 m s^{-1}) at these conditions corresponded to a plasma temperature of ~2 eV. Such estimation can be, to some extent, applied to the ablation of bulk graphite.

Meanwhile, the dynamics of plasma formation on the surface containing carbon-based multi-particle species embedded in epoxy resin is rather complicated. One can assume that, after heating by relatively long (nanosecond) pulses, the epoxy resin melts and evaporates earlier than these processes can occur in C_{60} and CNT. Thus the evaporated glue can take fullerenes and carbon nanotubes together with epoxy resin molecules out from the target surface. In that case, the concentration of fullerenes and carbon nanotubes in LIP seems hardly determined.

By any means, the harmonic yields from these plasmas were larger than from the ablated graphite. Notice, that this empirical observation does not point out the higher concentration of the above nanostructures in LIP compared with single atoms and ions of carbon during ablation of graphite. The specific properties of nanostructures leading to stronger nonlinear optical response during laser-NPs interaction can enhance the harmonic yield at rather smaller concentrations of NPs with regard to single atoms. In the case of NPs one has to take into account a large number of atoms in those structures. However, only those atoms, located on the outer parts of NPs, can participate in HHG, thus notably decreasing the number of potential participants of this nonlinear optical process. Concluding, it is hard to estimate the concentration of nanostructured materials in such plasmas during efficient HHG in "optimal" LIPs.

As for the mechanism of HHG, it follows the well-developed three-step model of harmonics generation [54]. The standard consideration including the tunnel ionization, as a first step, the acceleration of ejected electron, as a second step, and the recombination with parent particle, as a third step, well describes main experimentally determined parameters (harmonic cutoff, dependence of HHG yield on the wavelength of DP, etc.). As an example, theoretical studies of HHG from C_{60} involved extending the three-step model [55] and using dynamical simulations [56]. One can intuitively assume that large sizes of parent particles in the case of carbon NPs (for example, spherical nanoparticles, nanotubes, fullerenes, nanofibers, nanourchins, nanocubes, graphene, etc.) increase the cross-section of recombination of the accelerated electron with those particles. Moreover, the recombination can occur with either the same atom or other atoms on the surface of NP, which also increases the conversion efficiency of HHG, as has been suggested in numerous publications of HHG in gas clusters and plasma nanoparticles.

Meanwhile, the uncertainty in the exact mechanism of the HHG from NPs has previously been underlined in a few studies [57]. As was mentioned, among additional mechanisms, the ionization and recombination to the same ion, to the neighboring ions, and to the whole NP have been proposed. The experiments with gas and plasma NP have revealed some difficulties in disentangling the harmonics produced by different species (monomers and NPs of different sizes). Finally note that the comparative studies of HHG in the plasmas consisted of either NP or monomers have shown that, at optimal experimental conditions, the former species provide stronger harmonic yield, thus pointing out the advanced properties of the large emitters of harmonics in the carbon-containing plasmas, as was demonstrated in the discussed study.

5.3 Summary

In this Chapter, various processes during HHG in the plasmas containing carbon clusters and nanoparticles were discussed. The role of small carbon clusters, which inevitably appear during graphite ablation, on the efficiency of HHG was analyzed. The advantage of the proposed method for laser ablation and delay-dependent study of harmonic emission during ablation of graphite target is that plasma formation in double-laser configuration of HHG experiments could be efficiently applied for a variation of the delay between heating and driving pulses using electronic methods. It can also be applied for other clustered targets to determine the best conditions for HHG in such plasmas. We analyzed the advanced properties of the carbon plasma produced by nanosecond pulses as a medium for efficient low- and high-order harmonic generation. The role of C_5–C_{25} clusters in the enhancement of harmonic generation efficiency was discussed. HHG studies demonstrated that these carbon clusters appearing during ablation of graphite influence HHG efficiency at delays similar to those at which single carbon atoms and ions arrive in the area of femtosecond beam propagation. The discussed studies show that such clusters commonly appearing during time-of-flight mass spectroscopic studies of carbon ablation strongly influence the harmonics yield and thus could be considered as the sources of efficient harmonics in the 40–100 nm spectral range.

We also analyzed the high-order harmonics generation from the plasmas produced on the carbon-containing targets (graphite, fullerenes, and carbon nanotubes) using near-infrared laser radiation. The evolution of different plasma plumes using the extended delays from the beginning of ablation (up to a few tens microseconds) was discussed. We analyzed the role of three carbon-containing nanostructures in the frequency conversion of laser pulses at different conditions of plasma formation. The discussed research shows that all carbon clusters and nanostructures acquire from the very beginning the approximately same kinetic energy and spread out of the surface with the velocity approximately similar to the one that is possessed by a single carbon atom ablated from the bulk material. Additionally, we discussed the single-color pump (1030 nm) and two-color pump (1030 nm + 515 nm) of small carbon clusters and nanoparticles to determine the opportunity of resonance enhancement of even harmonics, as well as analyzed the role of epoxy resin in observed harmonic spectra from fullerenes and nanotubes.

References

1. S. Hädrich, M. Krebs, A. Hoffmann, A. Klenke, J. Rothhardt, J. Limpert, A. Tünnermann, Light: Sci. Appl. **4**, e320 (2015)
2. X. Geng, S. Zhong, G. Chen, W. Ling, X. He, Z. Wei, D.E. Kim, Opt. Express **26**, 3067 (2018)
3. R.A. Ganeev, V.V. Strelkov, C. Hutchison, A. Zaïr, D. Kilbane, M.A. Khokhlova, J.P. Marangos, Phys. Rev. A **85**, 023832 (2012)
4. A.M. Perego, S.K. Turitsyn, K. Staliunas, Light: Sci. Appl. **7**, 43 (2018)
5. B.D. Bruner, M. Krüger, O. Pedatzur, G. Orenstein, D. Azoury, N. Dudovich, Opt. Express **26**, 9310 (2018)
6. R.A. Ganeev, T. Witting, C. Hutchison, F. Frank, M. Tudorovskaya, M. Lein, W.A. Okell, A. Zaïr, J.P. Marangos, J.W.G. Tisch, Opt. Express **20**, 25239 (2012)
7. A. Bahabad, O. Cohen, M.M. Murnane, H.C. Kapteyn, Opt. Express **16**, 15923 (2008)
8. R.A. Ganeev, V. Tosa, K. Kovács, M. Suzuki, S. Yoneya, H. Kuroda, Phys. Rev. A **91**, 043823 (2015)
9. R.A. Ganeev, P.A. Naik, H. Singhal, J.A. Chakera, M. Kumar, M.P. Joshi, A.K. Srivastava, P.D. Gupta, Phys. Rev. A **83**, 013820 (2011)
10. B. Bódi, M. Aladi, P. Rácz, I.B. Földes, P. Dombi, Opt. Express **27**, 26721 (2019)
11. S. Ghimire, A.D. DiChiara, E. Sistrunk, P. Agostini, L.F. DiMauro, D.A. Reis, Nat. Phys. **7**, 138 (2011)
12. I.J. Kim, G.H. Lee, S.B. Park, Y.S. Lee, T.K. Kim, C.H. Nam, T. Mocek, K. Jakubczak, Appl. Phys. Lett. **92**, 021125 (2008)
13. I. Lopez-Quintas, M. Oujja, M. Sanz, M. Martin, R.A. Ganeev, M. Castillejo, Appl. Surf. Sci. **278**, 33 (2013)

14. R.A. Ganeev, G.S. Boltaev, K. Zhang, P.V. Redkin, S.K. Maurya, M. Venkatesh, Z. Yu, V.V. Kim, C. Guo, OSA Continuum **2**, 1510 (2019)
15. R.A. Ganeev, M. Suzuki, M. Baba, H. Kuroda, T. Ozaki, Opt. Lett. **31**, 1699 (2006)
16. V. Strelkov, Phys. Rev. Lett. **104**, 123901 (2010)
17. R.A. Ganeev, Z. Wang, P. Lan, P. Lu, M. Suzuki, H. Kuroda, Phys. Rev. A **93**, 043848 (2016)
18. H.T. Kim, I.J. Kim, D.G. Lee, K.-H. Hong, Y.S. Lee, V. Tosa, C.H. Nam, Phys. Rev. A **69**, 031805 (2004)
19. F. Brandi, F. Giammanco, W. Ubachs, Phys. Rev. Lett. **96**, 123904 (2006)
20. R.A. Ganeev, M. Suzuki, P.V. Redkin, M. Baba, H. Kuroda, Phys. Rev. A **76**, 023832 (2007)
21. A.M. Rubenchik, M.D. Feit, M.D. Perry, J.T. Larsen, Appl. Surf. Sci. **127–129**, 193 (1998)
22. H. Ruf, C. Handschin, R. Cireasa, N. Thire, A. Ferre, S. Petit, D. Descamps, E. Mevel, E. Constant, V. Blanchet, B. Fabre, Y. Mairesse, Phys. Rev. Lett. **110**, 083902 (2013)
23. V. Veniard, R. Taıeb, A. Maquet, Phys. Rev. A **65**, 013202 (2001)
24. F. Bourquard, A.-S. Loir, C. Donnet, F. Garrelie, Appl. Phys. Lett. **104**, 104101 (2014)
25. T.E. Itina, K. Gouriet, L.V. Zhigilei, S. Noël, J. Hermann, M. Sentis, Appl. Surf. Sci. **253**, 7656 (2007)
26. H.O. Jeschke, M.E. Garcia, K.H. Bennemann, Phys. Rev. Lett. **87**, 015003 (2001)
27. L.B. Elouga Bom, Y. Pertot, V.R. Bhardwaj, T. Ozaki, Opt. Express **19**, 3677 (2011)
28. M. López-Arias, M. Oujja, M. Sanz, R. de Nalda, R.A. Ganeev, M. Castillejo, Molec. Phys. **110**, 1651 (2012)
29. R.A. Ganeev, Opt. Spectrosc. **123**, 351 (2017)
30. M. Oujja, J.G. Izquierdo, L. Bañares, R. de Nalda, M. Castillejo, Phys. Chem. Chem. Phys. **20**, 16956 (2018)
31. R.A. Ganeev, C. Hutchison, T. Witting, F. Frank, W.A. Okell, A. Zaïr, S. Weber, P.V. Redkin, D.Y. Lei, T. Roschuk, S.A. Maier, I. López-Quintás, M. Martín, M. Castillejo, J.W.G. Tisch, J.P. Marangos, J. Phys. B **45**, 165402 (2012)
32. A. Kaplan, M. Lenner, C. Huchon, R.E. Palmer, Appl. Phys. A **92**, 999 (2008)
33. R. de Nalda, M. Lopez-Arias, M. Sanz, M. Oujja, M. Castillejo, Phys. Chem. Chem. Phys. **13**, 10755 (2011)
34. T.D. Donnelly, T. Ditmire, K. Neuman, M.D. Perry, R.W. Falcone, Phys. Rev. Lett. **76**, 2472 (1996)
35. R.A. Ganeev, G.S. Boltaev, V.V. Kim, M. Venkatesh, A.I. Zvyagin, M.S. Smirnov, O.V. Ovchinnikov, M. Wostmann, H. Zacharias, C. Guo, J. Appl. Phys. **126**, 193103 (2019)
36. J.W.G. Tisch, T. Ditmire, D.J. Fraser, N. Hay, M.B. Mason, E. Springate, J.P. Marangos, M.H.R. Hutchinson, J. Phys. B **30**, L709 (1997)
37. C. Vozzi, M. Nisoli, J.P. Caumes, G. Sansone, S. Stagira, S. De Silvestri, Appl. Phys. Lett. **86**, 111121 (2005)
38. J.W.G. Tisch, Phys. Rev. A **62**, 041802 (2000)
39. V. Veniard, R. Taieb, A. Maquet, Phys. Rev. A **65**, 013202 (2001)
40. J.R. Vazquez de Aldana, L. Roso, J. Opt. Soc. Am. B **18**, 325 (2001)
41. M. Kundu, S.V. Popruzhenko, D. Bauer, Phys. Rev. A **76**, 033201 (2007)
42. W.R. Creasy, J.T. Brenna, Chem. Phys. **126**, 453 (1988)
43. V.V. Kim, R.A. Ganeev, G.S. Boltaev, M. Iqbal, A.S. Alnaser, Phys. Plasmas **28**, 023111 (2021)
44. I.J. Kim, C.M. Kim, H.T. Kim, G.H. Lee, Y.S. Lee, J.Y. Park, D.J. Cho, C.H. Nam, Phys. Rev. Lett. **94**, 243901 (2005)
45. R.A. Ganeev, G.S. Boltaev, S.Y. Stremoukhov, V.V. Kim, A.V. Andreev, S. Alnaser, Eur. Phys. J. D **74**, 199 (2020)
46. Y. Akiyama, K. Midorikawa, Y. Matsunawa, Y. Nagata, M. Obara, H. Tashiro, K. Toyoda, Phys. Rev. Lett. **69**, 2176 (1992)
47. R.A. Ganeev, V.I. Redkorechev, T. Usmanov, Opt. Commun. **135**, 251 (1997)
48. K. Krushelnick, W. Tighe, S. Suckewer, J. Opt. Soc. Am. B **14**, 1687 (1997)
49. R.A. Ganeev, H. Singhal, P.A. Naik, V. Arora, U. Chakravarty, J.A. Chakera, R.A. Khan, P.V. Redkin, M. Raghuramaiah, P.D. Gupta, J. Opt. Soc. Am. B **23**, 2535 (2006)

50. R.A. Ganeev, M. Baba, M. Suzuki, H. Kuroda, J. Phys. B: At., Mol. Opt. Phys. **47**, 135401 (2014)
51. S. Noel, J. Hermann, T. Itina, Appl. Surf. Sci. **253**, 6310 (2007)
52. L.B. Elouga Bom, J.-C. Kieffer, R.A. Ganeev, M. Suzuki, H. Kuroda, T. Ozaki, Phys. Rev. A 75, 033804 (2007)
53. R.A. Ganeev, L.B. Elouga Bom, T. Ozaki, P.V. Redkin, J. Opt. Soc. Am. B **24**, 2770 (2007)
54. P.B. Corkum, Phys. Rev. Lett. **71**, 1994 (1993)
55. M.F. Ciappina, A. Becker, A. Jaron-Becker, Phys. Rev. A **76**, 063406 (2007)
56. G.P. Zhang, Phys. Rev. Lett. **95**, 047401 (2005)
57. P. Moreno, L. Plaja, L. Roso, Europhys. Lett. **28**, 629 (1994)

Chapter 6
HHG in Metal Nanoparticles

6.1 Application of Oxide Nanoparticles for HHG

6.1.1 High-Order Nonlinear Optical Studies of ZnO Nanoparticles

The interest in the study of the optical and structural properties of ZnO nanostructures is due to the nontoxicity and stability of the ZnO nanomaterials. The noncentrosymmetric structures of ZnO NPs can exhibit relatively strong second harmonic generation (SHG) [1]. Another promising application of nanosized ZnO crystals is a frequency conversion of laser pulses during high-order harmonics generation [2]. The enhancement of high-order harmonic generation (HHG) in the vicinity of ZnO nanoantennas was demonstrated in Ref. [3].

Here we discuss the high-order harmonic studies in the plasmas containing ZnO nanoparticles allowing the generation of harmonics up to the 27th order using single- and two-color pumps [4]. The high harmonic generation experiment was carried out in the laser plasmas produced on the surfaces of the pressed ZnO tablet. A laser-plasma created on the surfaces of ZnO tablet was probed using radiation of Ti:sapphire femtosecond laser. The driving femtosecond pulses (800 nm, 30 fs, 1 kHz) propagated through the plasma produced on the ZnO target at different delays from the beginning of target irradiation by nanosecond heating pulses (1064 nm, 5 ns, 10 Hz). The electronic delay system was used for the synchronization of two lasers. The maximal harmonic yield was achieved by adjusting the position of the target at a distance of ~200 μm above the target surface. The optimal delay between heating and driving pulses was equal to 300 ns. A nonlinear crystal (0.2-mm thick BBO crystal) was used to study the two-color pump scheme for simultaneous generation of even and odd harmonics in the laser-plasma plumes containing ZnO nanoparticles. The second harmonic conversion efficiency was measured to be ~4%. The harmonic spectrum was analyzed using the XUV spectrometer.

R. A. Ganeev, *High-Order Harmonics Generation in Plasmas*, Springer Series on Atomic, Optical, and Plasma Physics 122, https://doi.org/10.1007/978-3-031-09040-0_6

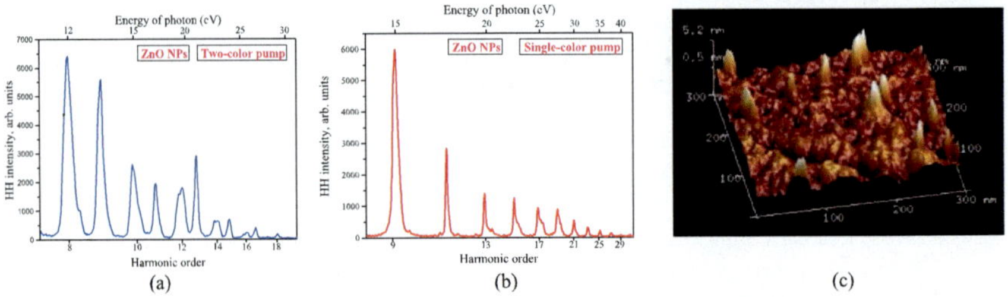

Fig. 6.1 Spectra of harmonics generated in the ZnO NPs plasma: **a** single-color pump scheme and **b** two-color pump scheme of high harmonic generation. **c** Microphotography of deposited ZnO nanoparticles on the surfaces of silicon wafer. Reproduced from [4] with permission from European Physical Journals

The ablated ZnO NPs as a nonlinear medium were analyzed for HHG of 800 nm, 30 fs laser pulses. Figure 6.1a, b shows the harmonic spectra generated in ZnO NPs-containing plasmas. The single- and two-color pump schemes were used for the generation of odd and odd + even harmonics. The observation of even harmonics does not indicate the anisotropy properties of ZnO NPs in laser-produced plasmas. This behavior can be considered to be due to the motion of an electron in the two-component electric field of laser pulses. The single-color pump showed the larger cutoff (up to the 29th harmonic) in ZnO plasma, compared to the two-color pump, which yielded a lower cutoff order (18th harmonic). The decrease of cutoff energy in the case of a two-color pump can be explained by the application of a shorter wavelength component and quadratic dependence of the cutoff energy on the wavelength of pump radiation. The AFM of deposited films showed the presence of ZnO NPs in the plasma plume during the propagation of driving pulses through the plasmas (Fig. 6.1c). The sizes of ablated nanoparticles were measured to be ~30 nm. Thus the presence of NPs in plasma coincided with the generation of strong harmonics.

6.1.2 Influence of Gadolinium Doping on High-Order Nonlinear Optical Properties of ZnO Nanomaterials

The dopants like gadolinium alter the structural parameters such as crystallite size, lattice constants, stress–strain mechanisms, etc., and hence related nonlinear optical properties. The nonlinear optical properties of Gd:ZnO thin films were analyzed by irradiating with an electron beam of various dosages, which lead to an increase in third-order nonlinear optical susceptibility [5]. Here we analyze the studies of HHG performed using laser ablation of pure ZnO and Gd-doped ZnO [6].

In order to synthesize zinc oxide and Gd doped ZnO nanocrystals (GZO), a glass microwave (MW) reactor vessel was loaded with the appropriate amount of corresponding components. The MW vessel was sealed, and the reactants were kept in a water bath at 60 °C for 30 min under continuous stirring, which helped the solids to get dissolved properly. The MW vessel was then inserted into the microwave cavity and heated to 280 °C temperature for 10 min. The solution was allowed to cool down to 50 °C, which was followed by precipitation of nanocrystals using the addition of methanol (~5 ml). The product was isolated by centrifugation and washed with toluene and acetone four times to remove organic residue. The resultant product was dried overnight under vacuum and stored in clean sealed plastic containers for measurements. Similarly, aluminum-doped zinc oxide (AZO) and Gd doped AZO (GAZO) nanocrystals were synthesized by adding Al salt powder into the aforementioned reaction mixture for the synthesis of ZnO and GZO by keeping other experimental parameters constant.

The studied samples were ablated by nanosecond pulses (1064 nm, 5 ns, 10 Hz) to create the plasma plume. The driving femtosecond pulses (800 nm, 30 fs, 1 kHz) propagated through the plasma at different delays after the beginning of irradiation of target by nanosecond heating pulses. The harmonics emission was directed to the XUV spectrometer.

Figure 6.2a shows the UV–visible spectra of ZnO and AZO as a function of the concentration of Gd. The ZnO, AZO, AZO + 1% Gd, AZO + 12.5% Gd and ZnO + 20% Gd samples have shown the absorption peaks at 354 (3.50 eV), 356 (3.48 eV), 358 (3.46 eV), 366 (3.38 eV) and 369 (3.36 eV) nm, respectively. Those measurements have shown that the bandgap of ZnO and AZO material is decreasing with the increase in Gd concentration. The addition of 20% Gd to ZnO shifted the absorption peak to larger wavelengths by 15 nm compared to pure ZnO. Thus the doping of ZnO and AZO by Gd led to the redshift of the absorption peak. Moreover, the doping of ZnO (i.e. AZO) by Al had no effect on shifting the absorption peak.

X-ray diffraction (XRD) patterns of ZnO, GZO, AZO, and GAZO nanocrystals (NCs) are presented in Fig. 6.2b. XRD results corresponding to AZO and GAZO were scaled by 1.5X and 6X, respectively, for better visualization. Diffraction peaks observed in the range of 30–80° were indexed for (100), (002), (101), (102), (110), (103), (112), (201) and (202) planes of zinc oxide with wurtzite crystal structure. The

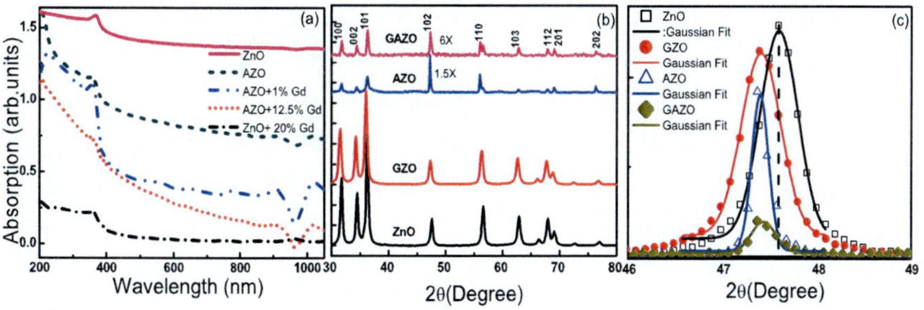

Fig. 6.2 **a** UV–visible spectra **b** XRD data of undoped Zinc oxide, 12.5% Gd doped zinc oxide (GZO), aluminum doped zinc oxide (AZO) and 12.5% Gd doped AZO. XRD intensity for AZO and GAZO is scaled by 1.5X and 6X respectively. **c** Zoomed view of (102) XRD peak for different zinc oxide samples without scaling. Reproduced from [6] with permission from Elsevier

diffraction peak corresponding to (102) plane has been fitted to Gaussian (Fig. 6.2c) in order to investigate the influence of doping on crystalline size and lattice constant of ZnO NCs.

Scanning electron microscopic investigation was carried out in order to analyze the influence of doping on the surface morphology of zinc oxide NCs. Figure 6.3 presents the SEM images of different zinc oxide samples. In all the cases it was found that the NCs in each sample have a great tendency to get self-assembled to make larger size crystals and aggregates. Gd doping had little influence on surface morphology, but aluminum doping significantly increased the rate of aggregation resulting in clusters with larger sizes. TEM images have shown that ZnO and GZO particles are almost spherical in shape with NPs sizes in the range of 10–20 nm. Small NPs get self-assembled to make larger clusters, but individual NPs can be easily identified. In contrast to this, Al doping produces non-spheroid particles and increases the rate of aggregation and self-assembly.

A search for new applications of different dopants of semiconductors is an important problem of optics. The interesting idea is to find the optimal conditions in the application of such multi-component structures as effective emitters of the high-order harmonics of femtosecond pulses for the development of efficient sources of coherent XUV radiation. Below we discuss the results of studies of the HHG in the

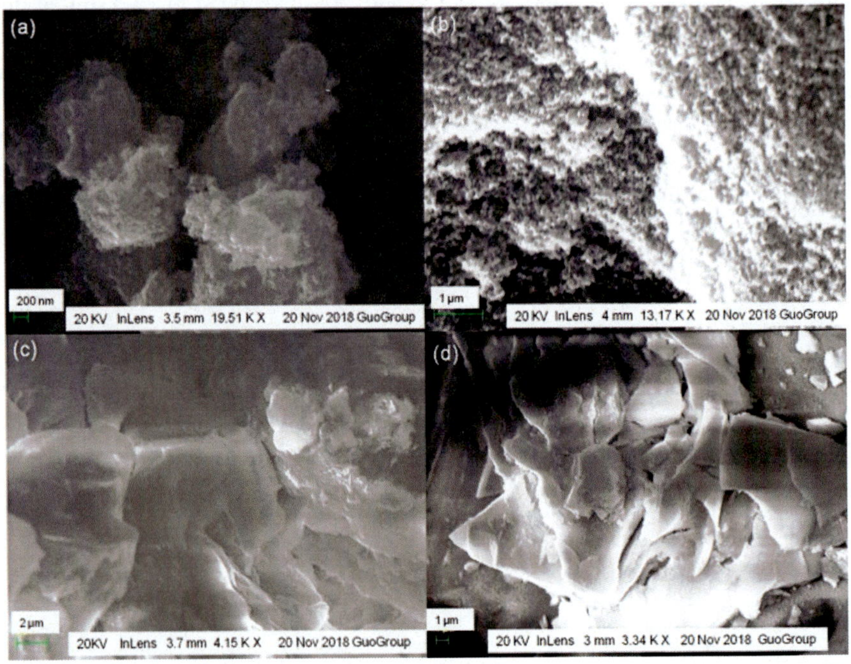

Fig. 6.3 SEM images of **a** ZnO (scale bar 200 nm) **b** GZO (scale bar 1 μm) **c** AZO (scale bar 2 μm) and **d** GAZO (scale bar 1 μm). Reproduced from [6] with permission from Elsevier

plasmas produced during ablation of bulk ZnO, ZnO nanocrystal (NC) powder, and Gd-doped ZnO NC powder using electronically driven delays between the heating and driving laser pulses.

Figure 6.4 shows high-order harmonics spectra generated from the plasmas produced on the surface of three targets [bulk ZnO, ZnO NC powder, and Gd (12% weight ratio)-doped ZnO NC powder] in the spectral range of 25–100 nm. HHG at different delays between the heating and driving pulses (from 100 to 3500 ns) is shown for pulse energies of 4 and 0.8 mJ, respectively. At the initial stages of plasma formation and spreading out from the target, i.e., at delays less than 50 ns, the concentration of particles (neutral atoms and molecules and singly charged ions as well as NCs) was insufficient for HHG, since the whole ensemble of particles possessing velocities of ~2×10^4 m s^{-1} cannot reach the spatial region of the driving beam. At larger delays (100 ns and longer), sufficient amounts of particles appeared in the path of the femtosecond beam that allowed the generation of harmonics. The optimal delay was ~200 ns for bulk ZnO, ~300 ns for powdered ZnO NCs, and ~500 ns for Gd-doped ZnO NC powder targets. Gadolinium (atomic weight 157.3 amu) can change the dynamics of delay-dependent harmonics spectra compared with the pure ZnO (81.4 amu) target. The harmonics cut-offs were similar for two powdered species (Fig. 6.4b, c], e.g., at ~23rd harmonic, while ablation of bulk ZnO showed the 29th harmonic as the highest one (Fig. 6.4a). An increase of the delay above the mentioned optimal values led to a gradual decrease in HHG efficiency. The harmonics almost disappeared at the delay of 3.5 μs. No harmonics were observed at delays up to 20 μs.

Experiments using bulk and powdered targets of the same materials led to similar dependencies of the harmonics yield on the delay between the heating and driving

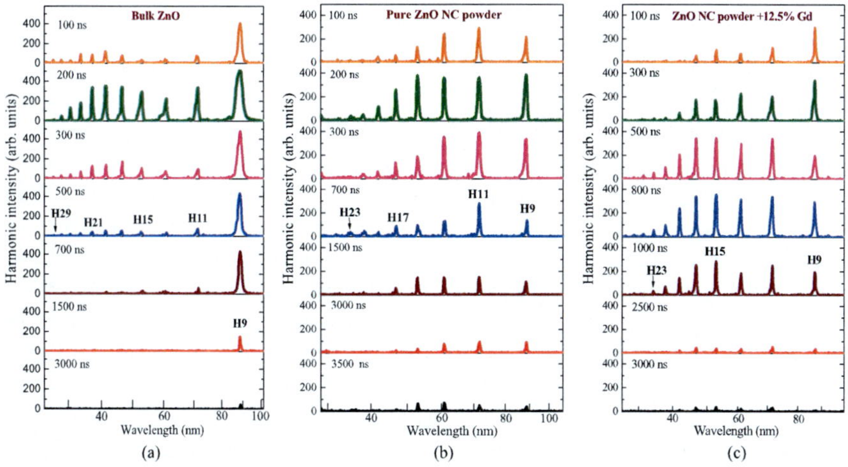

Fig. 6.4 Harmonics spectra from the ZnO plasmas produced on three targets at different delays between the heating and driving pulses. **a** Bulk ZnO target, **b** Powdered ZnO target. **c** Powdered ZnO doped with 12% weight parts of Gd. Reproduced from [6] with permission from Elsevier

pulses. The only difference was the notably strong 9th harmonic in the case of bulk ZnO (Fig. 6.4a). These experiments showed that independently on either molecules or powdered species of ZnO spread out from the target surface, most of them appear in the area of the femtosecond beam approximately at the same time (~200 ns).

The harmonics generation in such plasma can determine the role of the complex composition of ablated species in the former process. A signature revealing the nature of different emitters could be the variation of harmonics emission arising at some specific delay after the beginning of ablation. This delay may characterize the time of propagation of different components from the target's surface to the optical axis of ultrashort pulse propagation. The important peculiarity of this process in three plasmas is a difference in the variation of harmonics yield for different delays between the heating and driving pulses. Though the same components were used in the case of bulk ZnO and powdered ZnO NCs, the dynamics of plasma formation on these targets causes different harmonics spectra. One can see the strong 9th harmonic in the case of bulk ZnO, which is attributed to the resonance induced enhancement of a single harmonic in the vicinity of the ionic transition of Zn possessing large oscillator strength. The enhancement was due to the influence of Z II resonance (88.1 nm) [7].

Ablation of the powdered target requires a smaller fluence of heating pulse compared with the ablation of the bulk target of similar consistency, which changes the excitation conditions for the ions. Correspondingly, the role of the ionic transition responsible for the 9th harmonic enhancement in the case of the strongly excited bulk target has been diminished in the case of ZnO NC powder. In the case of powdered targets (Fig. 6.4b, c), the insignificant difference between the 9th and the 11th harmonics was observed, contrary to the case of a bulk target (Fig. 6.4a).

6.2 High-Order Harmonic Generation in Au Nanoparticle-Contained Plasmas

There is an urgent need to search for methods to generate intense harmonics and attosecond pulses. For these purposes, gas clusters and plasma produced from nanoparticle targets can be used to increase the intensity of harmonics. The motivation of these studies is related to the comparison of the high-order nonlinear optical response of the nanoparticles appearing from the targets during the ablation of NP-contained samples and bulk Au samples. The dynamics of plasma formation in the former case can be considered as follows. The material directly surrounding the NPs is a polymer (epoxy glue), which has a considerably lower ablation threshold than the metallic materials. The NPs absorb the laser pulse energy and pass the thermal energy to the surrounding media. Therefore, the NPs-carrying polymer begins to ablate at relatively low intensities, resulting in the lower laser fluence required for the preparation of the appropriate nonlinear medium for the HHG. This feature allowed for easier creation of the optimum plasma conditions, which resulted in better HHG conversion efficiency from the Au NPs-containing plume compared to the plasma from the bulk target.

The production of clusters in the laser-plasma during laser ablation of various targets has a high probability, while their presence and concentration in the plasma volume, where the frequency conversion occurs, is yet to be confirmed directly. Analysis of post-ablation conditions of the deposited debris can provide information on the nature of those species, despite the differences between the composition of the plasma formed at its early stages and the deposited material (due to the influence of aggregation on the substrate). Further evidence of the cluster contribution to the enhancement of the harmonic generation comes from investigations of intense laser ablation of a target, which would give the assumptions regarding the participation of in situ generated nanoparticles.

Gold nanomaterials have potential applications because their surface plasmon resonance affects the optical properties of materials. Here, we analyze the results of the study showing how harmonics obtained from Au NPs are affected by using the active substrate, which itself allows the generation of harmonics. In order to support the experimental measurements and to advance in understanding the fundamental physical processes underlying the laser-induced formation of Au NPs a series of computer simulations of laser ablation of the thick gold target at different irradiation parameters is performed. The used model is capable of addressing the kinetics of strongly non-equilibrium laser-induced phase transition processes at the atomic level with the Molecular Dynamics (MD) approach.

In this section, the analysis of systematic experimental and numerical studies of the NPs formation during laser ablation and the HHG from Au NPs, Au bulk, paper, and Au NPs deposited on paper and glass substrates were carried out [8]. These studies also allow understanding the effects of a single-color pump (SCP) and two-color pump (TCP) of plasma, laser chirping, laser pulse duration on the harmonic conversion efficiency from different ablated targets, and the choice of irradiation regimes for generation of NPs with predesigned optical and morphological properties.

6.2.1 Comparison of Harmonic Emission from Different Plasmas Containing Gold Nanoparticles

The 100 nm and 10 nmAu NPs were glued on the glass substrate and paper, respectively. Notice that ablated pure glue did not allow the generation of harmonics in laser-induced plasmas. The studied samples were ablated by nanosecond (1064 nm, 5 ns, 10 Hz), picosecond (800 nm, 200 ps, 200 Hz), and femtosecond (800 nm, 30 fs, 200 Hz) pulses. Both femtosecond (fs) and picosecond (ps) laser pulses were obtained from the same laser by separation of the part of uncompressed radiation

(200 ps) before entering into the compressor stage. The delay between heating radiation (i.e., fs or ps pulses) and driving fs pulses (800 nm, 30 fs, 200 Hz) was varied in the range of 0–120 ns by using the optical delay line. The delay between heating nanosecond (ns) pulses and driving fs pulses was varied electronically between 0 and 10^5 ns using a delay generator. LIP was created by heating pulses, while the driving pulses were focused inside the plasma using a 500 mm focal length spherical lens from the orthogonal direction with regard to ablating pulse to generate harmonics. The diameter of focused driving femtosecond pulses was 64 μm. The energy (intensity) of driving pulses employed in the experiment was 0.5 mJ (4×10^{14} W cm^{-2}). The HHG experiments using TCP of LIP were carried out using the 0.2-mm thick β-barium borate (BBO) crystal, which was kept inside the vacuum chamber on the path of 800 nm pulses to generate a second harmonic (400 nm). The harmonic spectra were analyzed using an XUV spectrometer and collected by a CCD camera.

The low- and high-order nonlinear response of gold NPs prepared by the chemical method was analyzed in Ref. [9]. Particularly, the harmonics up to the 27th order (H27) were generated during the ablation of thin Au NP-containing films. The difficulties encountered during the application of these films were related to their small thickness (100 nm) leading to evaporation during a single shot, which required the constant movement of the destroyed film. In the discussed study, the commercially available Au NP powder was used, which can be attached to the surface of different materials (glass, paper) to form the NP multilayer sample with a rather large thickness (~1–2 mm). The application of such samples allowed the maintenance of relatively stable harmonic emission for a longer period of ablation. This amendment in handling the NP target, in turn, allowed better optimization of HHG and achieving the conditions for the generation of higher-order harmonics (up to H39). Additionally, HHG was analyzed from the ablated bulk gold target. The simulations demonstrated the appearance of Au NPs in LIP at these conditions, which can also enhance the harmonic yield from such plasma.

The employed repetition rates of ns, fs, and ps laser were 10, 200, and 200 Hz, respectively. Every single HHG spectrum was obtained from the fresh sample. The irradiation of the same spot of targets at a high pulse repetition rate caused the crater formation and degradation of plasma thus decreasing the stability of harmonic yield. The most advanced method to increase the stability of harmonics emission was the application of the rotating target. The rotating speed in the range of 5–300 rpm did not influence the stability of harmonics. The application of rotation target and the movement of the spot of the heating beam along the height of this target being dragged up and down would allow further improvement of the stability of harmonic yield.

Figure 6.5 shows the harmonics in the spectral range of 20–130 nm generated from the plasmas produced on the surface of different targets (Au bulk target, Au 100 nm nanoparticles, and Au 10 nm nanoparticles glued on paper). In total, 5 ns, 1064 nm, 10 Hz, and 35 fs, 800 nm, 200 Hz laser pulses were used as the heating and driving pulses. These HHG spectra show the variations for different delays between the driving and heating pulses (from 100 to 7000 ns) at heating nanosecond and femtosecond pulse energies of 10 and 0.5 mJ, respectively. At smaller delays

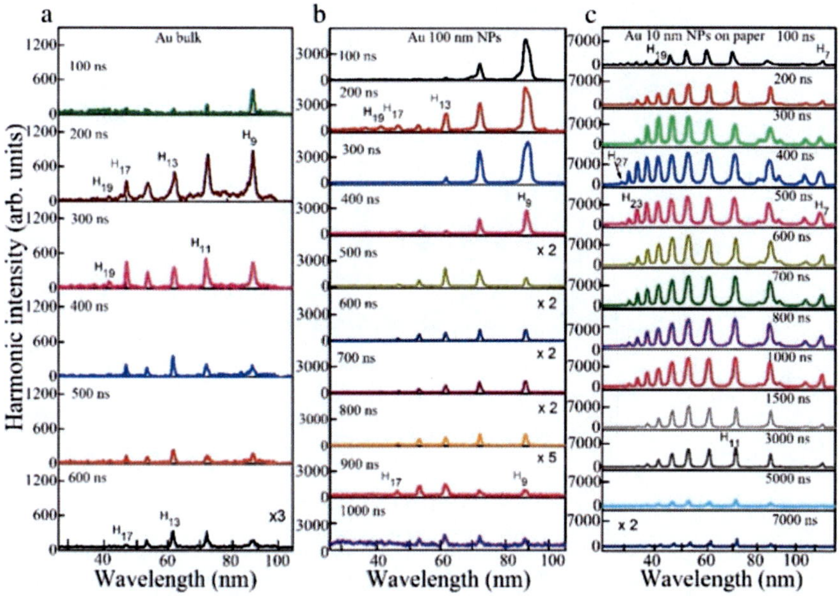

Fig. 6.5 Harmonics spectra from the Au plasmas produced on three targets at different delays between the heating and driving pulses. **a** Bulk Au, **b** Au 100 nm NPs, **c** Au 10 nm NPs glued on paper. Reproduced from [8] with permission from MDPI

(<20 ns), the concentration of particles (neutral atoms, molecules, single charged ions, and NPs) was insufficient for HHG because the whole cloud of ablated particles possessing velocities of ~2×10^4 m s^{-1} cannot reach the spatial region of the driving beam, which propagated at the distance of 0.5 mm above the surfaces of the targets. At larger delays ($\geq$100 ns), the concentration of the particles appearing on the path of the driving pulse became sufficient for the generation of harmonics.

The optimal delays allowing the generation of the maximal yield of harmonics were in the range of 150–400 ns for the ablated Au bulk, Au 100 nm NPs and Au 10 nm NPs on paper. At optimum delay, the harmonic range was extended up to H_{19} for Au bulk and Au 100 nm NPs, whereas it was H_{29} in the case of ablated Au 10 nm NPs on paper. An increase of the delay above the optimal values led to a gradual decrease in HHG efficiency. The harmonics were observed up to 600, 1000, and 7000 ns delays between the driving and heating pulses for the ablated Au bulk, Au 100 nm NPs, and Au 10 nm NPs on paper, respectively.

Figure 6.6 shows the comparative spectra of harmonics produced from the plasmas generated on the Au bulk, Au 100 nm NPs deposited on glass and Au 10 nm NPs on the paper at optimal delays between the heating and driving pulses. Some emission lines appeared in harmonic spectra and were attributed to the carbon presented in the paper-containing target.

At 200 ns delay, the harmonic intensity obtained from ablated Au 10 nm NPs on paper was approximately three and 14 times stronger with regard to Au NPs on glass

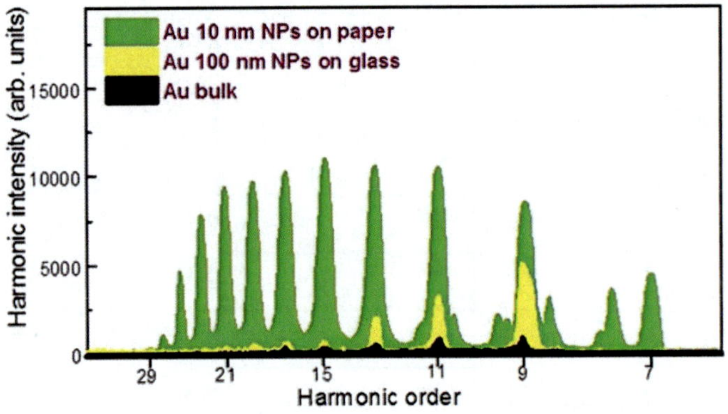

Fig. 6.6 Comparative harmonic spectra from ablated bulk Au (at 200 ns delay), Au 100 nm NPs (at 200 ns delay), and Au 10 nm NPs on paper (at 350 ns delay). Reproduced from [8] with permission from MDPI

and Au bulk, respectively. However, at 500 ns delay, the harmonic yield from ablated Au 10 nm NPs on paper was further enhanced up to five and 40 times compared to Au 100 nm NPs on glass and bulk Au, respectively. No harmonic was observed in the case of heating of the pure glass surfaces at similar fluencies of heating pulses.

The weak harmonics appearing from paper plasma were attributed to the presence of the carbon atoms and ions in LIP. The harmonics started from H11 and extended up to H29. The ablated Au 10 nm NPs on paper also demonstrated the harmonic range between H7 and H29, however, the intensities of H11, H13, H15, and H17 from the plasmas produced on the Au 10 nm NPs on paper were 40, three, four, and five times stronger with regard to those from the ablated paper. The significant enhancement of HHG efficiency from Au NPs on paper was attributed to the influence of the surface plasmon resonance of Au NPs which causes stronger absorption of incident laser light resulting in a larger amount of Au during propagation of the driving pulses compared to other samples.

This finding is supported by the experimental measurements of the complex dielectric function of gold [10]. Localized surface plasmon resonance increases the local electrical field around the metal nanoparticles, which can in principle lead to effective decreasing of ionization potential. According to the three-step model of HHG, this decreased ionization potential results in higher conversion efficiency for the low-order part of the HHG plateau and at the same time causes the shortening of the cutoff and plateau range. Additionally, local field enhancement is attributed to the collective motion of free electrons confined in narrowly localized regions, similar to that observed in colloidal nanoparticles exposed to an external electromagnetic field.

A method of HHG allowed exploiting the local field enhancement induced by plasmons within a metallic nanostructure consisting of bow-tie-shaped gold elements on a sapphire substrate [11, 12]. HHG resulting from the illumination of plasmonic nanostructures with a short laser pulse of a long-wavelength source was also studied in [13]. It was demonstrated that both the confinement of electron motion and the inhomogeneous character of the laser electric field play an important role in the HHG process and lead to a significant increase of the harmonic cutoff.

The harmonic intensity obtained from ablated Au 10 nm NPs on paper was approximately three and 14 times stronger with regard to Au NPs on glass and Au bulk, respectively. The conversion efficiency of samples was estimated using the comparison with known results from other plasma. The conversion efficiency from previous measurements of harmonic generation in the plasmas produced on the surface of bulk Ag was reported to be 8×10^{-6} [14]. In the case of Au bulk plasma at similar conditions, the HHG conversion efficiency was almost ¼ with regard to ablated bulk silver. Hence the conversion efficiency of in Au plasma was estimated to be 2×10^{-6}. Therefore, the HHG conversion efficiency in the plasma produced on the Au NPs glued on paper was determined to be 3×10^{-5}.

6.2.2 Role of Different Parameters of Driving and Heating Pulses on the HHG Efficiency in Au NP Containing Plasmas

The sample's absorption and evaporation show insignificant wavelength dependences of employed 1064 nm (ns pulses) and 800 nm (ps and fs pulses) laser sources due to the closeness of their wavelengths. Meanwhile, pulse duration strongly influences the ablation and harmonic emission due to different time scales of interaction with samples.

The effect of heating pulse duration on HHG from ablated Au bulk target is presented in Fig. 6.7a. Maximum harmonic intensity for ablated Au bulk target using ns heating pulses was observed at E(ns) = 10 mJ with harmonics extended up to H21 (upper panel). Harmonic yield decreased at E(ns) > 10 mJ due to the growth of free electrons density and phase mismatch between interacting waves. In the case of ps and fs heating pulses (two bottom panels of Fig. 6.7a), harmonics extended up to H33. The intensities of harmonics in the case of heating ps and fs pulses were 5 and 4 times larger compared to the ns pulses induced ablation, while the driving pulse and heating pulse fluencies were smaller. This variation of HHG yield confirms the fact that heating pulse duration affects the harmonic yield and cut-off. Those studies confirmed that LIP from bulk Au allows the generation of strong harmonics using short heating pulses. Laser ablation using ps and fs pulses allows the formation of relatively dense plasma, with regard to ns heating pulses while electron concentration was maintained at ~10% of the plasma concentration. To achieve a similar concentration of plasma with nanosecond heating pulse, one has to use stronger fluence, which causes the appearance of a notably larger number of free electrons. These electrons significantly suppress the conversion efficiency of harmonics [15].

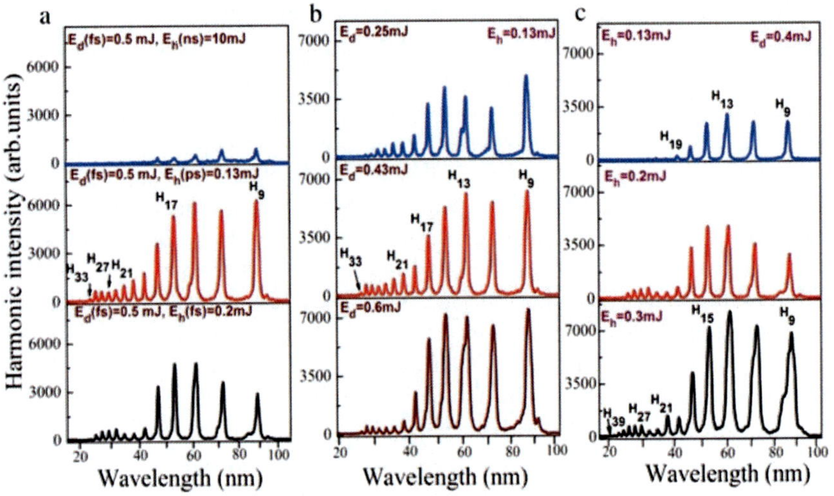

Fig. 6.7 Harmonics spectra from ablated bulk Au at **a** different energies of ns heating pulses, **b** different energies of fs driving pulses, and **c** different energies of ps heating pulses. Reproduced from [8] with permission from MDPI

At similar fluence, (i) concentration of free electrons is more in laser-plasma created by ns pulses with regard to fs and ps heating pulses, and (ii) the density characteristics of plasma (i.e., the concentration of neutrals and single charged atoms) are high at fs and ps ablation with regard to the ablation using ns pulses. A high concentration of generated free electrons leads to self-defocusing and self-modulation of the driving laser pulses resulting in the phase mismatch between the driving and harmonic waves. Hence, the possibility of phase-mismatch using ns heating pulses is larger with regard to ps and fs heating pulses. Therefore, the efficient generation of the highest harmonics from the Au plasma produced by fs and ps heating pulse is attributed to better phase matching conditions between the driving pulses and harmonic waves compared with the ns-induced LIP.

Figure 6.7b, c show the variable harmonic intensities with respect to variation in the energies of heating and driving laser pulses. The harmonic intensity and harmonic range were increased with the growth in driving and heating pulse fluencies. These results illustrate that harmonics intensity and cutoff produced from the plasma of Au bulk target strongly depend on the driving and heating pulse energies.

With the increase in heating pulse and driving pulse energies, the particle density and number of photons available for particle acceleration increase leading to the growth of harmonic yield and extension of harmonic cutoff. The decrease in harmonic efficiency after crossing certain energy of heating pulses is due to the growth of free electrons density, which leads to the self-defocusing and self-modulation of fs pulse resulting in phase mismatch.

The diameter ($2w_0$) of the focused radiation was 64 μm. The corresponding Rayleigh length was $z_o = k(w_0)^2/2 = 4$ mm. Here, k is the wavenumber and w_0

is the beam waist radius. For the plasma length of 0.3 mm, the driving beam interacted with the plasma plume at the conditions of the plain wave propagation. The coherence length for mid-term harmonic ($q = 21$) ($L_{coh}(mm) \approx 1.4 \times 10^{18}(q \times N_e)^{-1}$ [16]) was equal to 2 mm at the concentration of free electrons $N_e = 3 \times 10^{16}$ cm^{-1} corresponding to 10% of plasma concentration (3×10^{17} cm^{-1}). At these conditions, no phase mismatch occurs, since the coherence length is larger than the size of the nonlinear medium. The over-excitation of the gold target causes the growth of free electrons concentration until the conditions when almost all particles became ionized. In that case, the coherence length decreases down to 0.2 mm, which caused the strong phase-mismatch between interacting waves inside the 0.3 mm long medium.

Figure 6.8a shows the harmonic spectra obtained from ablated Au bulk target using SCP of LIP and different heating laser pulse duration. The obtained harmonics were extended up to H33 in the case of the shortest heating pulse duration. The harmonic cut-off was decreased in the case of TCP (H23, Fig. 6.8b) as compared to SCP (H31) due to the application of a shorter wavelength component of pumping radiation. The harmonic intensities obtained using TCP were two times stronger with regard to SCP, despite the small conversion efficiency of second-harmonic radiation in BBO (~2.5%) and a small ratio of 800 and 400 nm energies (1:40). TCP generated broader harmonics with regard to SCP due to self-modulation of the driving pulses in the plasma area (Fig. 6.8b). Figure 6.8c shows the harmonic spectra obtained from ablated Au bulk and Au 100 nm NPs using TCP geometry and ns heating pulses. One can see that ablated Au 100 nm NPs produced stronger harmonics as compared to ablated Au bulk target in the two-color configuration as well.

Figure 6.8d shows the effect of laser chirp on the generation efficiency of harmonics. The reduced harmonic cutoff in the case of chirped laser pulses is

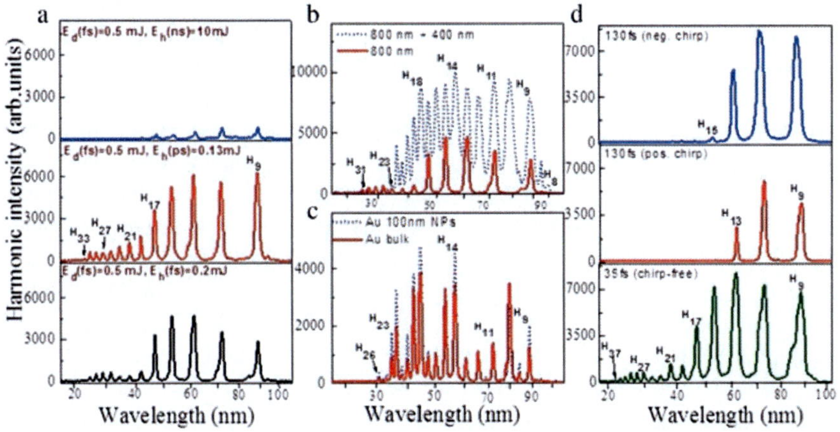

Fig. 6.8 Harmonics spectra from the plasma of bulk Au. **a** single-color driving pulses (800 nm) and different pulse durations of heating pulses, **b** single- and two-color (800 + 400 nm) driving pulses while using 0.2 mJfemtosecond heating pulses, **c** TCP of ablated bulk Au and Au 100 nm NPs, and **d** application of chirped laser pulses. Reproduced from [8] with permission from MDPI

attributed to the reduced intensity at longer pulse duration. The harmonics were insignificantly red- and blue-shifted in the case of positively and negatively chirped 130 fs laser pulses. The variation of the sign of laser chirp also affected the conversion efficiency of harmonics. The low-order harmonics (e.g., H9 and H11) were larger in the case of negatively chirped 130 fs pulses with regard to positively chirped pulses. The neutrals and ions present in plasma can be responsible for additional chirp-induced blue- and redshifts. Particularly, self-phase modulation of laser pulses during propagation of the leading part of the laser pulse through the plasma containing neutral and ions might be the reason for the observed red-shift of harmonics.

6.2.3 *Numerical Simulations of Au Nanoparticles Formation During Laser Ablation*

In order to support the experimental data and investigate the mechanism of NPs formation at different irradiation conditions, a series of MD-based modeling was performed. The classical MD method can describe the kinetics of fast non-equilibrium phase transition processes at atomic precision. This method, however, does not have free electrons included explicitly and, therefore, cannot address the processes of laser light absorption, laser-induced electron–phonon nonequilibrium, and fast electron heat conduction. The last three phenomena are playing a determinant role during short laser pulse interactions with metals and can be described within the frames of the Two Temperature Model (TTM) [17], where the effect of free carriers is accounted for via the electron–phonon temperature dynamics. In the hybrid MD-TTM method, the advantages of both MD and TTM methods can be combined so that all the above-mentioned processes can be described with the scale of a single computational approach.

The basics of the atomistic-continuum MD-TTM approach are described in [18]. The schematic view of the total computational cell for the simulation of Au NPs formation in a vacuum is shown in Fig. 6.9. The spot size of the laser beam focused on the metal surface is considered large enough (0.3 mm) as compared to the lateral size of the modeling cell. The lateral size of the computational box also suits the requirements for the identical comparison of the simulation results with the experiment data.

Based on the above assumptions, the total supercell consisting of ~90,000,000 atoms was taken with dimensions of 65 × 65 × 400 nm in X, Y, and Z directions respectively. While Periodic Boundary conditions were imposed in X and Y directions, to avoid unnecessary and expensive MD integrations in the deep bulk of the material one can impose the Non-Reflective Boundary conditions in Z at the depth of 400 nm from the metal surface. The total computational box was divided by a number of cells and each of them was processed by an individual processor core. At the same time, the applied model was solved in 3D mesh inside of each processor cell (Fig. 6.9).

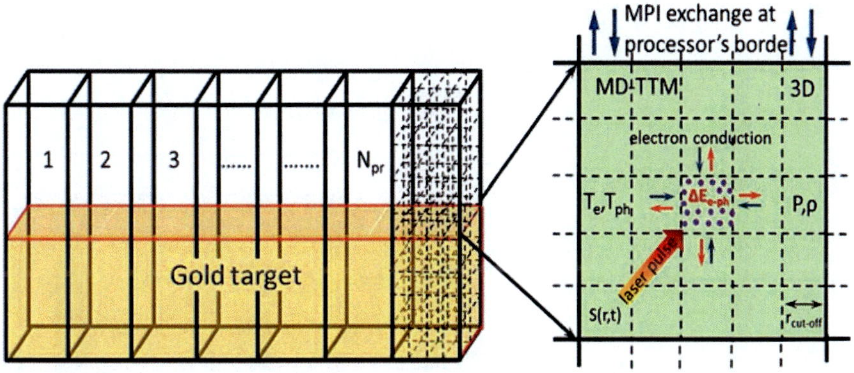

Fig. 6.9 Computational box for simulations of laser-induced formation of Au NPs during the laser ablation process on the experimental scale. The box is divided to the number of subcells N_x and N_y in X and Y directions correspondingly to be processed by N_x*N_y processor cores (left) in multiprocessing regime with the MPI library. Each core is divided to 3D mesh and shown in 2D view (right) with an illustration of the MD-TTM model scheme. The effect of free carriers via the electron, T_e, and phonon, T_{ph} temperatures dynamics are considered to describe the laser light absorption, S(r,t), the process of thermal energy exchange between electrons and phonons, $\Delta E_{e\text{-}ph}$, and fast electron heat conduction. Reproduced from [8] with permission from MDPI

For an equilibrium crystal at P = 0 GPa and T = 0 K the interatomic potential for Au gives −367.609 kJ/mole for the cohesive energy, 179.4 GPa for the bulk modulus, and 0.4065 nm for the lattice constant. Furthermore, this potential represents the experimental thermophysical properties of the modeled material (such as equilibrium melting temperature, heat capacity, the volume of melting, and linear thermal expansion coefficient) with an accuracy of more than 99.5%. For instance, such an important property in the discussed model as the melting temperature T_m = 1343 K was computed in a series of liquid–crystal coexistence simulations and showed a good agreement with its experimental value of T_{exp} = 1337 K. In order to investigate the target's response to a laser pulse at different regimes, three simulations were performed for the pulse durations of 300 fs, 4 ps, and 100 ps that are correspondingly much shorter, comparable, and much longer than the characteristic electron–phonon equilibration time $\tau_{e\text{-}ph}$ in gold (~10 ps, [19]).

The results of modeling of NPs formation due to a laser pulse focused on a thick Au target can be seen in Fig. 6.10 for three pulse durations. In Fig. 6.10a one can see a general view of the ablation plume caught at the time of 500 ps due to a 300 fs laser pulse (λ = 800 nm) ablation at the incident fluence of 2 J cm^{-2}. Here one can clearly identify three zones: "1", "2", and "3", where NPs form with different sizes and shapes. These three zones are zoomed for a more detailed observation in Fig. 6.10b, where the atoms are colored by Central Symmetry Parameter (CSP) for identification of their local structure: solid < 0.08 < defects < 0.12 < liquid < 0.25 < surface < 0.50 < vapors. The top zone "3" contains the smallest NPs with a rather spherical shape and the characteristic size of ~2–3 nm. Zone "2" has particles of rougher form, but of bigger size (~5–15 nm). And finally, the area "1" shows the part

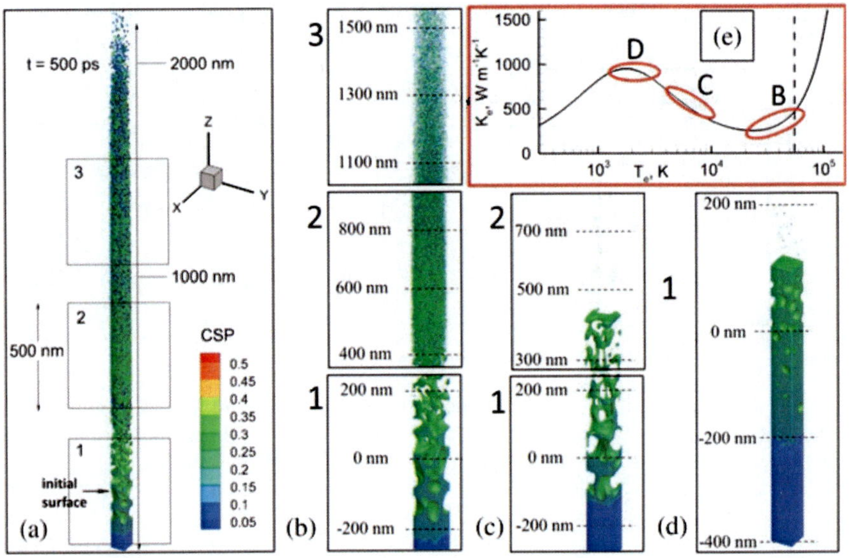

Fig. 6.10 The atomic snapshots are shown for a general view on the ablation plume formed in vacuum. **a** at the time of 500 ps. The three zones of the NPs formations are shown for the case of 0.3 ps pulse ($\lambda = 800$ nm, $F_{inc} = 2$ J cm^{-2}) in (**b**). Two zones of NPs formations are shown for the case of 4.0 ps pulse ($\lambda = 800$ nm, $F_{inc} = 2$ J cm^{-2}) in (**c**). One zone of the material foaming process is seen for the case of 100 ps laser pulse ($\lambda = 1064$ nm, $F_{inc} = 10$ J cm^{-2}) in (**d**). The atoms are colored by CSP for identification of the local crystal structures as follow: solid < 0.08 < defects < 0.12 < liquid < 0.25 < surface < 0.50 < vapor. The free (volatile) particles are blanked here for a better visualization of the NPs formation process. The electron heat conductivity k_e as a function of electronic temperature T_e is shown in the red rectangle (**e**) with red ovals correspondingly indicating the characteristic conductivity values for all three pulses (B) 0.3 ps, **c** 4.0 ps, and **d** 100 ps. Reproduced from [8] with permission from MDPI

of the material, where the formation of NPs is still in progress due to the spallation mechanism, with the forming clusters of ~30–40 nm characteristic size.

For the case of 0.3 ps pulse duration (Fig. 6.10b), the rate of target heating is determined by the characteristic time of the electron–phonon relaxation process, which is on the order of 20 ps. Providing the efficient laser heating depth large enough (~200 nm), the mechanical relaxation rate is weaker and the high compressive stresses are built up in the vicinity of the material surface. These conditions are frequently referred to as the regime of internal stress confinement, which provides the mechanical damage of the target with the subsequent formation of large NPs. However, due to the relatively large values of the electronic temperature developed during the pulse (up to 45,000 K), the conductivity values (k_e) of the exciting free carriers is dynamically changing from $k_e \sim T_i/T_e$ dependence when T_e is low to $k_e \sim T_e/(T_i + T_e^2)$ dependence for higher values of T_e. This situation leads to the accumulation of the laser deposited energy in the proximity of the surface and referred to as thermal confinement regime [20], is indicated in Fig. 6.10e in the red rectangle, where the conductivity of free electrons in gold is plotted as a function of electronic temperature at the fixed value of the lattice temperature, and marked as *B*.

While simulating the process of NPs formation with the MD-TTM model, one can monitor all thermophysical properties of matter (pressure, density, temperature) and also density and temperature of the electrons (due to the TTM part of the combined MD-TTM model). Therefore, the electron temperature at any moment of simulation time is known as well, and its maximum value during the 0.3 ps laser pulse interaction with thick Au sample was measured during the simulation. Note that the thermal confinement regime results in the lattice temperature at the surface reaching its critical value (~7000 K) and the explosive boiling process serves as the driving mechanism for the material ejection. The whole ablation process for the case of 0.3 ps pulse, therefore, is a mixture of the thermal damage at the surface (i.e., ablation, area "3" in Fig. 6.10b), and the mechanical damage of deeper parts of the material (i.e., spallation, area "1" in Fig. 6.10b).

The material ejection process due to the explosive boiling mechanism results in a high velocity of the small NPs, is shown in the region "3" of Fig. 6.10, where they move with the speed of ~6500 m s^{-1}. At the same time, the intensive evaporation process efficiently cools down the material's surface due to the transfer of thermal energy to the enthalpy of vaporization. Thus, the NPs with the lower kinetic energy, but with large size, are shown in the region "2", where they move at a slower speed of ~3500 m s^{-1}. Finally, when the material's surface loses most of its thermal energy, the ordinary material ejection process due to the spallation mechanism forming therefore the region "1" and moving roughly at the speed of ~750 m s^{-1} was observed. Further evolution of this ablation plume will result in more pronounced segregation of the generated NPs in accordance with their size and the lift-off velocity.

The result of 4.0 ps laser pulse interaction with gold at the incident fluence of $F_{inc} = 2$ J cm^{-2} ($\lambda = 800$ nm) is shown in Fig. 6.10c. Unlike the previous situation, the formation of large droplets of the material without noticeable contribution of the small clusters and free atoms shown in zone "2" was observed. The process of foam formation due to spallation is seen in zone "1". This time, since the pulse is longer, the laser intensity at the peak of the pulse is much lower as compared to the case of 0.3 ps pulse. The elevation of the electronic temperature, therefore, is just 15,000 K, which corresponds to the scaling of electron conductivity function as $k_e \sim T_e/T_i$, and reflected in Fig. 6.10e in the red rectangle by the marker *C*. Therefore, the deposited laser energy will efficiently penetrate the bulk of the material resulting in the establishment of temperature and pressure gradients by the time of the electron–phonon equilibrium on a spatial scale of ~200 nm.

The result of 100 ps laser pulse interaction with gold at the incident fluence of $F_{inc} = 10$ J cm^{-2} is shown in Fig. 6.10d, where the only foaming process in zone "1" is observed. The 100 ps laser pulse, being much longer than the characteristic electron–phonon relaxation time, cannot induce strong electron–phonon non-equilibrium conditions. The electronic temperature, therefore, does not reach high values and is limited roughly within 5000 K, where the laser deposited energy dissipation channel through the electron heat conduction is strong, but the induced lattice heating is weak. This situation is indicated in Fig. 6.10e in the red oval marked *D*.

From the performed simulation, one can thus conclude that, due to a specific dependence of the electron heat conductivity function of gold from the induced electronic temperature, the shorter (fs) laser pulses or the pulses at sufficiently high incident energy can result in an explosive boiling mechanism of the material removal process and NPs generation of few nm in size and of high yield. This corresponds to the experimental data above and can be related to the measurements presented in Fig. 6.7.

To compare the simulations and experimental data and to prove the presence of clusters in gold-ablated plasmas under optimal conditions of harmonic generation, the morphology of the deposited debris from Au plasma created during target surface ablation by laser pulses was analyzed. Laser ablation of solid material is a widely accepted technique for the generation of nanoparticles. However, this process has previously been studied without taking into account the role of free electrons and highly excited ions, which violates the optimal conditions for phase-matched HHG.

SEM measurements of the deposited debris of 10 nmAu NPs (Fig. 6.11a) were carried out under laser ablation conditions corresponding to the optimal plasma formation for the highest yield of harmonics. The substrates (glass plates and silicon wafers) used to collect the deposited material were placed at a distance of 40 mm in front of the ablation area, and the debris was further analyzed by SEM. It was determined that the mean size of deposited NPs (12 nm) was close to the NPs sizes measured prior to ablation (8–15 nm).

During ablation of a bulk target, under weak gold plasma formation conditions, the SEM images did not reveal the presence of nanoparticles in deposited debris with sizes above the microscope detection limit (3 nm). This was probably due to the small fluence (0.5 J cm^{-2}) of the heating 200 ps pulses on the target surface. Another pattern was observed upon ablation of the target using higher heating fluence (2 J cm^{-2}), which caused the appearance of small nanoparticles deposited onto a nearby substrate. At these conditions, the NPs appeared in the SEM images of the deposits, with their mean size of ~5 nm, while some larger NPs of a mean size of

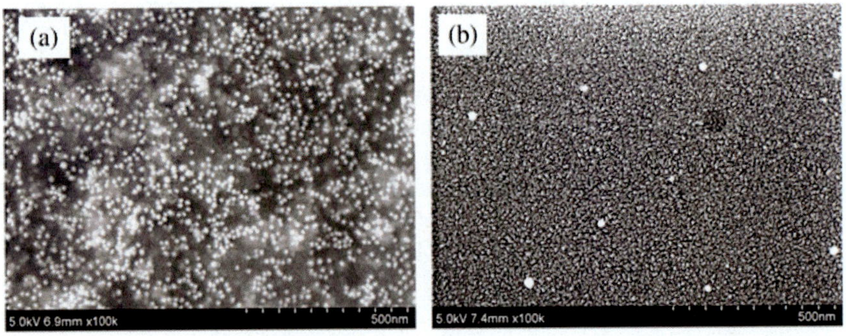

Fig. 6.11 SEM images of **a** deposited 10 nm NPs after ablation, and **b** deposited NPs synthesized during ablation of Au bulk target by 200 ps pulses. Reproduced from [8] with permission from MDPI

about 20 nm and higher were also seen (Fig. 6.11b). One has to reiterate that these characteristics of debris were measured once obtaining them at the maximum HHG conversion efficiency. These morphological studies confirmed the presence of a large number of tiny NPs and a small amount of large NPs simultaneously deposited on the substrates under conditions of 'optimal' laser ablation. This observation points out the presence of synthesized NPs at the moment of femtosecond pulses propagation.

The concentration of these small NPs increased with the growth of heating pulse intensity. The production of nanoparticles by laser ablation of metallic targets is a well-studied phenomenon. However, the use of high ablation fluencies allowing the synthesis of a large amount of Au NPs resulted in the growth of free-electron concentration, which is a most detrimental factor in HHG due to the contribution to the phase mismatch between the driving and harmonic waves. This explains why, under ablation conditions leading to NP formation in the plume of bulk metallic targets, the HHG signals are weaker with regard to the ablation of already existing Au NPs on the target surface. In that case, the presence of NPs in the plasma does not compensate for the deteriorated phase mismatch conditions caused by over-ionization and the presence of a large number of electrons.

6.2.4 Discussion

The characterization of the gold nanoparticles being presented in LIP is an important component of morphology and HHG studies, though it is very difficult to properly determine some of those parameters. Below we address a few issues related to the above experiments using Au NPs and try to answer some questions regarding these studies.

(1) Can the laser radiation destroy or modify Au NPs? Yes, it can. NPs can be disintegrated by the radiation of the heating pulses. We analyzed the debris of deposited Au NPs under the irradiation by ns, ps, and fs pulses. The mean size of NP debris was dominantly the same as of the initial NPs (10 and 100 nm), though the presence of the smaller-sized wing in the histogram of size distribution pointed out the appearance of disintegrated NPs. Their ratio was insignificant since we ablated Au NPs at the fluencies allowing evaporation of those particles without the notable modification of their structure. Stronger irradiation of targets led to the appearance of large aggregates and small NPs in the nearby substrates, alongside the decrease of HHG conversion efficiency attributed to a large concentration of free electrons.

(2) How much of them remained from pulse to pulse? As for the remained particles on the target surface, their concentration did not change since only an insignificant part of NPs was ejected from the surface during a single shot. The problem of crater formation was resolved by moving the target surface either by dragging the sample up and down or by using the rotating targets or by both above methods.

(3) How many particles escaped into the plasma? To properly analyze and determine the amount of NPs ablated and appeared in the plasma area one has to carry out the accurate measurements of the weight of NP powder before and after ablation by a large number of pulses and then calculate how many particles escaped into the plasma during a single shot. Even this information is insufficient for the determination of the amount of NPs participating in HHG during the interaction with fs driving pulses since the plasma cloud had relatively large sizes (a few mm, in accordance with the observations of the emission of incoherent radiation of the plasma moving out from the surface) and an only small portion of them met the fs driving beam. How much NPs were inside the "tube" with the diameter ~64 μm (i.e., sizes of the focused driving beam) and length ~0.4 mm (diameter of spreading plasma at the distance of ~0.5 mm from the target surface) at the moment when the largest amount of them reaches the axis of propagation of the fs pulse remains unknown.

(4) What was their concentration? Here we can talk only about the averaged concentration of NPs in the plasma cloud at the moment of propagation of the driving fs pulses. Earlier, estimates and calculations using code HYADES of the density of ablated particles at the conditions suitable for efficient HHG were reported on the 2×10^{17} cm^{-3} concentrations of silver atoms and ions in the case of efficient HHG in Ag plasma [21]. The discussed studies were carried out at other experimental conditions (i.e., easier ablation and lower laser heating fluencies, formation of NP cloud, presence of different groups of NPs, etc.). It is also difficult to compare the concentrations of the homogeneous plasma containing separated atomic and ionic species and of the plasma containing ultrasmall "solid" species like NPs.

One can easily see that the density of the formed NPs is close to the density of the solid. Furthermore, while modeling the NPs formation process with the MD-TTM model, one can see that the formation of the NPs takes place via the establishing of metallic bonding between the atoms, that only possible (via the interatomic function construction) if the atomic density is close (within 10%) to that of the solid.

An explanation for intense harmonic generation from Au NPs could be the higher concentration of neutral atoms due to the presence of nanoparticles. Unlike single atoms and ions, whose density quickly decreases due to plasma expansion, the NPs retain densities that are close to their solid-state, while the distance between NPs in plasma jet is notably larger as compared with the distance between particles in atoms-containing LIPs. The concentration of atoms in solid species depends on its density and varies in the range of 10^{22}–10^{23} cm^{-3}. Combined with the higher harmonic efficiency of neutral atoms compared with their ions, the neutral atoms within the NPs could generate high-order harmonics efficiently.

How many atoms in the nanoparticle become optimal for efficient generation of coherent extreme ultraviolet radiation using a whole ensemble of particles, which allow increasing the number of photons of high-order harmonics, remains a puzzle despite the fact that to date numerous experiments using the ablated NPs were conducted [22–24]. Qualitative assessments predict that the presence of the particles containing a few hundred to a few thousand atoms in the area of interaction with a strong laser field may lead to the maximal growth of generated harmonics. The comparative analysis of HHG spectra generated in the plasmas excited on the surfaces of pure paper and Au NPs-contained paper show the advantages of ablated NPs-containing targets and the influence of these species on the enhancement of harmonic yield at the same conditions of the experiment.

There are still a lot of issues that have to be clarified in the case of the application of the used Au NPs for HHG. It remains a puzzle how the composition and spatial scales of the NPs affect the HHG efficiency and the cutoff frequency. There remain also other questions regarding the charge state of this plasma medium, which could be resolved only by using time-of-flight mass spectrometry. Particularly, are the Au NPs were charged before the driving pulse comes? How many electrons are ionized under the high-intensity pump pulse from each NP? What are the exact ionization potentials of the chosen Au NPs and how do they affect the HHG cutoff? Neither of those questions still found the proper answer. It is difficult to resolve numerous issues while dealing with NPs in plasmas due to the availability of only indirect methods of the measurements of plasma parameters. Some additional abovementioned issues also still wait to be resolved.

6.3 Summary

In this chapter, we discussed the conditions when the plasma plumes created on the surfaces of ZnO thin films in vacuum conditions can be used as the efficient emitters of high-order harmonics of 800 nm, 30 fs laser pulses. We also analyzed the studies of the influence of dopants in nanocrystal-based media on the variation of high-order harmonics spectra produced during propagation of ultrashort laser pulses through the laser-produced plasmas containing ZnO and Gd. The strong 9th harmonic was observed in the case of plasmas produced on the bulk ZnO due to the resonance-induced enhancement of single harmonic, whereas in powered targets (powdered ZnO NCs and ZnO NCs + 12.5% Gd) insignificant difference between the 9th and the 11th harmonics was observed. We analyzed the influence of dopant on the harmonics spectra at different delays between the heating and driving pulses. Those studies confirmed that doping of ZnO materials with Gd caused variation of transient, morphological, and high-order nonlinear optical properties of ZnO.

The 5- and 40-fold enhancement of harmonic yield was analyzed from ablated Au NPs on paper as compared with ablated Au NPs on glass and Au bulk targets. The harmonic cutoffs obtained from ablated Au NPs on paper, Au NPs on glass, and Au bulk target were 29th, 21st, and 21st orders, respectively, in the case of nanosecond heating pulses. In addition, the effects of heating pulse duration, TCP, and laser chirp on harmonic yield were discussed in Au NP-contained LPPs. The enhanced harmonic intensity was obtained from ablated Au bulk target by decreasing the heating pulse duration. Those studies showed that the application of picosecond and femtosecond heating pulses for plasma formation allowed the generation of 5 and 4 times stronger harmonics with regard to the nanosecond heating pulses. The harmonic cutoffs were

33rd and 39th orders in the case of picosecond and femtosecond heating pulses, respectively. The enhanced harmonic intensity was demonstrated from ablated Au bulk target by applying TCP (800 nm + 400 nm) of LPP. Two-fold enhancement of harmonics was observed using TCP of ablated Au bulk target with regard to SCP. Negatively chirped 130 fs pulses enhanced 9th and 11th harmonics with regard to chirp-free (35 fs) and positively chirped 130 fs pulses. Performed MD-based simulations of NPs formation at different pulse durations and the incident fluencies supported the experimental measurements and revealed the main driving mechanisms responsible for the size and morphology of the generated particles.

References

1. H. Linnenbank, Y. Grynko, J.Förstner, S. Linden, Light: Sci. Appl. **5**, e16013 (2016)
2. S. Ghimire, D.A. Dichiara, E. Sistrunk, P. Agostini, L.F. Dimauro, D.A. Reis, Nature Phys. **7**, 138 (2011)
3. K. Imasaka, T. Kaji, T. Shimura, S. Ashihara, Opt. Express **26**, 21364 (2018)
4. A. Rout, G.S. Boltaev, R. Ganeev, K.S. Rao, D. Fu, R. Rakhimov, S. Kurbanov, S. Urolov, Z. Shaymardanov, C. Guo, Eur. Phys. J. D **73**, 235 (2019)
5. K. Spoorthi, S. Pramodini, I.V. Kityk, M. Abd-Lefdil, M. Sekkati, A.E. Fakir, R. Ashok, S. Ganesh, P. Poornesh, Laser Phys. **27**, 065403 (2017)
6. M. Venkatesh, R.A. Ganeev, K.S. Rao, G.S. Boltaev, K. Zhang, A. Srivastava, J.K. Bindra, S. Singh, V.V. Kim, S.K. Maurya, G.F. Strouse, N.S. Dalal, C. Guo, Opt. Mater. **95**, 109241 (2019)
7. A.M. Crooker, K.A. Dick, Canad. J. Phys. **46**, 1241 (1968)
8. M. Venkatesh, R.A. Ganeev, D.S. Ivanov, G.S. Boltaev, V.V. Kim, L. Jingguang, I.N. Zavestovskaya, M.E. Garcia, C. Guo, Nanomater. **10**, 234 (2020)
9. A. Rout, G.S. Boltaev, R.A. Ganeev, Y. Fu, S.K. Maurya, V.V. Kim, K.S. Rao, C. Guo, Nanomater. **9**, 291 (2019)
10. R.L. Olmon, B. Slovick, T.W. Johnson, D. Shelton, S.-H. Oh, G.D. Boreman, M.B. Raschke, Phys. Rev. B **86**, 235147 (2012)
11. S. Kim, J. Jin, Y.-J. Kim, I.-Y. Park, Y. Kim, S.-W. Kim, Nature **453**, 757 (2008)
12. I.-Y. Park, S. Kim, J. Choi, D.-H. Lee, Y.-J. Kim, M.F. Kling, M.I. Stockman, S.-W. Kim, Nature Phot. **5**, 677 (2011)
13. M.F. Ciappina, S.S. Aćimović, T. Shaaran, J. Biegert, R. Quidant, M. Lewenstein, Opt. Express **20**, 26261 (2021)
14. R.A. Ganeev, M. Baba, M. Suzuki, H. Kuroda, Phys. Lett. A **339**, 103 (2005)
15. R.A. Ganeev, G.S. Boltaev, V.V. Kim, C. Guo, Nanomater. **9**, 572 (2019)
16. C. Kan, N.H. Burnett, C.E. Capjack, R. Rankin, Phys. Rev. Lett. **79**, 2971 (1997)
17. S.I. Anisimov, B.L. Kapeliovich, T.L. Perel'man, Zh. Eksp. Teor. Fiz. **66**, 776 (1974)
18. D.S. Ivanov, L.V. Zhigilei, Phys. Rev. B **68**, 064114 (2003)
19. P.B. Corkum, F. Brunel, N. Sherman, T.J. Srinivasan-Rao, Phys. Rev. Lett. **61**, 2886 (1988)
20. D.S. Ivanov, B. Rethfeld, Appl. Surf. Sci. **255**, 9724 (2009)
21. L.B. Elouga Bom, J.-C. Kieffer, R.A. Ganeev, M. Suzuki, H. Kuroda, T. Ozaki, Phys. Rev. A **75**, 033804 (2007)
22. R.A. Ganeev, C. Hutchison, M. Castillejo, I. Lopez-Quintas, F. McGrath, D.Y. Lei, J.P. Marangos, Phys. Rev. A **88**, 033803 (2013)
23. M. Wöstmann, P.V. Redkin, J. Zheng, H. Witte, R.A. Ganeev, H. Zacharias, Appl. Phys. B **120**, 17 (2015)
24. R.A. Ganeev, M. Suzuki, H. Kuroda, Eur. Phys. J. **70**, 21 (2016)

Chapter 7
Two-Color Pump of Laser-Induced Plasmas

7.1 Comparative Studies of High-Order Harmonic Generation in Argon Gas and Different Laser-Produced Plasmas Using Single-Color and Two-Color Pumps

7.1.1 Consideration of Different Media for HHG

The advanced properties of LIP over gas media have been demonstrated in the case of carbon plasma [1–3]. The efficiencies of high-order harmonics generation in carbon-containing plasmas and gases were compared under closely matched experimental conditions [1]. Those measurements have shown that, in the 14–25 eV photon energy range, carbon plumes are more efficient than argon suggesting that the high-order nonlinear response of carbon nanoparticles is significantly enhanced. The studies have demonstrated higher conversion efficiency for the low-order harmonics generated in graphite plasma compared with argon gas in the case of 10-ns ablating pulses. Among the main factors influencing HHG conversion efficiency in medium, density and length play an important role. The other experimental conditions of HHG (confocal parameter of driving radiation, pulse duration, and energy, etc.) were maintained equal for both media. In both cases, weak focusing conditions, with the confocal parameter much longer than the medium, were used.

In the past, comparative studies of carbon plasma and argon gas have shown the advantages of the former medium once one became aimed at stronger harmonic emission in the 50–100 nm spectral range of XUV. One limitation of the intense carbon harmonic source driven by a 0.8 μm wavelength Ti:sapphire laser has been the low cutoff around ~32 eV. The application of relatively longer-wavelength pulses allows the generation of harmonics from carbon plasma up to ~70 eV by increasing the driving laser wavelength to 1.71 μm [4]. As it was revealed in [4], the carbon harmonic intensity is found to be high despite the long-wavelength driving laser. Experiments have shown only a 30% decrease in the harmonic intensity when changing the driving laser wavelength from 0.8 μm to 1.71 μm. Meanwhile, notice

R. A. Ganeev, *High-Order Harmonics Generation in Plasmas*, Springer Series on Atomic, Optical, and Plasma Physics 122, https://doi.org/10.1007/978-3-031-09040-0_7

that, at the shortest wavelength region, the HHG conversion efficiency from these lasers was notably low, which may diminish the advantages of the application of different concepts of harmonic enhancement, like quasi-phase-matching (QPM), in the case of this plasma even using the near-infrared sources. Such a conclusion has been revealed during QPM studies of the harmonics generated from the tunable (1100–2400 nm) femtosecond source [5]. A search for suitable plasma media showing preferable features compared with gases thus became an interesting topic of study. To analyze the potentially efficient plasma media chosen some of them were chosen taking into account earlier reported data on their ability in generating higher-order harmonics of 800 nm driving laser, compared with the carbon plasma.

The comparison of harmonic yields from argon and other plasma media has yet been reported. Meanwhile, many samples to be ablated have shown relatively strong harmonics in the XUV region shorter than 40 nm. Among them, silver and silver sulfide plasmas are distinguished with the strongest yield down to the cutoff region at around 20 nm and beyond. Another advantage in the application of these and other plasmas for comparison with gas-induced harmonics is the analysis of the behavior of XUV yield in the case of different geometries of experiments. In particular, the tuning of LIP with regard to gas jet allows defining the optimal conditions of joint influence of two sources of harmonics. LIP being installed before the gas jet can provide the harmonics distinguished by divergence, cutoff, and yield with regard to those produced from the plasmas coincided with the gas jet or installed after the gas jet.

Silver has earlier shown the excellent properties of harmonic generation. The calibration of the XUV registration system using different approaches has demonstrated large conversion efficiency of the harmonics in the range of H20–H40 (~10^{-5} [6]). However, there have never been back-to-back measurements that compare the efficiencies of gas and plasma harmonics, using the same pump laser and under the same conditions. Such comparison is important since the harmonic emission in the case of HHG in LIPs is extremely sensitive to various experimental conditions, including the heating pulse intensity and the distance of the driving laser from the target surface.

In this section, we analyze and compare the harmonic yields from the argon gas, silver plasma, silver sulfide plasma, and stainless steel plasma [7]. The single- and two-color pumps of gas and plasma media allow enhancement of odd and appearance of strong even harmonics during propagation through the LIPs and gases, even at the very small ratio between second and first waves (~1:30). We analyze the variation of the divergence of harmonics at different positions of LIP with regard to gas jet as well as resonance enhancement of single harmonic. The discussed studies demonstrated a 3- to a 20-fold enhancement of harmonics from LLPs compared with a gas medium at similar concentrations and lengths of gas and plasma.

HHG experiments were carried out using 800-nm, 35-fs, 1-kHz, 1.5-mJ laser pulses (Spitfire Ace, SpectraPhysics), which were focused using a 400 mm focal length spherical lens in the vacuum chamber contained gas jet and LIP (Fig. 7.1a). The gas jet (diameter of nozzle 0.25 mm) was placed at the focal plane of the focused radiation and operated with a continuous Ar flow. LIP was produced using either 35 fs, 800 nm pulses or 200 ps, 800 nm pulses from the uncompressed radiation

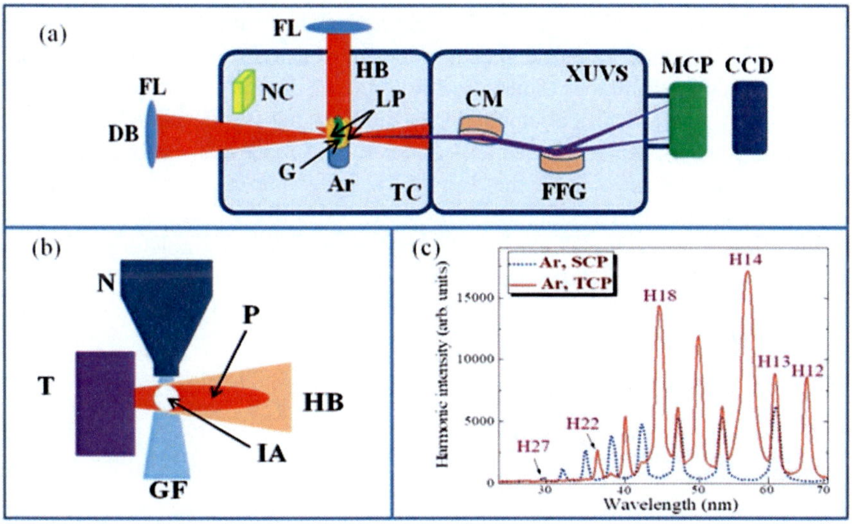

Fig. 7.1 **a** HHG setup. DB, driving beam; HB, heating beam; FL, focusing lens; NC, nonlinear crystal (BBO); Ar, argon gas inlet; TC, target chamber; LP, laser plasma; G, gas flow; XUVS, extreme ultraviolet spectrometer; CM, cylindrical mirror; FFG, flat field grating; MCP, microchannel plate; CCD, CCD camera. **b** Area of laser-gas-plasma interaction. N, target for ablation; HB, heating beam; P, laser-produced plasma; N, nozzle of gas tube; GF, gas flow; IA, interaction area of spreading gas and plasma, though which the driving femtosecond pulse propagated in the direction normal to the picture. **c** Harmonic intensities distribution of harmonics in the cases of single-color pump (800 nm, dotted curve) and two-color pump (800 nm + 400 nm, solid curve) of Ar gas. Reproduced from [7] with permission from the Optical Society of America

of the same laser on the surfaces of bulk silver sulfide and silver targets placed at different positions with regard to the gas jet. The plasma was also formed on the surface of the stainless steel nozzle in the vicinity of gas flow. The debris from the ablation coats the window from which the ablating pulse enters the vacuum chamber, thus reducing the fluence of heating radiation on the target with time. To exclude this effect, the protective microscope slide was used. It was installed between the target and the entrance window and could be easily replaced from time to time.

The targets were installed either at the focal plane of focused radiation or in front of and after the gas jet at a distance of ~5 mm. Thus the positions of LIPs were within the confocal parameter of the focused driving beam (b = 14 mm). The maximal used intensity of femtosecond pulses in the focal plane was 5×10^{14} W cm^{-2}. In most cases, the single-color pump (SCP) of gas or plasma was used for HHG when 800 nm pulses were used. To generate harmonics using a two-color pump (TCP), the 0.2-mm thick barium borate (BBO) crystal was inserted in the vacuum chamber on the path of the focused 800 nm radiation. The conversion efficiency of 400 nm radiation was 3%.

The concentration of gas particles was estimated to be 5×10^{17} cm^{-3}, while plasma concentration was estimated to be 3×10^{17} cm^{-3}. The diameter of the nozzle providing gas flow was 0.25 mm, while the sizes of the heating pulse on the target surface were ~0.3 mm. The driving beam propagated at a distance of ~0.5 mm from the nozzle and target surfaces. Thus one can assume a similarity in the concentrations of harmonic emitters and the lengths of gas and plasma media through which the 35 fs pulse was propagated to generate harmonics. Particularly, the active lengths of the media at the distances of 0.5 mm from the target and nozzle were assumed to be 0.3 and 0.4 mm for the gas and plasma medium, respectively.

Gas and plasma jets were intercepted with each other once the ablating target was placed close to the nozzle of the gas jet (Fig. 7.1b). In most experiments, the LIP was positioned either in front of or behind the gas jet. Harmonic emission was analyzed using an XUV spectrometer.

7.1.2 Comparative Studies of Gas and Plasma Harmonics

In most of the cases, a strong first field (800 nm) and a weak second field (400 nm) were used to pump the gas plasma plume. These studies demonstrated that, even at a 30:1 ratio between these two fields, dramatic changes in the spectrum of harmonics occur, which point out the decisive influence of a weak second field on the whole process of HHG. A similar influence could be expected once one considers the variation of cut-off for TCP versus SCP schemes. The decrease of cutoff did not follow the quadratic dependence on the wavelength, though the almost two-fold decrease of maximally generated harmonic order and stronger odd harmonics was observed in the case of TCP of both gas and plasma media.

Strong harmonic generation in the case of TCP is possible due to the formation of a quasi-linear field, selection of a short quantum path component, which has a denser electron wave packet, and higher ionization rate compared with SCP [8]. The orthogonally polarized second field also participates in the modification of the trajectory of an accelerated electron from being two-dimensional to three-dimensional, which may lead to the removal of the medium symmetry. With suitable control of the relative phase between the fundamental and the second harmonic radiation, the latter field enhances the short path contribution, resulting in a clean spectrum of harmonics.

The TCP-induced enhancement of harmonics has earlier been realized through generation of 400-nm radiation in a separate channel with further mixing with 800-nm radiation in gases, as well as through direct generation of second harmonic (H2) in thin BBO crystals followed with focusing of two co-propagating beams in gases [9, 10]. The following studies of TCP in LIPs [11–13] have also shown the advantages of the TCP of LIP with regard to the SCP using both 800 nm and 1300 nm fundamental sources and their H2.

Figure 7.1c shows the SCP- and TCP-induced HHG spectra from the argon gas. The efficiency of HHG was larger in the case of TCP with regard to SCP. Also, strong even harmonics were observed. Even harmonics were two times stronger than odd harmonics, while the cutoff was decreased in the case of TCP with regard o SCP (H27 and H22 respectively).

Harmonic generation in gas and silver plasma jets at different positions of LIP was carried out using SCP at a similar intensity of driving pulse and approximately equal concentrations of gas and plasma media. The procedure of these experiments was as follows. Firstly, the best harmonic yield from the gas jet was achieved, and then the heating pulse was scanned either on the target placed close to the needle, through which gas flowed to the chamber, or targets placed prior to or after the needle.

The upper panel of Fig. 7.2a shows the harmonic spectrum from argon. The H23 cutoff was observed in the case of gas emitters. The second, third, and fourth panels of this figure show the HHG in the cases of plasmas produced in front of the gas jet, at the position of the gas jet, and behind the gas jet, respectively. Correspondingly, the distance between LIP and gas jets along Z-axis was −5, 0, and +5 mm with regard to the focal plane of the driving beam, which corresponded to the position of plasma within the confocal parameter of focused radiation. Figure 7.1a shows the interaction of gas and plasma jets in the case of their coincidence along the Z-axis. Gas jet was placed in the focal plane of the focusing lens. One can see both the growth of harmonic yield and cutoff extension in the case of gas + plasma configuration.

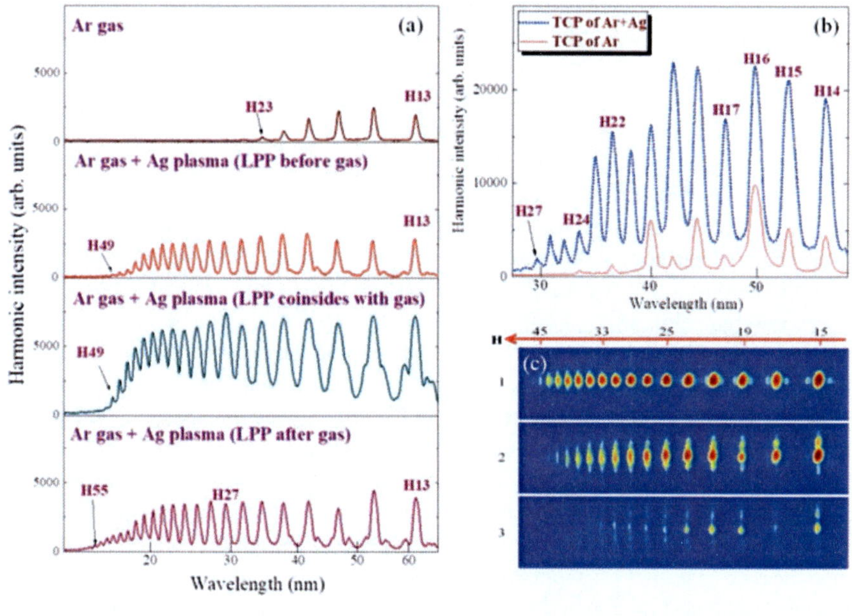

Fig. 7.2 Comparison of harmonic spectra generating in pure Ar gas and mixture of Ar gas and Ag plasma in the case of **a** SCP and **b** TCP. **a** LIP was placed before gas jet (second panel from top), after gas jet (fourth panel), as well as was coincided with the gas jet (third panel). **b** LIP was placed after gas jet. The energies of driving and heating femtosecond pulses 0.9 and 0.4 mJ. The delay between pulses was maintained at 73 ns. **c** Images of harmonic spectra from Ar + Ag medium observed by CCD camera. Panel 1 corresponds to the position of LIP before gas jet. Panel 2 corresponds to the position of LIP after gas jet. Panel 3 shows the harmonic spectrum during shorter collection time to distinguish the details of the divergence of harmonics and the appearance of interference pattern along the vertical axis. Reproduced from [7] with permission from the Optical Society of America

The strongest harmonics were observed in the case of the coincidence of gas and plasma jets (third panel from the top of Fig. 7.2a).

A comparison of harmonic spectra generated in a pure Ar gas and a combination of Ar gas and Ag plasma in the case of TCP is shown in Fig. 7.2b. LIP in that case was placed at 5 mm after the gas jet. The application of thin (0.2 mm) BBO crystal allowed the generation of approximately equal odd and even harmonics from the gas + plasma medium. Three- to five-fold growth of even harmonic yield in the case of gas + plasma compared with a gas medium at similar experimental conditions was observed. The enhancement factor was larger in the case of odd harmonics (compare the yields of H15, H17, and H19 in the cases of gas and gas + plasma media).

The raw images of harmonic spectra from different combinations of generating media in the case of replacement of cylindrical mirror by plane one are presented in Fig. 7.2c. The goal of those studies was the qualitative definition of the conditions of different gas and plasma formations on the variation of harmonic divergence and harmonic distribution in the XUV range, rather than the quantitative measurements of the gain and conversion efficiency of HHG. Visual presentation of spectra and observation of the difference between the divergences of harmonics in the cases of plasma + gas and gas + plasma have motivated to present the images of generating spectra appearing on the screen of the monitor. The saturated images of harmonics are intentionally chosen to present the spectra for better viewing. In the meantime, for the line-outs of the HHG spectra, the unsaturated images were used.

Figure 7.2c demonstrates the raw images of harmonic spectra obtained using SCP of plasma + gas (panel 1) and gas + plasma (panel 2) configurations. The harmonic spectrum generated in the argon gas and silver plasma using a smaller collection time of the CCD camera is also shown (panel 3). One can see the difference in the divergences of harmonics generation in the case of two different positions of plasma jet with regard to the gas jet. The divergence of harmonics arising from plasma + gas configuration was significantly smaller compared with the case of gas + plasma. Moreover, in the latter case the irregularity in the divergence resembling the interference pattern along the vertical axis for most of the harmonics was observed (panel 3). This interference-like distribution of divergence can be attributed to the interaction of harmonics originating from the gas jet with those from the plasma jet.

In the case of gas HHG, the installation of a gas jet prior to the focal plane causes the preferential involvement of short trajectory of an accelerated electron in harmonic generation leading to narrower divergence of coherent XUV emission. The opposite case (gas after focal plane) increased the role of a long trajectory of electrons thus increasing the divergence of generated harmonics. A similar difference in the divergences of harmonics was observed during plasma HHG studies.

Below we analyze the double media configuration for HHG. One can assume that weak harmonics from the first medium (gas jet) can further be amplified in the second medium (plasma jet). A similar concept of amplification has been analyzed in [14] when the first experimental demonstration of the parametric amplification of attosecond pulse trains at around 11 nm was demonstrated in the helium amplifier driven by intense laser pulses and seeded by high-order harmonics generated in a neon gas jet. Their measurements have suggested that amplification can take place only if the seed pulse trains are perfectly synchronized in time with the driving laser field in the amplifier. In the discussed case, the distance between gas and plasma media was maintained at ~5 mm. It is difficult to assume the amplification of weak harmonics from the gas jet in the following plasma jet at the used experimental conditions. The exclusion of gas harmonics, by closing the nozzle, led to an almost similar harmonic spectrum generated in the plasma medium. One of the possible scenarios of the interaction of gas harmonics and plasma harmonics in the case of gas + plasma configuration could be some interference between these two sources of harmonic emitters. A weak interference pattern was observed in that case (Fig. 7.2c, panel 3). Nevertheless, it would be an important step towards designing amplifiers for the realization of strong XUV pulses, provided the determination of the optimal conditions for enhancement from the first medium generating seeding pulses towards the second medium amplifying those XUV pulses. This method of harmonic amplification can also be useful for the determination of harmonic pulse duration, similarly to the technique demonstrated in [15]. Varying the delay with a sub-10-as temporal resolution, they were able to resolve electric field evolution within the attosecond pulse train, which allowed them to estimate the duration of the individual XUV pulses within the train to be about 0.2 fs.

Analogous studies of silver sulfide plasma were carried out by a similar manner to clarify SCP and TCP of gas and plasma media at different conditions of the experiment. Figure 7.3a shows the HHG using SCP in the case of plasma + gas configuration. One can see that in the case of LIP placed prior to gas jet the harmonic distribution was almost similar to the case when the gas nozzle was closed. The presence of gas in those experiments was insignificant. Another situation was observed when LIP was installed after the gas jet (Fig. 7.3b). The harmonic yield was notably stronger compared with the former case. The comparison of two HHG spectra (with and without gas, Fig. 7.3b) shows that the influence of gas harmonics can be attributed to the enhancement of lower-order harmonics. Similarly, stronger odd and even harmonics in the case of TCP of two configurations (plasma + gas and gas + plasma) revealed the preferable position of the plasma jet behind the gas jet (Fig. 7.3c).

Finally, the experiments were carried out by ablation of the nozzle of the gas jet. Figure 7.4 shows the harmonic distribution in the case of the argon gas and stainless steel plasma. The stainless steel needle was ablated in the area of the nozzle by picosecond pulses. Without the ablating nozzle by picosecond pulses, the harmonic spectrum represented the ordinary spectrum of gradually decreasing harmonics from the gas jet (red solid curve). The ablation of the nozzle led to the appearance of the components of stainless steel ablation in the axis of driving beam propagation. These components drastically changed the HHG spectrum (blue dotted curve). The harmonics from H13 to H25 generated in the mixture of gas and plasma were five times stronger than those generated in pure argon gas. The most intrigued pattern was observed in the area of harmonic cutoff. One can see that the 27th harmonic almost disappeared compared to the neighboring ones. Meanwhile, H29 was notably stronger than H25 and equal to H23.

Stainless steel contains a number of constituents that define its structure, hardness, and other physical and chemical properties. The elemental distribution in the used

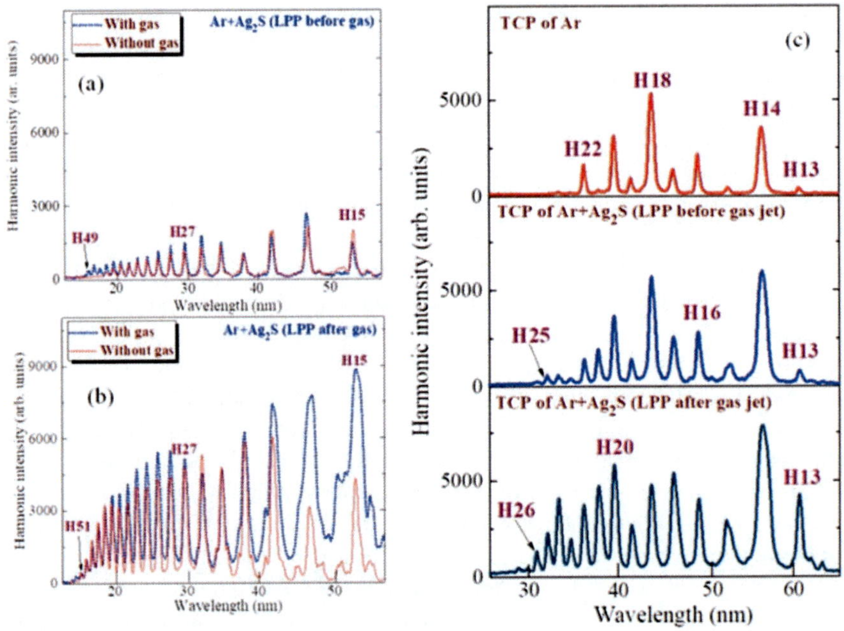

Fig. 7.3 Harmonic spectra generated in the Ar gas + Ag2S LIP medium in the case of presence (dotted curves) and absence (solid curve) of gas. **a** LIP was placed before gas jet. **b** LIP was placed after gas jet. **c** Two-color pump of argon gas and silver sulfide plasma. Upper panel: gas without plasma. Middle panel: gas after plasma. Bottom panel: gas before plasma. Reproduced from [7] with permission from the Optical Society of America

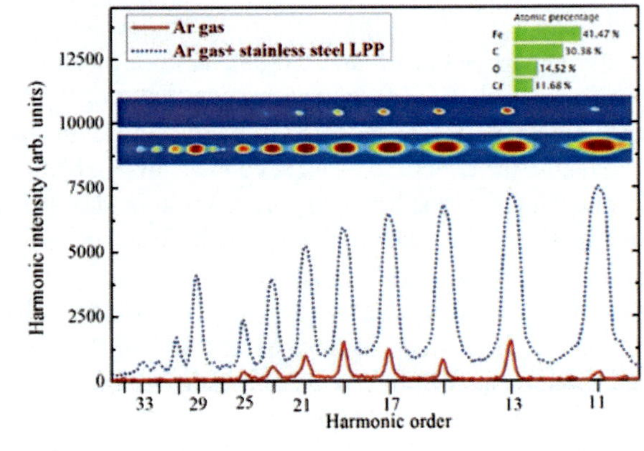

Fig. 7.4 Harmonic spectra distributions in the cases of argon gas (solid curves) and argon as + stainless steel LPP. Insets show the CCD images of harmonics spectra and elemental distribution of used stainless steel, which contained ~12% of chromium. Reproduced from [7] with permission from the Optical Society of America

sample is shown in the inset in Fig. 7.4. One can see that one of the important constituents was chromium (~12%). The role of Cr in the SS harmonic spectrum generated in stainless steel plasma has been early studied by using pure Cr and pure Fe as the targets for HHG experiments [15]. The former plasma has shown a considerable enhancement of the intensity of the 29th harmonic. Those observations clearly pointed out that the increase of the intensity of the 29th harmonic in the case of ablated stainless steel, although not as pronounced as in ablated Cr, was caused by the presence of the latter component in the ablated sample.

Previous studies of the photoabsorption and photoionization spectra of Cr plasma in the range of 41–42 eV [16] have demonstrated the presence of strong transitions, which could be responsible for a suppressed pattern of the harmonic spectrum in the wavelength region of 29.5–31 nm. The region of 'giant' 3p → 3d resonances (44–45 eV) of Cr II spectra was analyzed in [16, 17] and strong transitions that could enhance the nonlinear optical response of the plume were revealed. The harmonic wavelength can be resonant with the transition between the ground and the autoionizing state (the excited state embedded in the continuum) of the generating ion and that this transition exhibits high oscillator strength. In particular, in the four-step model describing the enhancement of resonance harmonics [18] the ionized and laser-accelerated electron is captured into the autoionizing state of the parent ion, and, in the final step, the radiative relaxation of this state to the ground state, a harmonic photon is emitted.

7.2 Single-Color and Two-Color Induced High-Harmonic Generation in Lanthanides–Containing Plasmas

Different metals, metal alloys, and semiconductors were analyzed by ablation and HHG in LIP. The map of the elements used in HHG studies is present in Fig. 7.5. Additionally, different molecular structures were used for plasma HHG [2, 7]. Some metal ions demonstrated the resonance enhancement of a single high-order harmonic of 800 nm pulses. HHG in nanoparticle-containing LIPs allowed the enhancement of harmonic yield with regard to the monomer-containing plasmas of the same elemental consistency. Quasi-phase-matching is another mechanism that allows for the enhancement of the group of harmonics in the plateau range. Some elements of the periodic table showed advanced properties as emitters of harmonics.

Among new potentially interesting elements of the periodic table, the group of non-radioactive solid materials with large atomic mass, such as lanthanides, can be considered. This also allows studying the high-order nonlinear optical properties of those elements through HHG in LIPs. The lanthanides comprise the 15 metallic chemical elements with atomic numbers 57–71 and are often known as rare earth elements. They are called lanthanides because the elements in the series are chemically similar to lanthanum.

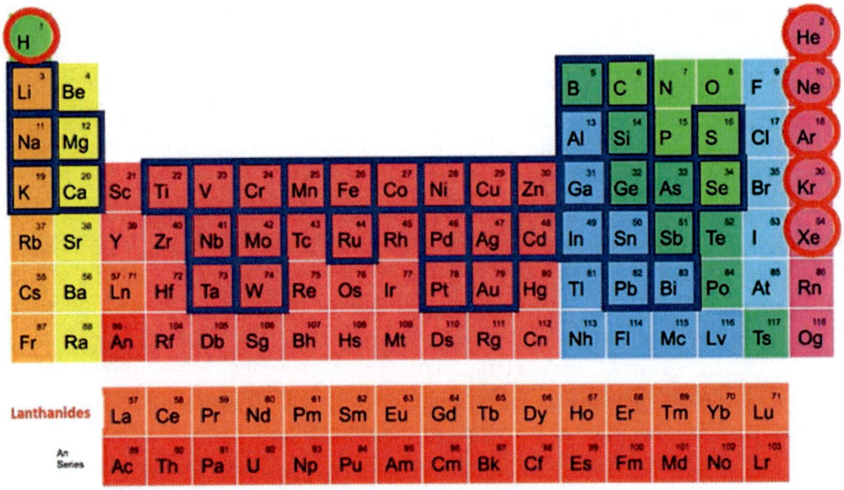

Fig. 7.5 Map of the solid elements of periodic table used in previous studies of HHG in plasmas (blue squares). Harmonic generation in gases is marked by red circles. Reproduced from [19] with permission from the American Institute of Physics

In the studies discussed in this section [19], four lanthanides (La, Yb, Pr_6O_{11}, and Tb_4O_7) were chosen for analysis of the high-order nonlinear optical properties of the LIPs produced on the surfaces of those targets. Here we analyze the generation of coherent XUV radiation in the plasmas containing lanthanide elements ablated by picosecond and femtosecond laser pulses. We also discuss HHG using femtosecond driving pulses in the case of the single- and two-color pump of LIPs.

Two lasers (central wavelength $\lambda = 800$ nm, 35 fs pulse duration, pulse energy $E \leq 1$ mJ, 1 kHz pulse repetition rate; $\lambda = 802$ nm, $t = 64$ fs, $E \leq 4$ mJ, 10 Hz) were used in these studies. The harmonic radiation was generated by focusing the femtosecond pulses in the plasmas formed on the surfaces of lanthanide pressed targets (in the case of lanthanide oxides) or lanthanide metals (in the case of La and Yb) using 370 ps, 200 ps, and 35 fs heating pulses (Fig. 7.6). The laser beam was split after regenerative amplifiers using the beam splitter at a 30/70 ratio. The larger part was compressed, while the lower one was used for LIP formation. The intensity of the heating pulses was varied in the range of 8×10^9 W cm^{-2}–2×10^{10} W cm^{-2} depending on the properties of ablating material.

The driving 35-fs, 0.5-mJ pulses or 64-fs, 1.7-mJ pulses were focused by a 500 mm focal length lens (single-color pump (SCP) scheme) inside the formed plasma to generate harmonics. The distance between the target and optical axis of femtosecond beam propagation was varied in the range of 0.2–0.5 mm. The intensity of driving pulses was $\leq 5 \times 10^{14}$ W cm^{-2}. For achieving the maximal harmonic yield the optical delay line between the driving and heating laser pulses was used. The delays were varied in the range of 30–100 ns. In the case of the two-color pump (TCP) scheme, the nonlinear crystal (0.2 or 0.3 mm thick barium borate (BBO) crystal, type I) was

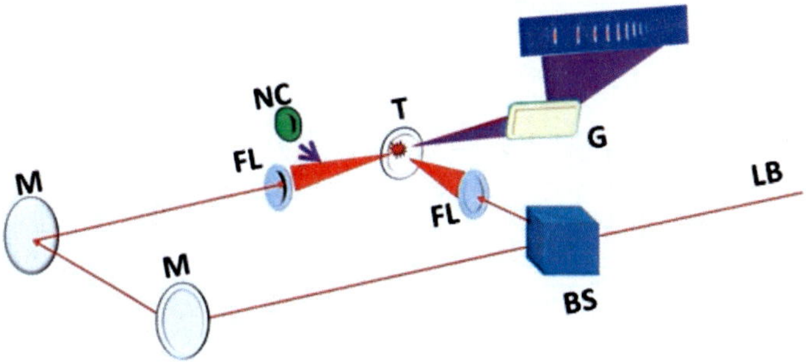

Fig. 7.6 The experimental setup of HHG in laser-produced plasma. LB: laser beam; BS: beam splitter; M: mirrors; FL: focusing lenses; NC: nonlinear crystal; T: target; G: flat field grating. Reproduced from [19] with permission from the American Institute of Physics

inserted inside the vacuum chamber along the propagation axis of the femtosecond driving pulses. The conversion efficiency of second harmonic radiation was 4%.

In the case of the application of 35-fs heating pulses, the beam was split after the compression stage. In that case, the same delay time between heating and driving pulses (70 ns) was used. During these experiments, the chirped pulses for HHG in LIPs were also used. To control pulse duration the optical autocorrelator for measuring the variable temporal parameters of the ultrafast pulses was applied. The vacuum in the target and XUV spectrometer chambers was kept at around 2.5 $\times$ 10^{-4} Pa. The generating short-wavelength spectrum was analyzed by an XUV spectrometer.

The pellet with 20 mm length and 7 mm diameter was used for the preparation of the target from powders. For the preparation of the pressed targets, which were composed of Pr_6O_{11} and Tb_4O_7 powders, the hydraulic press with 5 MPa pressure was applied. The prepared pressed tablets had a 7 mm diameter and were 5 mm thick. The elemental composition of the pellet targets was characterized using energy-dispersive X-ray spectroscopy. The oxides of lanthanides had the weight ratio of 79:21 for Pr and O and 84:16 for Tb and O respectively. Other targets were La and Yb plates cut with sizes of $5 \times 5 \times 1$ mm^3.

Figure 7.7a, b shows the harmonic spectra generated in the LIPs containing lanthanide oxides using the SCP scheme (800 nm). For comparative analysis of harmonic generation in lanthanides-contained plasmas, the 200 ps and 35 fs heating pulses were used. The harmonics of 35-fs, 800-nm driving pulses were observed from both plasmas, while higher yield was obtained in the case of the plasma produced by 35 fs heating pulses with regard to the HHG in the plasma produced by 200 ps heating pulses.

The intensity ratio of 19th harmonics (H19) generated from the plasmas excited using femtosecond and picosecond pulses was equal to 7 in the case of ablated Pr_6O_{11} tablet (Fig. 7.7a) and 5 in the case of Tb_4O_7 plasma (Fig. 7.7b). These ratios varied between 1.5 and 20 for the harmonics in the plateau region, thus demonstrating the advantages in the application of the shorter pulses for the formation of LIPs allowing

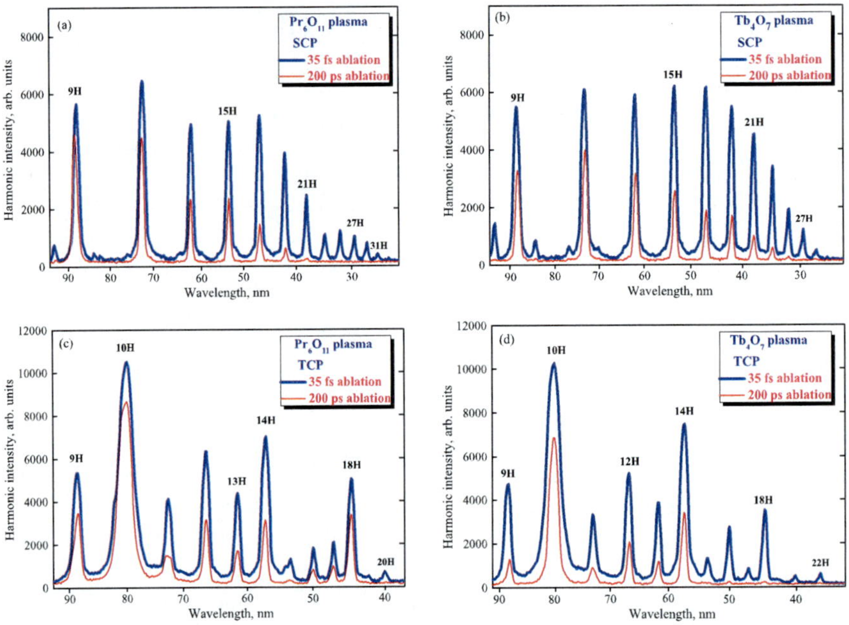

Fig. 7.7 Harmonic spectra generated in the plasmas containing **a** Pr_6O_{11} and **b** Tb_4O_7 and produced by 200 ps and 35 fs heating pulses at the wavelength of 800 nm (SCP scheme). **c**, **d** show the harmonic spectra generated in the Pr_6O_{11} and Tb_4O_7—containing LIPs produced by 200 ps and 35 fs heating pulses, respectively, and using the TCP scheme (800 nm + 400 nm). Reproduced from [19] with permission from the American Institute of Physics

higher conversion efficiency. The highest harmonic orders (H33 and H19) with cutoff energies of 51.1 and 29.4 eV in the case of femtosecond and picosecond ablation of Pr_6O_{11} target were achieved. In the case of Tb_4O_7 plasma, the harmonic cutoff energies of 48 eV (H31) and 38.7 eV (H25) were observed in the case of ablation using 35 fs and 200 ps heating pulses, respectively.

Figures 7.7c, d show the harmonic spectra generated using TCP of lanthanide oxide plasmas. The harmonic yields generated in that case were 1.5 to 5 times larger with regard to SCP depending on the order of harmonic. It was noticed that these spectra were obtained using the low conversion efficiency of second harmonic radiation (~4%) and small ratio of the energies of 400 and 800 nm orthogonally polarized pulses entering the plasma area (1:25). The harmonic cutoffs were decreased in the case of TCP (Fig. 7.7c, d) as compared to SCP (Fig. 7.7a, b) due to the application of a shorter wavelength component during the pump of plasma.

Below we analyze the application of the positively and negatively chirped femtosecond pulses for HHG in the lanthanide-containing LIPs. The variation of the chirp of driving laser radiation by adjusting the separation between two gratings of the pulse compressor allows compensating for the influence of the dynamic chirp of this pulse [20–22].

The pulse duration was properly adjusted between the minimal (35 fs) and maximal (130 fs) values of pulse duration used in these studies. Further increase of pulse duration by an additional separation between the gratings in the compressor led to a decrease of the fundamental intensity without modifying the chirp of the driving radiation. This was followed by a decrease of the cutoff energy without considerable modification of the brightness of the highest order harmonics. Figure 7.8 presents the dependences of harmonic intensity and cutoff energy on the chirp of fundamental driving pulses during TCP of two lanthanide LIPs. It was shown that the negatively chirped pulses caused a larger decrease of harmonic intensity and cutoff energy with regard to the application of the positively chirped pulses.

The relatively strong 10th harmonic was obtained in the Pr_6O_{11} plasma produced by femtosecond and picosecond heating pulses during propagation of both chirp-free 35 fs pulses, or positively and negatively chirped 70 and 130 fs pulses (Fig. 7.8a, b). The anticipated decrease of harmonic cutoff was observed in the case of both positively and negatively chirped pulses propagated through the plasma produced by 200 ps heating pulses (Fig. 7.4a) due to a decrease of the intensity of driving pulses, while similar experiments in the plasma produced by 35-fs pulses showed lesser variations of the harmonic cutoff for different driving pulses (Fig. 7.8b).

Figure 7.9a, b show the harmonic spectra generated in the Tb_4O_7 plasma produced by 200 ps and 35 fs heating pulses. In that case, similar features of the variable cutoffs and harmonic yields like in the case of the Pr_6O_{11} plasma were observed. The

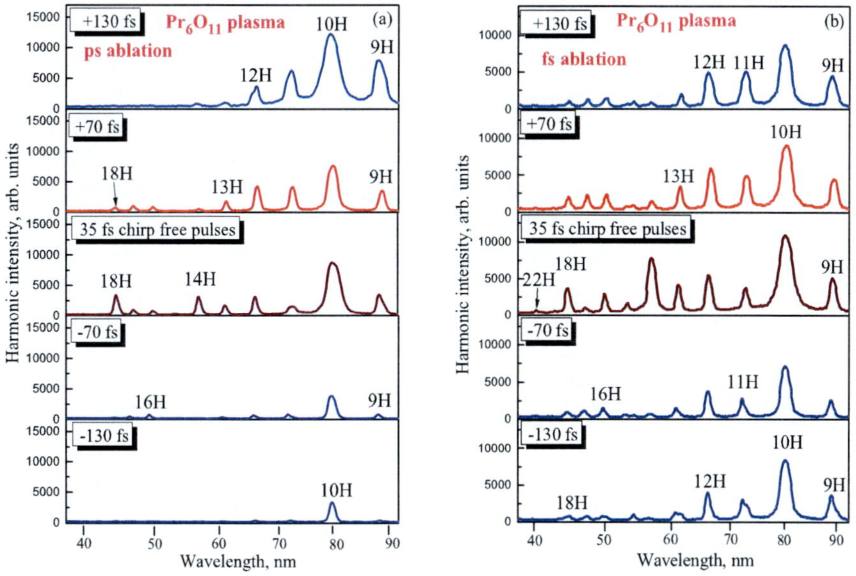

Fig. 7.8 Harmonic spectra generated in Pr_6O_{11} plasmas produced by **a** 200 ps and **b** 35 fs heating pulses during TCP using chirped and chirp-free IR pulses. Positive and negative values of pulse durations correspond to the positively and negatively chirped pulses. Reproduced from [19] with permission from the American Institute of Physics

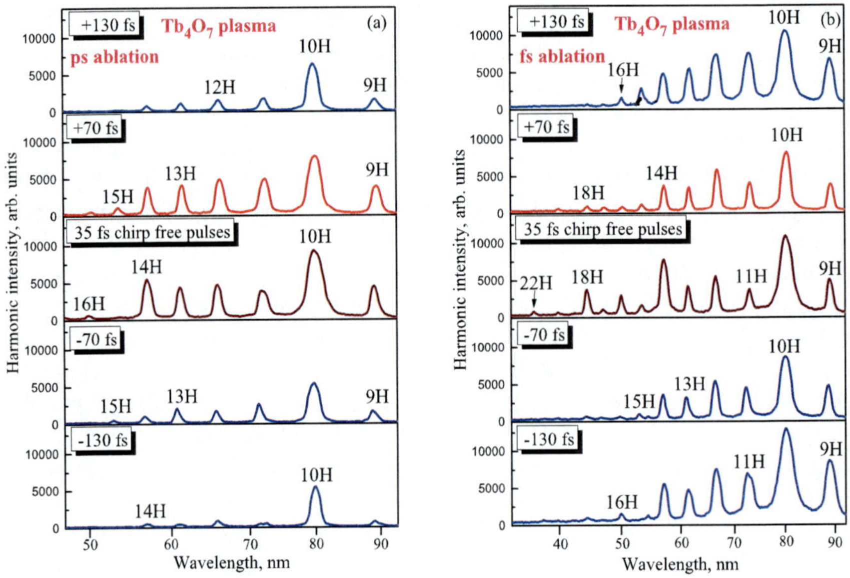

Fig. 7.9 Harmonic spectra generated in Tb_4O_7 plasmas produced by **a** 200 ps and **b** 35 fs heating pulses during TCP using chirped and chirp-free IR pulses. Reproduced from [19] with permission from the American Institute of Physics

decrease of harmonic yield can be induced, apart from other factors, by the influence of plasma concentration. Probably, in the case of femtosecond pulses induced LIP, the plasma concentration was larger at optimal conditions of HHG with regard to the picosecond pulses induced LIPs. At these conditions, the variation of the chirp of driving pulses leading to the decrease of their intensity was less influential due to some saturation of HHG at chirp-free conditions. In that case the application of the former plasma and chirped pulses led to a lesser decrease of harmonic yield compared with the picosecond plasma. The mid-order harmonics showed approximately the same brightness in the cases of chirped and chirp-free pulses. At longer distances between the gratings (i.e., far from the chirp-free position, when the absolute value of the chirp does not further change, but only the pulse duration increases), the decrease of fundamental intensity leads to the decrease of harmonic emission. This feature was clearly seen in the case of the negatively chirped 130 fs pulses propagating through the two studied plasmas.

Ablation of bulk La and Yb metals by 370 fs, 800 nm pulses allowed the formation of LIPs suitable for harmonic generation using 64 fs pulses. In Fig. 6a, the upper panel shows the spectral range close to the harmonic cutoff in the case of HHG in La plasma (H31, $E = 48$ eV, $I_{fs} = 2 \times 10^{14}$ W cm^{-2}, $I_{ps} = 1 \times 10^{10}$ W cm^{-2}) using SCP scheme. The growth of heating pulse fluence and driving pulse intensity did not increase the cutoff energy of the harmonic spectrum, thus pointing out the saturation of HHG. Stronger heating of La target caused the emission of intense 9.1 nm radiation. The bottom panel of Fig. 7.10a shows the appearance of a strong

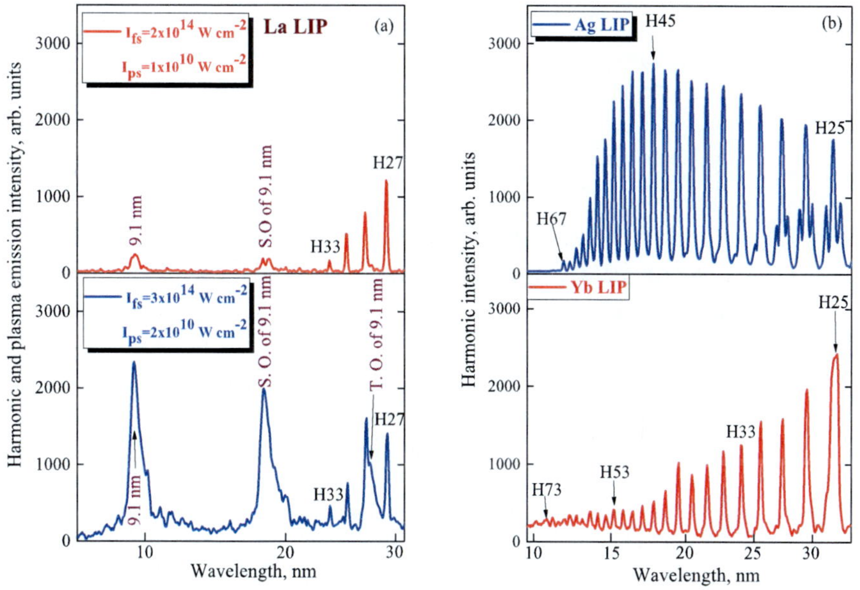

Fig. 7.10 **a** Harmonic and plasma emission spectra from La LIP in the case of optimal ($I_{ps} = 1 \times 10^{10}$ W cm^{-2}, upper panel) ablation and stronger ablation ($I_{ps} = 2 \times 10^{10}$ W cm^{-2}, bottom panel) of target by 370 ps pulses. **b** Comparative spectra of harmonics generated in silver (upper panel) and ytterbium (bottom panel) LIPs. Reproduced from [19] with permission from the American Institute of Physics

emission line (9.1 nm) and its second and third-order ghosts, alongside the moderate continuum and harmonic emission at $I_{fs} = 3 \times 10^{14}$ W cm^{-2}, $I_{ps} = 2 \times 10^{10}$ W cm^{-2}). The application of TCP decreased the harmonic cutoff.

Analogous dependences were observed in the case of ablation of the Yb bulk target. The increase of heating pulse fluence caused the growth of incoherent emission and a decrease in harmonic yield. Meanwhile, at optimal conditions of Yb plasma formation, the HHG in this medium showed the highest harmonic cutoff (H73) among other studied lanthanides (Fig. 7.10b, bottom panel). For comparison, the harmonic distribution generated in silver plasma is shown (Fig. 7.10b, upper panel). The latter plasma is one of the best media for harmonic generation in the shorter-wavelength range of XUV. One can see that the harmonic cutoff in the case of Yb plasma was larger than in the case of Ag plasma (H67, Fig. 7.10b, upper panel).

Regarding the plasma concentration, one has to estimate the concentration of neutrals, ions, and electrons to define their influence on the process of HHG. To understand this phenomenon, the simulations using the hydrodynamic code HYADES have been reported [23]. The expansion of a silver slab interacting with a laser pulse was simulated and then the electron density, ionization level, and ion density were determined as the functions of the heating pulse intensity at a fixed distance from the target surface. Based on the above assumptions and taking into account the studies reported in [23, 24], the plasma density (2×10^{17} cm^{-3}) was characterized in the case of picosecond pulses induced ablation of solid lanthanides at the moment of propagation of the driving beam through the plasma cloud at the conditions of highest HHG efficiency.

7.3 High-Order Harmonic Generation During Different Overlaps of Two-Colored Pulses in Laser-Produced Plasmas and Gases

A consistent theoretical study of the nonlinear interactions between complex structures (extended plasmas, multi-jet plasmas, quantum dots, etc.) and multi-component laser fields constitute a challenge for modern theoretical approaches. In the course of such studies, it is not enough to solve the problem at the microscopic (atomic) level of interaction [25, 26], but it is critical to consider the peculiarities of the laser field propagation in the media.

In this section, we analyze the conditions when a small difference in the thickness of two barium borate (BBO) crystals used for the formation of TCP can dramatically change the harmonic output and distribution regardless of other experimental conditions (choice of nonlinear medium, pulse duration and polarization of 800 nm or 1030 nm pulses, plasma structure, emitter sizes, and the effect of ion transitions) [27]. These results are confirmed and explained by the non-perturbative theory of single atoms and the modeling of interference media.

The HHG scheme was analogous to those used in the above-described experiments. Silver, gold, manganese, chromium, carbon, indium, antimony, cadmium, zinc plates ($5 \times 5 \times 1$ mm^3), as well as some nanoparticles and quantum dots, were used as targets for ablation and laser-produced plasmas formation. The pressure in the chamber was ~3.5×10^{-3} mbar during the experiments, while the XUV spectrometer chamber was maintained at ~3×10^{-5} mbar. A gas jet with a circular nozzle of 1.5 mm diameter was used as a target. The inner diameter of the gas nozzle (1 mm) was considered to be the effective length of the nonlinear media. The pressure of gases was maintained at ~70 mbar at the area of interaction with femtosecond pulses.

7.3.1 Analysis of Different Temporal Overlaps of Driving Pulses in Laser-Produced Plasmas and Gases

In the case of TCP containing 806 and 403 nm waves, the IR pulse was delayed by 38 and 76 fs relative to the second harmonic pulse when using a 0.2 or 0.4 mm crystal due to the group velocity dispersion in BBO crystal. The mismatch between the driving pulses in the laser-plasma region did not prevent them from being converted to odd and even harmonics in the case of a 0.2 mm thick crystal. Meanwhile, in the case of a 0.4 mm crystal, the HHG spectrum consisted of odd harmonics of ω and 2ω waves, while the even harmonics determined by the ratio $H = 4(n + 1)$ were considerably weaker due to the insufficient overlap of 800 and 400 nm waves in the plasma region. At the same time, the appearance of all even harmonics in the case of the thinner (0.2 mm) crystal revealed sufficient overlap of 800 nm and 400 nm pulses in the laser-produced plasmas.

Figure 7.11 shows a few examples of harmonic spectra confirming the above statement. The application of 0.2 mm thick BBO allowed the generation of enhanced harmonics from cadmium plasma (Fig. 7.11a). One can see a gradual decrease of odd and even harmonics. The former harmonics were approximately equal to each other and showed decay in the harmonic yield towards the shorter-wavelength region. Lower orders of harmonics generated using TCP were two to three times higher than those using SCP. Even a higher ratio between the lower-order harmonics induced by TCP and SCP was observed in carbon plasma at similar conditions of experiments (Fig. 7.11b). The HHG efficiency, in that case, was ~10 times higher when the 0.2 mm BBO was for second field formation, though the harmonic cutoff was smaller in the case of TCP of C LIP compared with the SCP case. One can again see the equality in odd and even harmonics through the whole range of harmonic emission.

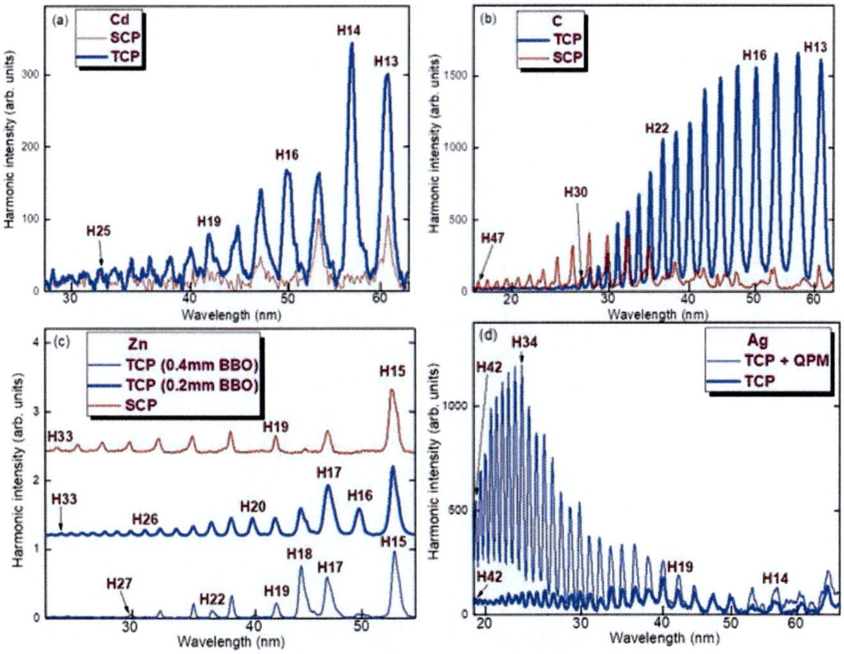

Fig. 7.11 Harmonic spectra from different plasmas in the case of SCP and TCP. **a** cadmium LIP, **b** carbon LIP, **c** zinc LIP, **d** extended silver LIP. Reproduced from [27] with permission from the European Physical Journals

To distinguish the HHG spectra in the case of different overlaps of driving pulses in the plasma area, comparative studies using 0.2 and 0.4 mm thick crystals were carried out. Figure 7.11c shows the results of these studies when, alongside the SCP-induced spectra (upper red panel), two groups of harmonic emission using TCP (middle and bottom panels corresponding to the second harmonic generation in 0.2 and 0.4 mm thick BBO crystals) are presented. The intensities of harmonics were normalized to 1 for the strongest lowest-order harmonic. SCP-induced HHG spectrum represents a plateau-like distribution of harmonics up to the 39th order (not shown in the upper curve). TCP in the case of 0.2 mm thick crystal also showed almost the plateau-like shape of harmonics distribution up to the 35th harmonic (middle panel). Meanwhile, the application of 0.4 mm BBO caused a drastic change in the harmonics distribution. In spite of the higher energy of the second wave, the harmonic distribution clearly shows the insufficient overlap between the 806 and 403 nm pulses in the plasma area resulting in the generation of 2(2n + 1) harmonics (H18, H22), while other even harmonics corresponding to 4(n + 1) orders (H16, H20, H24) were significantly suppressed.

These studies showed that the efficiencies of even and odd harmonics were comparable only in the case of a 0.2 mm thick BBO crystal. The harmonic cutoff for TCP was decreased compared to SCP of laser-produced plasmas, and the efficiency of HHG conversion in the former case was increased. This phenomenon was observed in almost all other comparative experiments conducted under similar conditions. Once the 0.4 mm BBO has been inserted into the target chamber in the path of the 800 nm radiation (bottom panel of Fig. 7.11c), the even harmonics corresponding to the odd harmonics of the second harmonic wave, i.e. the 9th and 11th harmonics of 400 nm radiation, which correspond to H18, H22 waves of 800 nm pulses, were dominant in the XUV region. One can see the almost complete absence of 4(n + 1) harmonics due to the aforementioned delay between the two driving fields. The 800 nm wave barely overlaps with the 400 nm wave in the plasma region, while each generates its own harmonics. Some slight overlap between these two pumps still exists, and weak 4(n + 1) harmonics can be generated under these adverse conditions.

The efficiency of harmonics produced by shorter wavelength sources becomes higher due to the strong wavelength-dependent harmonic yield ($I_{\text{harm}} \propto \lambda^{-5}$) ($I_{\text{harm}}$ is the harmonic intensity and λ is the driving field wavelength [28]). The $I_{\text{harm}} \propto \lambda^{-5}$ rule leads to a significant increase of harmonic yield in the case of shorter-wavelength sources compared with the 806 nm pump. The application of twice smaller wavelength of driving pulses at other similar conditions should result in a 32-fold growth of harmonic emission. This rule becomes diminished due to the smaller intensity of the second harmonic field. Nevertheless, in many cases, the intensity of even harmonics becomes stronger than those of odd harmonics. There is an explanation stipulating that a strong harmonic generation in the case of the two-color pump is possible due to the formation of a quasi-linear field, selection of a short quantum path component, which has a denser electron wave packet, and higher ionization rate compared with the SCP. With suitable control of the relative phase between the fundamental and the second harmonic radiation, the latter field enhances the short path contribution, resulting in a clean spectrum of harmonics.

Stronger excitation of the targets can increase the number of harmonic emitters and thus compensate for, to some extent, the weak interaction between the two waves. However, under these conditions, the large concentration of free electrons in LIPs strongly affects the phase matching between the driving and harmonic waves. Phase mismatch under these conditions plays a crucial role in suppressing the entire coherent extreme ultraviolet radiation generation. In addition, incoherent plasma emission becomes noticeably stronger than the expected harmonics.

To overpass this obstacle, one can use the quasi-phase-matching concept [29–31] to diminish the phase mismatch in these conditions. Application of multi-jet plasma instead of extended unstructured laser-produced plasmas allows quasi-phase-matching conditions to be achieved using SCP pump [32], with emphasized growth of the group of harmonics in the short-wavelength region of extreme ultraviolet. Here we analyze the results of the quasi-phase-matching experiments at the conditions of TCP conducted using a multi-slit mask consisting of 0.2 mm long slits separated from each other by a distance of 0.2 mm and placed between the heating pulses focused by cylindrical lens and silver target.

The thick curve in Fig. 7.11d shows the almost homogeneous distribution of odd and even harmonics generating in the multi-jet Ag LIP in the case of using 0.2 mm BBO. The insertion of the multi-slit mask on the path of the heating pulse drastically changed the harmonic distribution. A thin blue curve shows the TCP-induced harmonic spectrum generated in perforated Ag plasma produced using a multi-slit mask. The group of harmonics centered around H34 was significantly stronger compared with the case of extended homogeneous plasma.

The comparative SCP and TCP experiments were also carried out using nanoparticles- and quantum dot-containing plasmas. Similar features of harmonics enhancement were observed in the case of the application of two overlapping fields. The conversion efficiency of even harmonics in the case of TCP was higher than of odd harmonics in the case of SCP, despite the low pulse energy of the 403 nm radiation with respect to the 806 nm one. Even at a very small ratio of 403 and 806 nm pulse energies, the influence of the weak field was sufficient to strongly modify the whole harmonic spectrum from multiparticle plasmas compared to the single color pumping case.

Notice the relatively weaker $4(n + 1)$ orders compared to $2(2n + 1)$ ones in the case of using a thicker nonlinear crystal for second harmonic generation. Similar to the above cases with the plasmas produced on the bulk targets, this heterogeneity in the even harmonic yields was caused by group velocity dispersion in the 0.4 mm thick BBO crystal, which causes a notable delay between the 400 and 800 nm pulses at the output of the crystal in the case of 50 fs pulses. Only a small part of the second harmonic overlapped with the driving fundamental 806 nm pulse. This causes a further decrease in the influence of the second harmonic field on the whole pattern of harmonic distribution. On the other hand, this second harmonic pulse generated relatively strong low-order odd harmonics.

The application of a 100 kHzlaser for TCP in plasma also led to significant growth of harmonic yield compared to SCP. The role of the second field in the variations of the harmonic intensity generated in the plasmas produced on rotating targets was analyzed during those discussed studies. BBO crystals with different thicknesses and rotating the solid targets at different velocities were used to optimize the harmonic yields. Two silver samples were used: a silver bulk rod and thin silver film wrapped around a stainless steel rod to analyze whether heat transfer inside the target depends on the thermal conductivity of samples. SCP of the plasma produced on the bulk rod rotated at 30 rpm led to the generation of harmonics only up to the 17th order. The application of 0.1 mm thick BBO allowed the generation of a few even orders during the TCP of silver plasma (up to H14). Though the overlap between 1030 and 515 nm pulses in the area of LIP was perfect in the case of using this thin crystal, the small ratio of the amplitudes of overlapped 515 nm and 1030 nm pulses did not allow the generation of strong odd and even harmonics.

The efficiency of HHG from the 1030 nm pump was strongly enhanced in the case of 0.4 mm thick BBO. A significant improvement of odd and even harmonics compared with the application of 0.1 mm thick BBO was observed. SHG efficiencies in these comparative experiments were 1.5% (in the case of 0.1 mm thick BBO) and 4.7% (in the case of 0.4 mm thick BBO). Harmonics up to the 28th order were routinely obtained during this set of studies. The application of thicker crystal for SHG can cause a temporal separation of two pumps (fundamental and second harmonic waves) due to the group velocity dispersion in the BBO crystal. This separation did not allow efficient generation of even harmonics in previous studies using 800 nm fundamental radiation from Ti: sapphire lasers. Meanwhile, in the case of longer wavelength driving sources, like Yb-doped fiber laser, the difference in group velocity dispersion of ω and 2ω waves was smaller, which allowed the application of thicker crystal to produce a larger amount of second harmonic photons while maintaining sufficient overlap with the fundamental radiation in the area of LIP.

Similar to the case of metal plasmas, the weak orthogonally- polarized second field did not improve the whole yield of harmonics and allowed the generation of only a few even harmonics, which were notably weaker than the odd ones. However, even at these unfavorable conditions, the HHG yield of the lower order harmonics was two to four times stronger compared with metal LIPs. The reasons for the appearance of enhanced harmonics in the case of carbon LIP could be attributed to the formation of the nanoparticles in plasma, which increased the recombination cross-section for the accelerated electrons and correspondingly amended the conditions of harmonic generation. The presence of nanoparticles during these experiments was confirmed during an analysis of the debris from ablated graphite deposited on the nearby substrates.

Figure 7.12 shows the raw images of harmonic spectra in the case of SCP and TCP (using two crystals) of gold LIP. The raw images are presented for a better vision of the difference in harmonics distribution. One can clearly see the enhanced harmonics in the case of the 0.4 mm crystal (bottom panel). The even harmonics were equal to the odd ones, contrary to the case of using 0.1 mm (middle panel), which points out the good overlap between the pump pulses in LIP. Notice the increase of cutoff energy in this case compared with SCP of Au plasma.

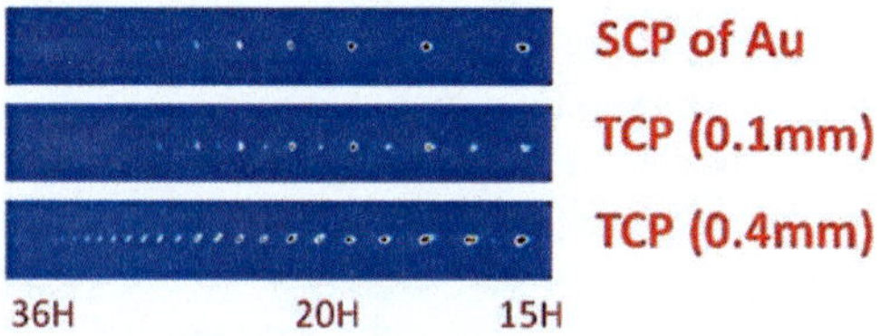

Fig. 7.12 Raw images of harmonic spectra from gold plasma in the case of single- color pump (1030 nm, upper panel), two-color pump using 0.1 mm BBO (1030 nm + 515 nm, 60:1 pulse ratio, second panel), and two-color pump using 0.4 mm BBO (1030 nm + 515 nm, 20:1 pulse ratio, bottom panel). Reproduced from [27] with permission from the European Physical Journals

As in the case of the metal plasma pumped by 806 nm + 403 nm pulses, a weak orthogonally-polarized second field (515 nm) did not improve the entire output of harmonics from carbon plasma and allowed only a few even harmonics to be generated in the case of using 0.1 mm BBO, which were noticeably weaker than the odd ones. However, even under these adverse conditions, the output of lower-order harmonics from carbon plasma was two to four times stronger than the case of using 806 nm + 403 nm pulses for metal LIPs. The reasons for the appearance of stronger harmonics in the case of carbon LIP can be explained by the formation of nanoparticles in the plasma, which increased the recombination cross-section for accelerated electrons and, accordingly, changed the conditions for generating harmonics. The presence of nanoparticles in these experiments was confirmed by analyzing fragments of ablated graphite deposited on nearby substrates.

Argon gas, being pumped by 806 nm + 403 nm pulses, demonstrated similar characteristics of the generated harmonic spectrum. The generated even harmonics were noticeably stronger compared to the odd harmonics in the case of TCP using 0.2 mm thick BBO. These experiments were conducted using a narrow gas jet from a stainless steel needle with an internal diameter of 0.3 mm. Similar studies using 1030 nm + 515 nm pump of molecular gases, showed the advantages of applying thicker BBO (0.4 mm) compared to the thinner one (0.1 mm). The dependence of highest order harmonic on the thickness of nonlinear crystal was analyzed at $I_{\omega} = 1.4 \times 10^{14}$ W cm^{-2} and $I_{2\omega} = 1.3 \times 10^{13}$ W cm^{-2} intensities of fundamental radiation and its second harmonic, respectively, when a change in the highest order of harmonics in N_2 and O_2 gases is observed. The highest harmonic order (H41) was observed in N_2 gas. Using TCP with 0.1 mm thick BBO led to an increase of the highest order harmonic (H46) compared to the SCP at the same intensity of the fundamental driving pulse. One can get only $2(2n + 1)$ orders in the case of TCP of gases when the temporal overlap of these pulses becomes insufficient. In this case, the odd orders of the 2ω pump dominantly appear in the harmonic spectra. This feature was demonstrated by using BBO crystals of different thicknesses when the temporal overlap of two laser pulses becomes worsened while using the thick crystal. Particularly, in the case of a 1.5 mm thick crystal, the lower intensities of some $4(n + 1)$ orders (H12, H16, H20) and stronger $2(2n + 1)$ orders of the second harmonic pump (H14, H18, H22) were observed. Thus the imperfect temporal overlap of two pumps in the gas medium leads to a decrease of mixing efficiency of two waves resulting in weaker 4(n + 1) harmonics compared to the 2(2n + 1) orders, similar to the above-described experiments of harmonics generation in plasmas.

The evidence of the spatial overlap of the pump beams in the gas and plasma media is the symmetrical distribution of the odd and even harmonic orders with respect to the axis of XUV radiation dispersion. The raw images of the harmonic spectra also showed the almost perfect spatial overlap of two laser beams in the area of interaction with gas atoms resulting in almost equal groups of odd and even harmonics positioned on the same axis of dispersion of the flat field grating. Notice that the application of BBO on the path of focused radiation inside the vacuum chamber diminished the separation of the two pumps.

7.3.2 Resonance-Induced Enhancement of Harmonics During SCP and TCP of Mn, Sb, Cr, and in LIPs

Most of the previous studies related to the resonance-induced processes during HHG in gases and plasmas were carried out using SCP [33–38]. Here we analyze strong enhancement of single even and odd harmonics, which is attributed to the influence of different transitions of ions in the increase of harmonic efficiency.

TCP at the conditions of relatively weak and strong overlap of two driving fields in plasmas allows determining the prevalence of resonance processes over temporal overlapping of two driving fields. The example of such interaction of two fields is shown in Fig. 7.13a where the harmonic spectra in the case of SCP (bottom red curve) and TCP at the conditions of weak overlap (upper blue curve) of 806 and 403 nm pulses in manganese plasma are shown. Manganese plasma is distinguished among other plasmas to produce resonance-induced enhancement of a harmonic centered at around 50–51 eV and arising from intense $3p \rightarrow 3d$ inner-shell transitions in the photoionization cross-sections of those ions [39–41]. The enhanced order in Mn plasma corresponded to the 31st harmonic of 780 nm laser ($\lambda = 25.16$ nm) [40]. Obviously, the use of another driving wavelength allows for observing another harmonic order. In the discussed case, the enhancement was observed in the case of 33rd order of 806 nm radiation ($\lambda = 24.42$ nm; bottom red curve). The application of 0.4 mm BBO for second harmonic generation caused insufficient overlap of two waves in the plasma area. In that case, the 2(2n + 1) orders (H22, H26, H30) dominated in the longer-wavelength harmonic spectra (upper blue curve). Meanwhile, the following harmonic (H34) showed exceptional amplification among other harmonics. The next orders (H38, H42) were also stronger than the odd harmonics. Thus the resonance enhancement in the vicinity of $3p \rightarrow 3d$ transitions allowed observation of the even harmonic (H34), which a few times exceeded the enhanced 33rd harmonic.

Antimony plasma has also demonstrated the enhancement of single (H21) harmonic (Fig. 7.13b, H21, bottom red curve) in the case of SCP. In the case of TCP, when thinner (0.2 mm) crystal was used for the formation of second pump

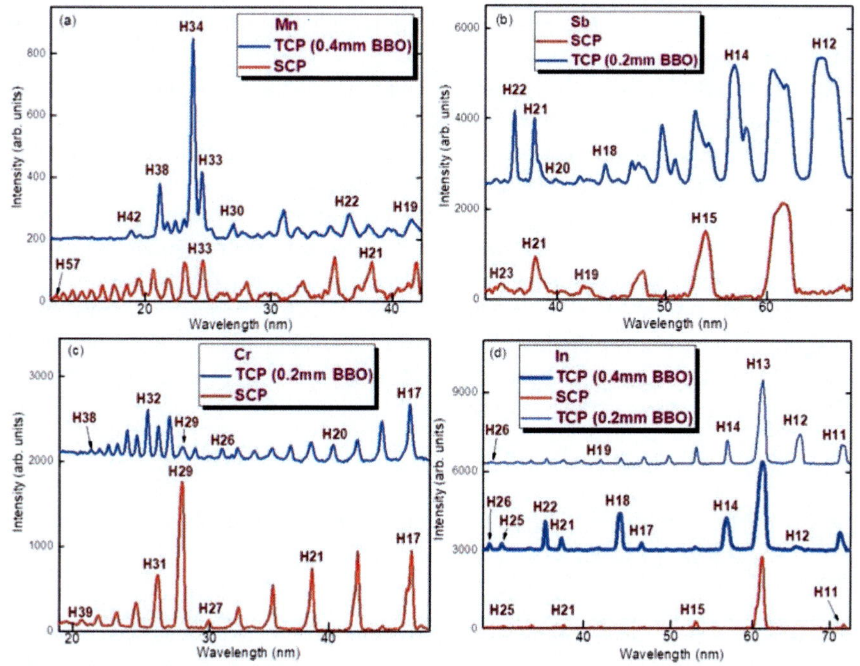

Fig. 7.13 Harmonic spectra using SCP and TCP of different plasmas demonstrating resonance enhancement of harmonics. **a** Mn, **b** Sb, **c** Cr, **d** In. Reproduced from [27] with permission from the European Physical Journals

wave, both odd and even harmonics were approximately equal to each other (upper blue curve), except for the 22nd order ($\lambda = 36.6$ nm) located in the vicinity of the strong radiative transition of the Sb II ions ($4d^{10}5s^22p^3P_2$-$4d^95s^25p^3(^2D)^3D_3$ and $4d^{10}5s^22p^1D_2$-$4d^95s^25p^3(^2D)^3F_3$ at the wavelengths of 37.82 nm and 37.55 nm, respectively, have been reported and analyzed [42]. Thus the enhancement of H22 cannot be attributed to belonging to the $2(2n + 1)$ group of even harmonics but rather to the influence of strong ionic resonance of antimony.

The application of thinner crystal in the case of HHG in chromium plasma also led to the gradual decrease of approximately equal odd and even harmonics (Fig. 7.13c, upper blue curve) until the region when strong single harmonic emission (H29) was observed in the case of SCP (Fig. 7.13c, bottom red curve). Previous studies of the photoabsorption and photoionization spectra of Cr plasma in the range of 41–42 eV [16] have demonstrated the presence of strong transitions, which could be responsible for a modification of the harmonic spectrum in the wavelength region of 29.5–31 nm. The region of 'giant'$3p \rightarrow 3d$ resonances (44–45 eV) of Cr II spectra was analyzed in [43] and the strong transitions that could enhance the nonlinear optical response of the plume were revealed. Notice the decrease of the enhancement factor of H29 in the case of TCP. The following harmonics in the upper curve also demonstrate the equality of odd and even orders, thus confirming the good overlap of two pump pulses in Cr plasma.

Two TCP-induced harmonic spectra from indium plasma distinguished from each other in the case of using thick and thin BBO crystals are shown in Fig. 7.13d. The thick and thin blue curves correspond to the use of thick and thin crystals for the formation of a 403 nm field. The Upper blue curve shows the equal odd and even harmonics along the whole range of harmonics generation, except for the H13 (λ = 62 nm), which was enhanced using both SCP (bottom red curve) and TCP (two other curves) due to the influence of strong ionic transition ($4d^{10}5s^2\,{}^1S_0$-$4d^95s_25p$ $(^2D)\,{}^1P_1$) transition of In II at the wavelength of 62.2 nm, which has an oscillator strength of 1.11 [44]. Meanwhile, the middle curve demonstrates the preferential enhancement of the group of 2(2n + 1) even harmonics due to the poor overlap of 806 and 403 nm waves in the indium plasma.

7.3.3 *Calculations of Harmonic Spectra Generated in Gases and Plasmas Using SCP and TCP*

To verify such unpredicted spectrum obtained in TCP in argon gas (i.e. presence of more strong even harmonics in comparison with the odd ones at the given ratio of ω/2ω intensities) this interaction was simulated using non-perturbative quantum mechanical theory [26] and the interference model [45].

Figures 7.14a, b show the spectra of SCP and TCP of argon atoms calculated quantum-mechanically for the experimental conditions (5×10^{14} W cm^{-2}, 800 nm, 50 fs for SCP in addition to ~2% of 400 nm wave delayed by 38 fs due to the group velocity dispersion in 0.2-mm thick BBO used for TCP). The calculations showed the harmonic spectrum in the case of SCP demonstrating the presence of odd harmonics (Fig. 7.14a), while the spectrum calculated for TCP demonstrated the presence of odd

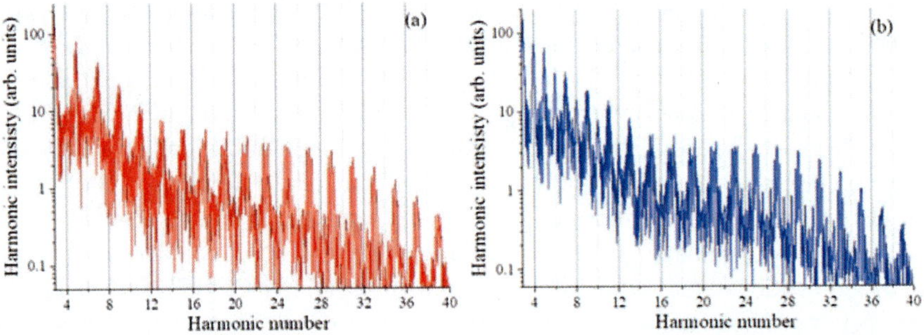

Fig. 7.14 **a** Calculated HHG spectrum from Ar obtained for the SCP. **b** Calculated HHG spectrum from Ar obtained for the TCP. The harmonic intensity axes are presented in logarithmic scale. Reproduced from [27] with permission from the European Physical Journals

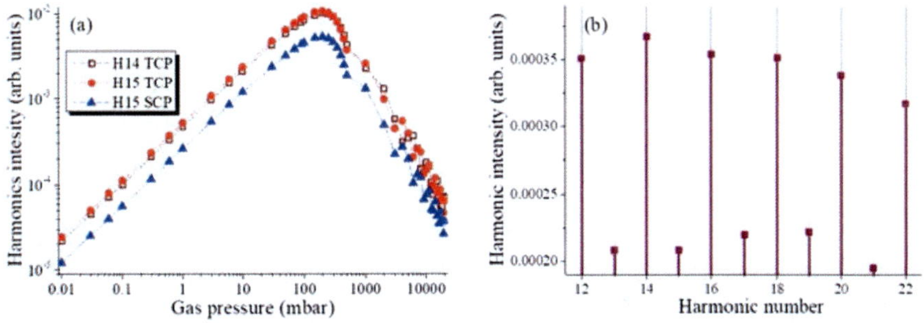

Fig. 7.15 **a** Dependence of the harmonics efficiency as a function of Ar gas pressure calculated for the 25 mm long gas jet interacting with TCP for H14 (black curve with squares) and H15 (red curve with circles). The dependence of the H15 efficiency as a function of gas pressure (blue curve with triangles) calculated at the same cases but for SCP is presented for comparison. The harmonic intensity axis is presented in logarithmic scale. **b** Harmonic spectra of extended gas response calculated for the case of higher than optimal Ar gas pressure for TCP case. Reproduced from [27] with permission from the European Physical Journals

and even harmonics (Fig. 7.14b), especially for the low-order part of the spectrum. The latter is due to the weak overlap of ω and 2ω pulses.

Then the response from 0.25 mm long gas jet was studied under the action of TCP. The impact of the free electrons and neutrals was calculated in accordance with Ref. [46]. Figure 7.15a demonstrates the dependences of the harmonics efficiency as a function of gas pressure. It is clearly seen that there is a specific value for the pressure to optimize the generation of a given harmonic, and this "optimal" pressure depends on the number of the harmonic [45]. The difference between the two above figures is related to the response from a single atom (Fig. 7.14) determined through the quantum mechanical calculations and from the group of atoms (Fig. 7.15), which is determined by the interference model.

At the same time, the values of odd harmonics around this "optimal" pressure are higher than the even ones, which is in contradiction with experimentally measured behavior. This can be due to the fact that experimental conditions were far from optimal. Indeed, for higher than "optimal" pressure there are some cases when even harmonics become stronger than the odd ones (see Fig. 7.15a where only one odd (H15) and one even (H14) dependencies are presented for simplicity). Spectrum calculated for the case of such range of gas pressure is presented in Fig. 7.15b. It is clearly seen that the even harmonics are stronger than the odd ones. An additional destructive factor, which can also break the "traditional" ratio between odd end even harmonics efficiencies, can be the use of relatively high intensities of the laser field used in experiments. This factor increases the value of atomic ionization and increases the destructive role of electrons in the harmonics phase matching. Here the 98:2 ratio of the intensities of 800 and 400 nm waves was used.

Using the quantum–mechanical theory with the interference model the silver plasma response on the action of SCP was simulated (Fig. 7.16a), TCP for the case of imperforated plasma (Fig. 7.16b), and perforated plasma consisting of 5 jets 0.5 mm

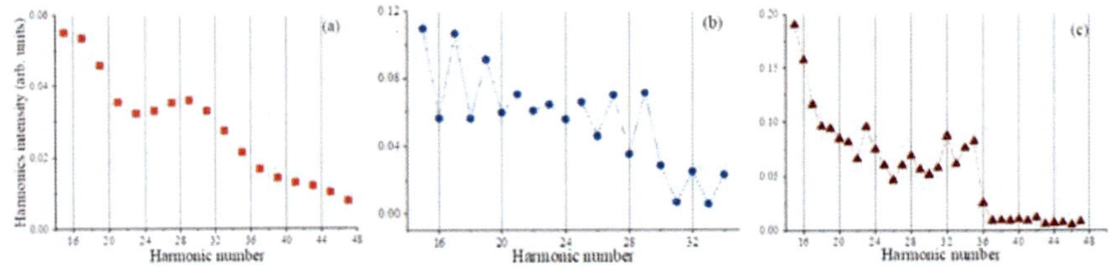

Fig. 7.16 The calculated response of silver plasma with **a** SCP, **b** TCP in the case of 0.2-mm thick BBO converter and **c** TCP-induced harmonic spectrum in the case of quasi-phase matching in multi-jet plasma. Reproduced from [27] with permission from the European Physical Journals

long each (Fig. 7.16c). The response of the single Ag atoms was calculated in accordance with the methods of the quasi-phase-matching and HHG in plasma previously reported in [47]. From the quantum–mechanical calculations, the amplitudes and phases for all generated harmonics were extracted. These calculations were carried out in accordance with the variations of the laser parameters during propagation in a plasma medium when the Gouy phase and Kerr effect contributions were taken into account [48]. Then the response of the extended plasma media was calculated. The calculations carried out for the experimental parameters of laser-plasma interaction showed a good agreement with the experimental results. It is clearly seen that due to the quasi-phase-matching the part of the spectrum increases while some other harmonics generated from different atoms are interfered destructively (Fig. 7.15c).

7.4 Summary

In this chapter, we analyzed different cases of the application of a two-color pump for efficient HHG in LIPs. A search for suitable schemes for the two-color pump of different isotropic media resulting in efficient HHG is of special interest to the laser community. Originally, gases and laser-produced plasmas were the main subjects of studies for the conversion of ultrashort near-infrared laser pulses towards the XUV region. The comparative studies of carbon plasma and argon gas have shown in the past the advantages of the former medium once one became aimed in stronger harmonic emission in the 40–100 nm spectral region of XUV. Carbon plasma did not allow the generation of strong harmonic in the shorter wavelength region. A search of suitable plasma media showing preferable features compared with gases thus became an interesting topic of study. To analyze the potentially efficient plasma media some of them were chosen taking into account earlier reported data on their ability in generating higher-order harmonics, compared with carbon plasma,

We discussed stronger HHG in the case of different plasmas (Ag, Ag_2S, stainless steel) with regard to the harmonics generating in argon gas in the cases of SCP and TCP. We analyzed the harmonic yields from gas and gas + plasma media under closely matched experimental conditions. Three- to five-fold growth of harmonic yield in the case of LIP compared with the gas medium at similar experimental conditions was achieved. Those studies have shown that, in the spectral range of 15–50 nm, the used plumes are more efficient than argon. Different schemes of HHG, such as single- and two-color pumps, as well as tunable positions of LIP with regard to gas jet, were analyzed, which allowed determining the best configuration of gas + plasma applications for a harmonic generation.

We analyzed the high-order harmonic spectra produced during propagation of the femtosecond pulses through the lanthanide-contained laser-produced plasmas. The species under investigation (praseodymium, terbium, lanthanum, and ytter bium) are among the heaviest non-radioactive solid elements of the periodic table. We discussed HHG in the plasmas produced on the surfaces of these lanthanides and their oxides (La, Yb, Pr_6O_{11}, Tb_4O_7) using different techniques (two-color pump, application of chirped pulses, variation of the parameters of heating and driving pulses, etc.). The ablation of samples using femtosecond pulses has proven to be more efficient from the point of view of harmonic yield with regard to the ablation using picosecond pulses in the case of a two-color pump of plasma. It was shown that Yb plasma is the efficient medium for harmonic generation up to the 73rd order, which is one of the largest harmonic cutoffs generated in the laser-produced plasmas.

The novelty of the discussed studies is a wavelength-dependent demonstration of the prevailing role of the partial overlap of two orthogonally polarized waves in laser-produced plasmas and gases over the larger ratio of the second and fundamental photons produced in a thicker nonlinear crystal. We analyzed the influence of two pulses overlapping in a nonlinear medium on the high-order harmonics generation. The overlap of pulses strongly depended on the group velocity dispersion in the nonlinear crystal used for the formation of a second weak field. We discussed the conditions when the application of thicker crystal leading to a higher yield of second harmonic emission causes a different effect in the case of different sources of laser radiation. Even a small difference in the variations of group velocities in BBO for the 806 and 1030 nm laser sources led to the lesser influence of this effect in the case of the latter emission. Correspondingly, the delay between 806 and 403 nm pulses introduced in thick (0.4 mm) BBO resulted in the generation of 2(2n + 1) even harmonics and suppression of 4(n + 1) harmonics in spite of higher ratio of $I_{\omega}/I_{2\omega}$ compared with the case of thinner (0.2 mm) crystal. Thinner crystal was a preferable choice since it allowed the generation of almost equal odd and even harmonics.

Another pattern was observed in the case of the 1030 nm + 515 nm pump of laser-produced plasmas and gases. The less influential role of group velocity dispersion using a longer-wavelength laser allowed achieving sufficient overlap of fundamental and second harmonic pulses in the nonlinear media used for HHG while applying relatively thick (0.4 mm) BBO crystal. Best conditions of HHG were achieved using this crystal thus showing that the higher conversion efficiency towards the second harmonic ($I_{2\omega(0.4\text{mmBBO})}/I_{2\omega(0.1\text{ mm BBO})} \approx 3$) at this driving wavelength overpasses the negative influence of lesser overlap of converting pulses in plasmas and gases in the former case.

We discussed different mechanisms of harmonics enhancement in the case of 806 nm driving pulses and second harmonic pulses (quasi-phase matching and resonance enhancement of harmonics using TCP of LIPs). In the discussed studies using 806 nm radiation, the enhancement of single even harmonics was observed in Cr, Mn, and Sb plasmas. Quasi-phase matching of odd and even harmonics in multi-jet Ag plasma allowed shifting the group of maximally enhanced harmonics towards the 24 nm range (H34). Atomic Ar and molecular N_2 and O_2 showed similar features as plasmas once we considered the overlap effects in gaseous media.

References

1. R.A. Ganeev, T. Witting, C. Hutchison, F. Frank, P.V. Redkin, W.A. Okell, D.Y. Lei, T. Roschuk, S.A. Maier, J.P. Marangos, J.W.G. Tisch, Phys. Rev. A **85**, 015807 (2012)
2. Y. Pertot, S. Chen, S.D. Khan, L.B.E. Bom, T. Ozaki, Z. Chang, J. Phys. B **45**, 074017 (2012)
3. M. Wöstmann, P.V. Redkin, J. Zheng, H. Witte, R.A. Ganeev, H. Zacharias, Appl. Phys. B **120**, 17 (2015)
4. M.A. Fareed, N. Thiré, S. Mondal, B.E. Schmidt, F. Légaré, T. Ozaki, Appl. Phys. Lett. **108**, 124104 (2016)
5. R.A. Ganeev, A. Husakou, M. Suzuki, H. Kuroda, Opt. Express **24**, 3414 (2016)
6. R.A. Ganeev, M. Baba, M. Suzuki, H. Kuroda, Phys. Lett. A **339**, 103 (2005)
7. R.A. Ganeev, G.S. Boltaev, V.V. Kim, M. Venkatesh, C. Guo, OSA Continuum **2**, 2381 (2019)
8. I.J. Kim, G.H. Lee, S.B. Park, Y.S. Lee, T.K. Kim, C.H. Nam, T. Mocek, K. Jakubczak, Appl. Phys. Lett. **92**, 021125 (2008)
9. I.J. Kim, C.M. Kim, H.T. Kim, G.H. Lee, Y.S. Lee, J.Y. Park, D.J. Cho, C.H. Nam, Phys. Rev. Lett. **94**, 243901 (2005)
10. J. Mauritsson, P. Johnsson, E. Gustafsson, A. L'Huillier, K.J. Schafer, M.B. Gaarde, Phys. Rev. Lett. **97**, 013001 (2006)
11. R. Ganeev, H. Singhal, P. Naik, I. Kulagin, P. Redkin, J. Chakera, M. Tayyab, R. Khan, P. Gupta, Phys. Rev. A **80**, 033845 (2009)
12. R.A. Ganeev, C. Hutchison, A. Zaïr, T. Witting, F. Frank, W.A. Okell, J.W.G. Tisch, J.P. Marangos, Opt. Express **20**, 90 (2012)
13. R.A. Ganeev, M. Suzuki, H. Kuroda, J. Opt. Soc. Am. B **31**, 911 (2014)
14. J. Seres, E. Seres, B. Landgraf, B. Ecker, B. Aurand, A. Hoffmann, G. Winkler, S. Namba, T. Kuehl, C. Spielmann, Scient. Reports **4**, 4254 (2014)
15. R.A. Ganeev, P.A. Naik, H. Singhal, J.A. Chakera, P.D. Gupta, Opt. Lett. **32**, 65 (2007)
16. C. McGuinness, M. Martins, P. Wernet, B. F. Sonntag, P. v. Kampen, J. P. Mosnier, E. T. Kennedy, J.T. Costello, J. Phys. B **32**, L583 (1999)
17. G. Duffy, P. v. Kampen, P. Dunne, J. Phys. B **34**, 3171 (2001)
18. V. Strelkov, Phys. Rev. Lett. **104**, 123901 (2010)
19. G.S. Boltaev, R.A. Ganeev, V.V. Kim, A.S. Alnaser, AIP Adv. **10**, 015231 (2020)
20. V. Tosa, H.T. Kim, I.J. Kim, C.H. Nam, Phys. Rev. A **71**, 063808 (2005)
21. R.A. Ganeev, H. Singhal, P.A. Naik, V. Arora, U. Chakravarty, J.A. Chakera, R.A. Khan, I.A. Kulagin, P.V. Redkin, M. Raghuramaiah, P.D. Gupta, Phys. Rev. A **74**, 063824 (2006)
22. R.A. Ganeev, M. Suzuki, P.V. Redkin, M. Baba, H. Kuroda, Phys. Rev. A **76**, 023832 (2007)
23. L.B. Elouga Bom, J.-C. Kieffer, R.A. Ganeev, M. Suzuki, H. Kuroda, T. Ozaki, Phys. Rev. A 75, 033804 (2007)
24. R.A. Ganeev, C. Hutchison, T. Witting, F. Frank, W.A. Okell, A. Zaïr, S. Weber, P.V. Redkin, D.Y. Lei, T. Roschuk, S.A. Maier, I. López-Quintás, M. Martín, M. Castillejo, J.W.G. Tisch, J.P. Marangos, J. Phys. B **45**, 165402 (2012)

25. P.B. Corkum, Phys. Rev. Lett. **71**, 1994 (1993)
26. A.V. Andreev, S.Yu. Stremoukhov, O.A. Shoutova, Eur. Phys. J. D **66**, 6 (2012)
27. R.A. Ganeev, G.S. Boltaev, S.Y. Stremoukhov, V.V. Kim, A.V. Andreev, A.S. Alnaser, Eur. Phys. J. D **74**, 199 (2020)
28. P. Lan, E. Takahashi, K. Midorikawa, Phys. Rev. A **81**, 061802 (2010)
29. A. Pirri, C. Corsi, M. Bellini, Phys. Rev. A **78**, 011801 (2008)
30. R.A. Ganeev, M. Suzuki, H. Kuroda, Phys. Rev. A **89**, 033821 (2014)
31. R.A. Ganeev, G.S. Boltaev, V.V. Kim, P.V. Redkin, C. Guo, J. Appl. Phys. **125**, 153101 (2019)
32. R.A. Ganeev, V. Tosa, K. Kovács, M. Suzuki, S. Yoneya, H. Kuroda, Phys. Rev. A **91**, 043823 (2015)
33. E.S. Toma, P. Antoine, A. de Bohan, H.G. Muller, J. Phys. B **32**, 5843 (1999)
34. M.B. Gaarde, K.J. Schafer, Phys. Rev. A **64**, 013820 (2001)
35. Z. Zeng, R. Li, Y. Cheng, W. Yu, Z. Xu, Phys. Scr. **66**, 321 (2002)
36. R. Taieb, V. Veniard, J. Wassaf, A. Maquet, Phys. Rev. A **68**, 033403 (2003)
37. R.A. Ganeev, M. Suzuki, T. Ozaki, M. Baba, H. Kuroda, Opt. Lett. **31**, 1699 (2006)
38. R.A. Ganeev, V.V. Strelkov, C. Hutchison, A. Zaïr, D. Kilbane, M.A. Khokhlova, J.P. Marangos, Phys. Rev. A **85**, 023832 (2012)
39. M.V. Frolov, N.L. Manakov, A.F. Starace, Phys. Rev. A **82**, 023424 (2010)
40. R.A. Ganeev, T. Witting, C. Hutchison, F. Frank, M. Tudorovskaya, M. Lein, W.A. Okell, A. Zaïr, J.P. Marangos, J.W.G. Tisch, Opt. Express **20**, 25239 (2012)
41. M.A. Fareed, V.V. Strelkov, M. Singh, N. Thiré, S. Mondal, B.E. Schmidt, F. Légaré, T. Ozaki, Phys. Rev. Lett. **121**, 023201 (2018)
42. R. D'Arcy, J.T. Costello, C. McGuinnes, G. O'Sullivan, J. Phys. B **32**, 4859 (1999)
43. J.B. West, J.E. Hansen, B. Kristensen, F. Folkmann, H. Kjeldsen, J. Phys. B **36**, L327 (2003)
44. G. Duffy, P. Dunne, J. Phys. B **34**, L173 (2001)
45. S. Yu, Stremoukhov and A. V. Andreev, Laser Phys. **28**, 035403 (2018)
46. R.A. Ganeev, S.Y. Stremoukhov, A.V. Andreev, A.S. Alnaser, Appl. Sci. **9**, 1701 (2019)
47. A.V. Andreev, R.A. Ganeev, H. Kuroda, S.Yu. Stremoukhov, O.A. Shoutova, Eur. Phys. J. D **67**, 22 (2013)
48. A. Couairon, A. Mysyrowicz, Phys. Rep. **441**, 47 (2007)

Chapter 8
Resonance-Induced Enhancement of Single Harmonic

8.1 Controlling Single Harmonic Enhancement in Laser-Produced Plasmas

Generation of coherent short-wavelength radiation is the attractive field of optics allowing various scientific and practical applications. The most advanced method here is the high-order harmonic generation of laser radiation that has long been considered a promising tool to retrieve the structural and spectral information through analysis of the harmonics [1]. The role of atomic resonances in increasing the laser radiation conversion efficiency towards shorter-wavelength region was actively discussed in the framework of perturbation theory at the early stages of the studies of low-order harmonic generation [2]. Resonance enhancement introduces an additional possibility of increasing the conversion efficiency of a specific harmonic order by more than one order of magnitude. The resonance-induced enhancement of the harmonics of ultrashort pulses is one of the attractive features of coherent radiation frequency conversion in laser-induced plasmas (LIP), with the maximal enhancement factor of a single harmonic approaching 80 in the case of indium plasma [3]. The use of LIP allowed the analysis of ionic transitions with high oscillator strengths (i.e. high *gf* values, which are the product of the oscillator strength *f* of an atomic transition and the statistical weight *g* of the lower level).

Notice that no experiments with gases showing the resonance enhancement of single harmonics have been reported so far. The only study where the partial enhancement of a narrow component of harmonic was demonstrated during HHG in Ar gas was related to the influence of Fano resonances [4]. Why have resonant harmonics not been observed in gas media? The answer to this fundamental question is related to the basic principles of the role of ionic resonances in the enhancement of harmonics. Various factors prevented the observation of resonance enhancement. Among them are the self-absorption near the ionic transitions, which emit strong radiation in the extreme ultraviolet range, and the weak oscillator strength of these transitions. Furthermore, it is obvious that the choice of gases, mostly limited to a few noble gases, is dramatically narrowyк compared with that of solids, which comprise most part

R. A. Ganeev, *High-Order Harmonics Generation in Plasmas*, Springer Series on Atomic, Optical, and Plasma Physics 122, https://doi.org/10.1007/978-3-031-09040-0_8

of the non-radioactive elements of the periodic table. Consequently, the probability of finding the appropriate resonance with high oscillator strength, which matches with some harmonics, is significantly smaller in the case of HHG in gases compared with HHG in plasmas. These conditions are rarely fulfilled, and if fulfilled they most probably occur in media produced during the ablation of thousands of solids rather than in a few gases. That is why the analysis of this phenomenon in plasma media has more chances of success.

In this section, we study a few plasma samples, which showed numerous cases of the coincidence of harmonics with some emission lines of corresponding ions [5]. We show that ionic transitions demonstrating high oscillator strengths are responsible for enhanced emission of a single order or group of harmonics. Meanwhile, plasma emission plays rather a destructive role during HHG in the plasma plumes.

8.1.1 Experimental Arrangements

Experimental setup comprised the Ti:sapphire laser (λ = 810 nm, E = 1.53 eV), traveling-wave optical parametric amplifier (OPA), and HHG scheme using the propagation of amplified signal radiation from OPA or 810 nm radiation through the extended LIP. Part of amplified uncompressed radiation (370 ps pulse duration, 5 mJ pulse energy) was separated from a whole beam and used as a heating pulse for homogeneous extended plasma formation using the 200 mm focal length cylindrical lens installed in front of the extended targets placed in the vacuum chamber. The 5-mm-long Sb, Sn, Mn, Zn, Se, In, Cr, Ag, Au, Mg, Cd, and graphite samples were used as the targets for ablation and formation of extended homogeneous plasma.

The compressed radiation of Ti:sapphire laser (pulse energy 8 mJ, pulse duration 64 fs, 10 Hz pulse repetition rate, pulse bandwidth 17 nm, central wavelength 810 nm) pumped the OPA. Signal pulses from OPA allowed the tuning along the 1200–1600 nm range. Most of the experiments were carried out using the 1 mJ, 70 fs signal pulses or 3 mJ, 64 fs, 810 nm pulses. The intensity of focused 1310 nm pulses (E = 1.53 eV) inside the LIP was 2×10^{14} W cm^{-2}. The 810 nm pulses were also used as a single-color pump (SCP) separately from signal pulses for HHG. The harmonic emission was analyzed using an XUV spectrometer.

In most of the following studies, the two-color pump (TCP) scheme comprising near-infrared (NIR) pulses and second harmonic radiation as a second field was used. To add a second field, the 0.5-mm-thick BBO crystal (type I) was installed inside the vacuum chamber on the path of the focused signal pulse. The conversion efficiency of second harmonic pulses was 20%. The use of second wave is not a trick but rather a necessity to study the resonant enhancement of harmonics using a NIR radiation.

8.1.2 Featureless and Resonance-Enhanced Harmonic Distributions

In this subsection, the raw images of plasma and harmonic spectra from different LIPs are described. The reason in the presentation of experimental data as the images appearing on the screen of the computer rather than the groups of intensity distributions is caused by better viewing of obtained data and clear definition of the difference between the so-called featureless group of harmonics and the group of harmonics containing a single enhanced harmonic. One has to note that the goal of discussed studies was the qualitative definition of the influence of plasma emission on the harmonic conversion efficiency in different LIPs rather than the quantitative measurements of the gain of a single harmonic in the plasmas where this process could be realized. Also notice that visual analysis of raw images of harmonic and plasma emission allows distinguishing these sources of radiation by their divergences. The vertical dimensions of the raw images of plasma emission, which characterize its divergence, were a few times larger compared with those of harmonics.

The saturated images were intentionally chosen to present the spectra for better viewing. Note that for the line-outs of the HHG spectra, unsaturated images were used. The X-axes are shown in the figure on the basis of the calibration of the XUV spectrometer for better viewing of the distribution of harmonics along the short-wavelength region. The HHG spectrometer was calibrated using the plasma emission from the used ablated species, as well as other ablating targets.

The first set of studies was carried out in the plasmas where no resonance enhancement was observed. Figure 8.1a shows the plasma and harmonic spectra obtained during the ablation of silver and gold targets. The plasma emission spectra were observed at the conditions of ablation at which the HHG could be realized at unfavorable conditions. These plasma spectra, as well as all other emission spectra obtained during ablation of targets, were registered at the conditions when femtosecond pulses did not propagate through the LIPs. Meanwhile, the HHG was performed at the conditions of optimal LIP formation. The term "optimal LIP formation" refers to maximal harmonic yield during the propagation of femtosecond pulses through such plasmas. Both Ag and Au plasmas are attractive media for extended harmonic distribution. The harmonics up to H63 and H57 (not shown in Fig. 8.1a) were generated in the silver and gold plasmas during propagation of 810 nm, 64 fs pulses through these LIPs. These distributions did not reveal any enhancement of some specific harmonic order but rather represent the plateau-like shapes when the intensities of harmonics were approximately the same along the whole spectral range of generation with some gradual decrease up to the cutoff orders. One can see that, though H11 and H13 coincided with some emission lines of Ag and Au ions, no enhancement of these harmonics was observed.

A similar pattern was obtained in the case of Mg (Fig. 8.1b) and graphite (Fig. 8.1c) plasmas. One can see that at least H19, H27, and H37 of the single-color pump (1340 nm, 70 fs) propagating through the magnesium plasma coincided with the plasma emission lines. However, neither enhancement nor decrease of these

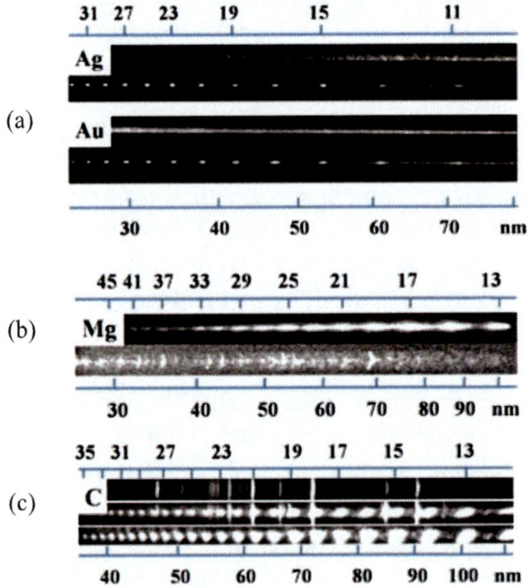

Fig. 8.1 Plasma and nonresonant harmonic spectra. **a** Raw images of plasma (upper panels) and harmonic (bottom panels) spectra obtained during ablation of Ag and Au targets and 810 nmfemtosecond pulses propagation. Harmonic orders and wavelengths in this and most of other figures are shown on the upper and bottom axes respectively. **b** Raw images of harmonic (upper panel) and plasma (bottom panel) spectra obtained during ablation of Mg target and 1340 nmfemtosecond pulses propagation. **c** Raw images of plasma (upper panel), plasma + harmonic (middle panel), and harmonic (bottom panel) spectra obtained during ablation of graphite target and 1340 nm + 670 nmfemtosecond pulses propagation. Reproduced from [5] with permission from the American Institute of Physics

harmonics was observed in the case of HHG in this plasma, but rather a gradual decrease of each next order of harmonics compared with the previous one.

The plasma emission from ablated graphite contained strong emission lines of carbon (upper panel of Fig. 8.1c). The propagation of the TCP (1340 nm + 670 nm) through such plasma allowed the generation of approximately equal odd and even harmonics of 1340 nm radiation up to the H43 (second panel). One can see that coincidence of a few harmonics with these emission lines did not cause the change of their intensity relative to neighbor orders. Moreover, as was mentioned above, a decrease of heating pulse energy on the target surface allowed the formation of optimal graphite plasma resulting in noticeable enhancement of all harmonics (bottom panel; compare with middle panel). Again as in previous cases (Fig. 8.1a and b), the featureless harmonic spectrum without the enhancement of a single harmonic was observed.

Most plasmas produced on the targets representing the solid elements of the periodic table allowed the generation of harmonic spectra similar to those shown in Fig. 8.1. Meanwhile, a few solid elements showed outstanding features being used for LIP formation and HHG. In Fig. 8.2, three samples of harmonic spectra

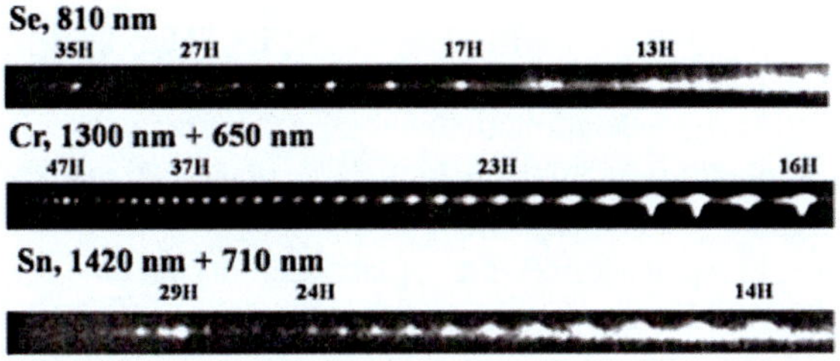

Fig. 8.2 Samples of resonance-induced enhancement of single harmonics in various plasmas. Upper, middle, and bottom panels correspond to the harmonic spectra generated in selenium, chromium, and tin plasmas using 810 nm, 1300 nm + 650 nm, and 1420 nm + 710 nm pumps respectively. One can see the enhanced H35 in the case of selenium plasma and 806 nm pump. In the case of chromium plasma, harmonics surrounded H47 were notably stronger compared with those in the range of H41–H45 in the case of 1300 nm + 650 nm pump. Similarly, tin plasma allowed observation of the enhanced harmonics in the vicinity of H29 of 1420 nm + 710 nm pump. Reproduced from [5] with permission from the American Institute of Physics

demonstrating the enhanced single harmonic, or group of harmonics, in different ranges of XUV using the selenium, chromium, and tin LIPs and shown.

The SCP of Se plasma using 810 nm radiation allowed odd harmonics generation up to H27 using the optimally formed LIP. The growth of heating pulse fluence on the surface of ablating target led to the appearance of plasma emission in the range of H10–H15 and the decrease of a whole yield of harmonics (Fig. 8.2, upper panel). Meanwhile, this over-excitation of target caused the growth of the population of some ionic states, which allowed the observation of an enhanced single harmonic (H35) far from the whole set of harmonics. This harmonic was visibly stronger than the lower-order ones (H23–H33), contrary to commonly observable patterns of lower intensities for the higher orders compared with lower-order ones. In fact, the neighboring harmonics around H35 were barely seen in this spectrum.

The HHG spectrum from Cr plasma also showed the gradual decrease of harmonic yield up to H42 using TCP (1300 nm + 650 nm). At higher excitation of the target, a group of harmonics centered on H47 was observed, apart from the set of lower-order harmonics (second panel).

The third sample of an unusual distribution of harmonic spectra was observed during experiments with tin plasma (third panel). The TCP (1420 nm + 710 nm) of over-excited plasma caused the appearance of both the mixture of harmonic and plasma emission in the longer wavelength range of XUV and the group of enhanced harmonics (H31, H30, and especially H29). Notice the absence of H25–H27 in this spectrum.

To address these results we compare them with the spectroscopic analysis of the properties of specific ionic transitions of Se, Cr, and Sn. The resonance-enhanced single harmonic observed in selenium plasma had the shortest wavelength (H35 of 802 nm pump, $\lambda = 22.9$ nm, $E = 54.1$ eV) reported so far among other resonance-enhanced processes [6]. The doubly ionized selenium atom has the ground configuration $4s^2 4p^2$ and the excited configurations $4s^4 p^3$, $4s^2 4pnd$ $(n > 4)$, and $4s^2 4pns$ $(n >$

5). The photon energy studied in previous spectroscopic experiments with Se plasma was ranged between 18.0 eV and 31.0 eV [7, 8]. However, they also reported the strong third-order line corresponding to the 54.62 eV energy (λ = 22.7 nm), without the deliberation of the origin of this transition. Note that this transition is close to the enhanced H35 in the discussed experiments. Probably, this transition can influence the nonlinear optical response of selenium plasma in this spectral region.

Previous studies of photoabsorption and photoionization spectra of Cr plasma in the range of 41–42 eV have demonstrated the presence of strong transitions, which could be responsible for such a pattern of the harmonic spectrum [9–11]. In particular, the region of the "giant" 3p $\rightarrow$ 3d resonance of Cr II spectrum was analyzed in Ref. [11]. The role of the $3d^5(^6S)$ state in determining the special position of Cr among the 3d elements was emphasized and the strong transitions, which could both enhance and diminish the optical and nonlinear optical response of the plume, were revealed. Some previous studies of chromium plasma revealed the enhancement of single harmonics in the case of different sources of driving pulses. Particularly, the H29 (λ = 27.6 nm) of the 800 nm pump was noticeably stronger than other harmonics [12].

In past work, the strong Sn II ion has been shown to possess a strong transition $4d^{10}5s^25p^2P_{3/2}$–$4d^95s^25p^2(^1D)^2D_{5/2}$ at the wavelength of 47.20 nm [13]. The *gf* value of this transition has been calculated to be 1.52, and this value is five times larger than other transitions from the ground state of Sn II. Therefore the enhancement of the H29 (λ = 48.9 nm) using the 1420 nm + 710 nm wavelength pump can be explained as being due to optimal closeness to the above transition. The resonance enhancement of single harmonic (H17 of 795 nm laser, $\lambda = 46.8$ nm) has previously been observed and attributed to the influence of the same transition [14].

These three samples of harmonic spectra point out the role of ablation conditions in the formation of the LIP allowing such an unusual distribution of harmonics appearing in the shorter wavelength range of XUV. During the last decade this phenomenon was observed in the case of some ionic transitions the *gf*s of which were significantly larger compared with similar parameters of other neighboring transitions in different spectral ranges [15–18]. Meanwhile, the role of plasma emission lines was yet examined. Below we analyze the comparative studies of plasma and harmonic spectra at the conditions of coincidence or mismatch of plasma and harmonic wavelengths.

8.1.3 Comparison of Plasma and Harmonic Spectra in the LIPs Allowing the Generation of Resonantly Enhanced Harmonics

To analyze the role of the ionic transitions responsible for plasma emission on the enhancement of a single harmonic one has to use a tunable source of femtosecond pulses. The pump wavelength was tuned using OPA, which allowed the variation of harmonic wavelengths in the vicinity of strong emission lines in the XUV range using Zn, Sb, Cd, and In LIPs. The fixed wavelength of pump radiation was also used to analyze the plasma and harmonic emission from Mn ablation. The data on the plasma emission of various elements were taken from the NIST Atomic Spectra Database [19].

Zinc plasma. The upper panel of Fig. 8.3a shows the raw image of zinc plasma emission where a few strong emission lines in the range of 65–90 nm dominated the XUV spectra. This spectrum was obtained at the unfavorable conditions for harmonic generation (i.e. at ~2 J cm^{-2} fluence of heating pulse), similarly to the LIP spectra shown in Fig. 8.1. At optimal conditions of zinc ablation when maximally

(a)
(b)

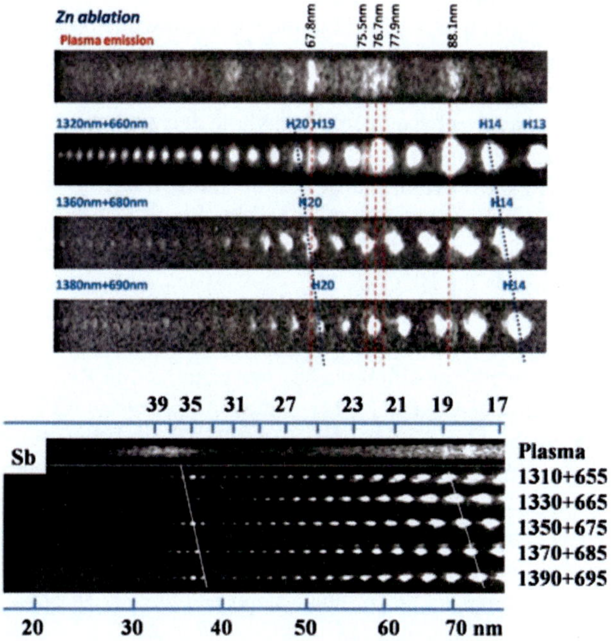

Fig. 8.3 **a** Raw images of plasma (upper panel) and harmonic (three other panels) spectra obtained during ablation of zinc. Harmonic spectra were tuned by changing the wavelength of driving NIR radiation and its second harmonic from 1320 nm + 660 nm (second panel from the top) to 1360 nm + 680 nm (third panel) and 1380 nm + 690 nm (fourth panel). Dashed lines show the positions of strong plasma emission lines in all four spectra. Dotted lines show the tuning of H14 and H20 at different pumps of plasma. The intensity of H20 remained approximately same during tuning through the strong resonance transition ($\lambda = 67.8$ nm), while H17 was stronger in the case of 1320 nm + 660 nm pump. **b** Raw images of plasma (upper panel) and harmonic (five other panels) spectra obtained during ablation of antimony. Harmonic spectra were tuned by changing the wavelength of driving NIR radiation and its second harmonic from 1310 nm + 665 nm (second panel from the top) to 1390 nm + 695 nm (sixth panel) with a step of 20 nm. Solid lines show the tuning of H36 and H19 at different pumps of plasma. The harmonic orders shown on the top of images correspond to those generated using the 1310 nm + 655 nm pump. Most of harmonics remain same during variation of pump wavelength, excluding those in the region of 37.5 nm. Reproduced from [5] with permission from the American Institute of Physics

efficient HHG was achieved a few plasma emission lines were still seen among the harmonics. Particularly, a strong 67.8 emission line of ZnIII attributed to the $3d^{10}$–$3d^{9}4p$ transition appeared between H19 and H20 of 1320 nm + 660 nm pump (second panel). The same can be said about the 75.5 nm emission line. The harmonic spectrum was extended up to H36. The H15 and H17, which coincided with 88.1 and 76.7 nm emission lines, appeared to be stronger than other neighboring harmonics. It was difficult to define whether this difference in harmonic intensities can be attributed to the influence of those transitions or to the naturally stronger odd harmonics compared to the even ones.

To define the role of plasma emission lines, the fluence of the heating pulse was decreased and the harmonics were tuned (third panel, 1360 nm + 680 nm pump). In that case, only 67.8 nm line remained in the spectrum while other plasma emission lines almost disappeared. The odd and even harmonics (at least up to H22) showed a gradual decrease of intensity for each next harmonic order, without any features indicating the resonance enhancement. The whole decrease of conversion efficiency compared with the 1320 + 660 nm pump was caused by the higher energy of the latter TCP compared with the previous pump. This decrease of HHG conversion efficiency is further seen in the case of the 1380 + 690 nm pump (fourth panel), which also confirmed the assumption about the negligible role of the ionic transitions responsible for plasma emission on the HHG.

The H17 of 1320 + 660 nm pump could be partially enhanced due to different processes. The involvement of the 76.7 nm transition of ZnII on the variations of harmonic yield near this transition was earlier examined in Ref. [20] by using the model potential introduced in Ref. [21]. The width of resonantly enhanced H17 in those studies was close to that of other harmonics, both in calculated and experimental results, similarly to the discussed data. Another case was observed in the case of the 67.8 nm transition. The enhancement region is narrower than the harmonic line width (third panel), so the shape of H20 is close to the one of surrounding harmonics but with a narrow maximum due to the possible resonant enhancement. A similar observation was reported using 775 nm, 4 fs probe pulses [22]. The observations of the strong emission of 18.3 eV transition (λ = 67.8 nm, $3d^{10}$–$3d^{9}4p$) of Zn III at the conditions of plasma excitation by a few cycles of broadband pulses centered at 775 nm was attributed to the enhancement of part of the 11th harmonic of this radiation, although the wavelength of this harmonic (70 nm) did not exactly match the wavelength of $3d^{10}$–$3d^{9}4p$ transition, which has an oscillator strength significantly greater than other lines in this spectral region [23]. In those studies, the narrowband-enhanced emission was similar to that observed here and could also be attributed to the influence of resonance and propagation processes.

Antimony plasma. Another example demonstrating the enhancement of single harmonic by tuning through the ionic transition possessing high *gf* rather than strong plasma emissions was demonstrated in the case of antimony plasma. This plasma has two strong ionic transitions. The XUV spectra of antimony plasma have been analyzed in Ref. [24]. The oscillator strengths of $4d^{10}5s^{2}2p^{3}P_{2}$–$4d^{9}5s^{2}5p^{33}D_{3}$ and $4d^{10}5s^{2}2p^{1}\ D_{2}$–$4d^{9}5s^{2}5p^{33}F_{3}$ transitions have been calculated to be 1.36 and 1.63, respectively, which were a few times larger than those of the neighboring transitions.

These strong Sb II transitions ($4d^{10}5s^{2}2p^{3}P_{2}$–$4d^{9}5s^{2}5p^{33}D_{3}$ and $4d^{10}5s^{2}2p^{1}$ D_{2}–$4d^{9}5s^{2}5p^{33}F_{3}$) at the wavelengths of 37.82 nm and 37.55 nm could be responsible for the enhancement of harmonics in the vicinity of these transitions.

The upper panel of Fig. 8.3b shows two groups of plasma emission comprising the continuum and emission lines. Notice that there are no peculiar emission lines in the vicinity of 37–38 nm. The strongest emission line was seen only at 32.5 nm, which is quite far from the spectral range of the expected enhancement of the nonlinear optical processes.

The application of TCP allowed a significant enhancement of harmonic yield compared with SCP. TCP was gradually tuned between 1310 nm + 655 nm to 1390 nm + 695 nm with a step of NIR pulses of 20 nm (second to sixth panels). The lower-order harmonics remained approximately the same during this procedure. The harmonic orders shown in the upper axis are attributed to those in the second panel. The white solid line at the right side of this picture shows the tuning of H19. No significant changes in the intensity of this harmonic were seen while tuning between the two plasma emission lines shown in the upper panel. The second group of harmonics appears in the 36–38 nm region. The white solid line at the left side of the picture shows the tuning of H36. One can see a significant change in the intensity of this harmonic, with maximal conversion efficiency at ~37.3 nm. Probably, the shorter-wavelength side of the two abovementioned resonances is responsible for better enhancement of the harmonics tuning in this region. The reasons for this peculiarity could be related to the propagation effect, when the phase-matching conditions became better for the system possessing lower difference between the wavenumbers of the pump and harmonic emissions due to anomalous dispersion in this side of resonances. These lines are not seen in the plasma emission spectrum (see the left group of lines and continuum appearing in the 28–36 nm range on the first panel). The efficiency of each neighboring harmonics in this region depended to the closeness with the $4d^{10}5s^{2}2p^{3}P_{2}$–$4d^{9}5s^{2}5p^{33}D_{3}$ and $4d^{10}5s^{2}2p^{1}$ D_{2}–$4d^{9}5s^{2}5p^{33}F_{3}$ transitions of SbII ion.

Cadmium plasma. The relative intensities of the H15 generating in Cd plasma were tuned by changing the SCP from 1360 to 1320 nm, as shown in Fig. 8.4a (two bottom panels). One can clearly see the relatively equal intensities of other harmonics in the case of 1320 and 1360 nm pumps. Meanwhile, the tuning of the pump and harmonic wavelengths towards the longer-wavelength region significantly changed the pattern of H15. H15 (λ = 90.7 nm) using a 1360 nm pump became significantly stronger than the H15 obtained using 1320 nm radiation. The emission lines from Cd plasma (upper panel) were far from the region of H15. Thus the influence of the transitions responsible for plasma emission was insignificant from the point of view of variations of the relative intensities of this harmonic.

Indium plasma. The strongest enhancement of single harmonic was observed in the case of In LIP (Fig. 8.4b, upper panel). H13 of 810 nm SCP was significantly stronger than other harmonics. The analysis of relative intensities showed that H13 was ~20 times higher compared with neighboring orders.

This harmonic generation process was carried out at a fixed wavelength (810 nm) of Ti:sapphire laser and could not be optimized from the point of view of further

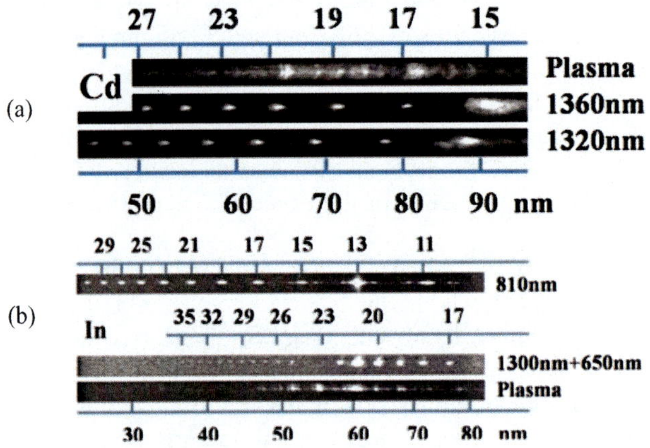

Fig. 8.4 **a** Raw images of plasma (upper panel) and harmonic (two bottom panels) spectra obtained during ablation of cadmium. Harmonic spectra were tuned by changing the wavelength of driving MIR radiation from 1320 nm (third panel from the top) to 1360 nm (second panel). The harmonic orders shown on the top of images correspond to those generated using the 1360 nm pump. **b** Raw images of harmonic (two upper panels) and plasma (bottom panel) spectra obtained during ablation of indium. One can see the strong H13 in the case of 810 nm pump (upper panel). The harmonics from 1300 nm + 650 nm pump were also maximally enhanced in the same spectral region (~61.5 nm, middle panel). The harmonic orders shown on the top of upper and middle images correspond to those generated using the 810 nm and 1300 nm + 650 nm pumps respectively. Reproduced from [5] with permission from the American Institute of Physics

enhancement of H13. This optimization assumes better matching of the H13 wavelength with the wavelength of the transition of InII responsible for such resonance-induced enhancement. The difference between the wavelengths of H11, H13, and H15 was large enough, so the ionic transition affected only single harmonic order (H13) and did not influence on other two neighboring harmonics. Another situation occurs in the case when the wavelength difference between the harmonics becomes smaller. To do so, the 1300 + 650 nm pump allowing the generation of equal odd and even harmonics in most plasmas was used (see for example Figs. 8.1c and 8.2, middle panel). In the case of In LIP, a significant deviation from the featureless homogeneous harmonic distribution was observed in the case of TCP. The middle panel of Fig. 8.4b shows a noticeable variation of harmonics in the region of 58–77 nm. The harmonic orders (from right to left) in that case start from the H16 (extremely weak emission) and then their intensity increases from H17 to H21, which is unusual for the HHG process. The strongest among them was the H21 of 1300 nm radiation ($\lambda = 61.9$ nm) almost coinciding with the H13 of 810 nm radiation ($\lambda = 62.3$ nm).

Occasionally, there was the emission line ($\lambda = 60.9$ nm, see bottom panel) in the plasma spectrum almost coinciding with the area of maximal enhancement. Notice that at weaker ablation (i.e. at a fluence of the heating pulse of ~1.3 J cm^{-2}), when most of the plasma emission lines disappear, the enhancement of harmonics in this region was even stronger compared with the above-described case (fluence of 1.6 J cm^{-2}).

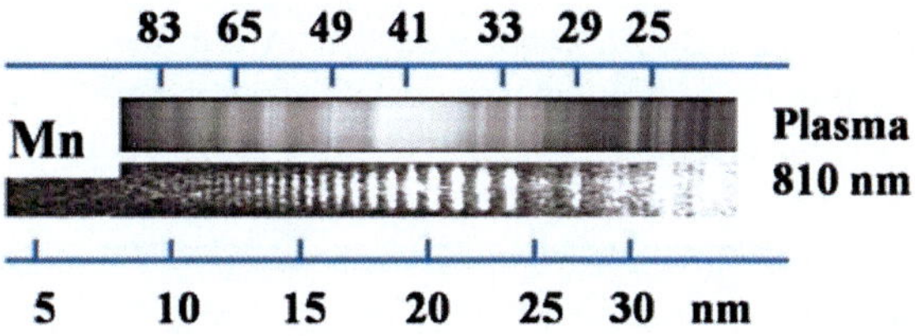

Fig. 8.5 Raw images of plasma (upper panel) and harmonic (bottom panel) spectra obtained during ablation of manganese. The intensity of H33 of the 810 nm pump was notably stronger than lower-order harmonics. Reproduced from [5] with permission from the American Institute of Physics

Meanwhile, other plasma lines shown in the bottom line did not affect the harmonic distribution using either 810 nm or 1300 nm + 650 nm pumps.

The enhancement of the 13th harmonic emission from the indium plasma observed in discussed studies is due to the influence of the radiative transitions between the $4d^{10}5s^{2}\,{}^{1}S_{0}$ ground state of In II and the low-lying $4d^{9}5s^{2}np$ transition array of In II. Among them, the transition of In II at 19.92 eV (62.24 nm, $4d^{10}5s^{2}\,{}^{1}S_{0} \rightarrow 4d^{9}5s^{2}5p\,{}^{1}P_{1}$) is exceptionally strong. The oscillator strength of this transition has been calculated to be $gf = 1.11$ [25] which is more than 12 times larger than that of any other transitions from the ground state of In II. This transition is energetically close to the H13 of 810 nm radiation and the H21 of 1300 nm radiation, thereby resonantly enhancing their intensity. Moreover, as one can see, this transition affects the harmonic yield of at least two neighboring harmonics H20 and H19, which was manifested by higher yields of these harmonics compared with lower-order one (H18, middle panel).

Manganese plasma. The final example of harmonic enhancement is shown in Fig. 8.5. The Mn plasma was emitted in a broad short-wavelength region (10 to 35 nm, upper panel) during over-excitation of target by 370 ps pulses (i.e. at a fluence of 1.9 J cm^{-2}). The harmonics at weaker ablation (1.2 J cm^{-2}) were limited by H29. The growth of heating pulse fluence led to the appearance of the extended group of harmonics starting from H33 and extending up to H83 (bottom panel). There was no relation between the emission lines and enhanced H33 and following harmonics. One can assume better conditions of enhancement for the H33 in the vicinity of the giant 3p → 3d resonances of Mn around the 51–52 eV range, where the metastable states of manganese are located [26]. Meanwhile there are no relations between the observed plasma emission lines (upper panel) and the harmonic enhancement in the region of ~24 nm (bottom panel).

8.1.4 Analysis of Experiments

Various theoretical approaches for the description of resonant HHG were introduced in Refs. [21, 27–35]. As one can see, plenty of theoretical studies were reported on the possibility of the resonance enhancement of harmonics to determine the availability of this process in HHG experiments. Below we briefly address another model of resonance enhancement. This model is based on multielectron resonance recombination.

According to this model the accelerated electrons while returning to the parent particle can transfer the energy to the electron, or a system of electrons, during the inelastic scattering in such a way that those electrons become excited on the resonant levels the structure of which is determined by the form of ionic potential. The excited resonance levels are independent of the Stark shift because these levels are not initially populated. Therefore, these resonances can be reasonably calculated. This model of multielectron resonance HHG model is a generalization of the four-step model. In the case of the multielectron system, the analytical equations derived in Ref. [21] are valid in the approximation that takes into account only one channel of excitation of the incident electron. Some of the principles of the discussed theory have already been described in Refs. [34, 36].

The difference in the character of the settlement between the resonance level proposed in Ref. [21] and the multielectron recombination resonance is that, in the former case, the excited state "captures" the accelerated electron, which after some time recombines to the ground state due to relaxation, whereas, in the model of multielectron resonance recombination, the capture of electron and the resonant nature of the process are determined by the maximum probability of electron energy transmission to the excited ion at the conditions of similarity of the electron kinetic energy and resonant energy. As for the simulation, this should be reflected in the fact that within a relatively short period of time the expected value of the induced dipole should remain constant or slightly oscillating, which allows one to test it on the basis of numerical methods. In the proposed model it is assumed that the use of multielectron methods allows the application of properly selected pseudopotentials, which avoid the artificial introduction of resonance levels to monitor the resonant HHG.

8.2 High-Order Harmonics Generation in Atomic and Molecular Zinc Plasmas

8.2.1 Introduction

Although the generation of high-order harmonics due to the interaction of intense laser pulses with matter provides a unique source of coherent femtosecond and attosecond pulses in the XUV range, the low efficiency of this process ($<10^{-6}$) is a serious disadvantage for its wide application. To increase the efficiency of HHG, various methods have been introduced, in particular, resonance-induced amplification, especially in the case of their generation in LIP formed during laser ablation of solids. The resonance-induced amplification of a single harmonic is attributed to the proximity of the autoionizing states (AIS) of single-charged ions and some specific harmonic orders. Resonant processes made it possible to increase the photon flux of such harmonics. This behavior was systematically analyzed using various materials. The use of LIP allows analyzing the ionic transitions with high oscillator strength (i.e., high *gf* values that are the product of the oscillator strength *f* of the atomic transition and the statistical weight *g* of the lower level).

A breakthrough in this area of research has been demonstrated in the indium plasma, where a strong resonant increase in the conversion efficiency for a single harmonic ($>10^{-5}$) from a low-ionized LIP was experimentally achieved. The main effort so far has been to increase the photon flux of resonant harmonics. However, only a few studies have been conducted to better understand the physics associated with this generation of resonant harmonics. These experiments in different plasmas showed that the intensity of the resonant harmonics strongly depends on the ellipticity of the driving laser and follows the phase-matching conditions.

Most studies of HHG from LIP have so far focused on the monatomic species, although HHG from diatomic molecules may offer some advantages due to their expanded structure and delocalized π electrons [37]. While single atoms with their relatively low ionization potentials probably limit HHG to low orders of generating harmonics, this is not obvious for the molecular species possessing higher ionization saturation intensities. Such molecular structures can have their own transitions, which can be used for resonant amplification of the non-linear optical response in plasmas. In addition, they can shift some of the transitions responsible for harmonic amplification in the LIP containing the components of these molecules. Demonstration of this phenomenon can show the advantages of high-order non-linear spectroscopic analysis of molecular transitions whose oscillator strengths have not yet been studied. The reason for this assumption is related to the modification of the properties of ionic transitions under consideration. In particular, a decrease of the oscillator strength of some transitions responsible for enhancement of single harmonic may correlate with the replacement of the atomic Zn-contained plasma (i.e., the one produced on the bulk Zn target) by the molecular Zn-contained plasma (i.e., the one produced on the bulk ZnSe target). The discussed study shows that this assumption is more reliable compared with the tuning of the ionic transition responsible for harmonic enhancement to explain the observed decay in single harmonic amplification in molecular plasma.

The relative role of atomic and molecular transitions in HHG resonance amplification is still unclear. To clarify this problem, it is necessary to study the generation of harmonics from the ionized molecules and atoms containing same element using different laser sources allowing the spectral tuning of the driving field. This approach is important from the point of view of the probability of molecular decay in laser plasma with subsequent participation of its components in HHG.

Changing the excitation of certain substances can lead to some additional excited states that can amplify neighboring harmonics. Among different types of ablated substances, zinc-containing plasma can be considered an interesting object of study, since it allows varying the excitation of various AIS depending on the ablation conditions of the bulk target. Target excitation makes it possible to analyze the nonlinear response of such LIPs by comparing the amplification of different harmonics, provided that they coincide or remain close to these autoionizing states. The crucial role of strong ion transitions in more complex systems than zinc atoms is still questionable. Here we discuss how the single harmonic enhanced in the case of atomic plasma completely lost this peculiarity of harmonic amplification once the zinc becomes a component of the molecular cloud produced during laser ablation [38]. We analyze the variations of single harmonic enhancement during single-color and two-color pumps of the zinc-containing plasmas using different laser sources.

8.2.2 Materials and Methods

Bulk zinc (Zn) metal and zinc selenide (ZnSe) targets were used in the discussed studies. To create plasma, the Ti:sapphire laser (central wavelength 806 nm, pulse duration of uncompressed radiation 370 ps, pulse energy 4 mJ, 10 Hz pulse repetition rate) was used. Part of the amplified uncompressed radiation was used as a heating pulse for LIP formation (Fig. 8.6). The intensity and fluence of the heating pulse on the target surface were varied up to $I = 4 \times 10^9$ W cm^{-2} and $F = 2$ J cm^{-2}, respectively. The ablation area was 0.6×0.6 mm^2. In the case of single-color pump (SCP) of LIP, the focused compressed 806 nm, 64 fs, 10 Hz pulses (with 120 μm focal spot sizes and intensity varying in the range of 1.5×10^{14} W cm^{-2} and 4×10^{14} W cm^{-2}) from this laser were aligned to propagate through plasma 0.2 mm above the ablating target. The 0.2-mm-thick beta-barium borate (BBO, type I) crystal was inserted into the vacuum chamber in the path of the 806 nm focused radiation (Fig. 8.6) to generate harmonics using a two-color pump (TCP, 806 nm + 403 nm).

Optical parametric amplifier (OPA) was used to analyze the tuning of NIR radiation during HHG in Zn-contained plasmas. Most of the experiments using this laser source were carried out using the 1 mJ, 70 fs signal pulses tunable in the range of 1280–1440 nm. OPA pulses were focused on the plasma from the orthogonal direction. The intensity of the 1310 nm pulses focused by 400 mm focal length lens inside

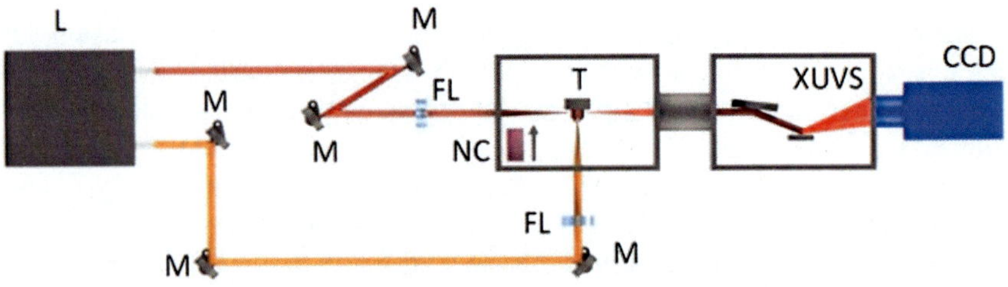

Fig. 8.6 Experimental setup for high-order harmonic generation (HHG) in zinc-contained plasmas. L: laser; M: mirrors; FL: focusing lenses; NC: nonlinear crystal; T: target; XUVS: extreme ultraviolet spectrometer; CCD: charge-coupled device. Reproduced from [38] with permission from MDPI

the plasma was 2×10^{14} W cm^{-2}. Most of the experiments using OPA were carried out using TCP of LIP. The 0.5 mm-thick BBO crystal (type I, $O = 21\circ$) was installed inside the vacuum chamber on the path of focused signal pulses. The conversion efficiency of H2 pulses was 12%.

Two orthogonally polarized pumps were used in the case of TCP of studied plasmas. Each of these pump components generate harmonics, while the mixture of pumps provided the 4(n + 1) orders (H12, H16, H20, H24, etc.). Application of second field (i.e., second harmonic of the main pump) even at the low ratio of 2ω and ω waves allows the enhancement of the overall conversion efficiency of harmonics, as well as generation of even orders. The mechanism of such enhancement in the case of TCP is due to selection of a short quantum path component, which has a denser electron wave packet, and higher ionization rate compared with the SCP. A suitable delay between fundamental and second fields in LIP was maintained by using thin BBO crystals.

An extreme ultraviolet spectrometer was used for the analysis of plasma and harmonic emission. The driving pulses and harmonics were separated by a flat field grating in the XUV spectrometer. Highest harmonic efficiency from LIP was observed at 100 or 250 ns delay from the beginning of Zn or ZnSe targets ablation. The variation of the optimal delay between heating and driving pulses was attributed to different velocities of spreading plasmas from the targets of different elemental weight.

8.2.3 Results

Previously, a study of the generation of high-order harmonics from ablated diatomic molecules (InSb, GaAs) showed that the discrepancy between the calculated molecular spectra and the observed resonantly amplified harmonics was due to the significantly greater influence of the strong electric field of laser radiation on the dynamic modification of molecular spectra [39]. The Stark shift in the energy levels of diatomic molecules can be considered the main reason for the decrease in the efficiency of single harmonics compared to monatomic LIPs (In, As, Sb), where the gain of single harmonics was explained by the proximity of the harmonic wavelength and the ion transitions with a strong oscillator strength. Below we address different aspects of this problem using single-atomic (Zn) and diatomic (ZnSe) plasma media.

Different fluencies of 806 nm pulses were used for the analysis of plasma and harmonic emission from an ablated Zn target. The application of fluence $F =$ 1.7 J cm^{-2} did not result in plasma emission in the XUV range. The growth of fluence ($F = 2$ J cm^{-2}) led to the appearance of emissions from two transitions of Zn II (88.1 and 89.3 nm) in the vicinity of the 9th harmonic of 806 nm pulses (89.6 nm).

Figure 8.7 shows the spectra observed during the generation of high-order harmonics of 806 nm, 64 fs, 10 Hz laser source in the case of ablation of zinc and zinc selenide bulk targets. In zinc plasma with optimal fluence of heating radiation ($F = 2$ J cm^{-2}, upper panel), a 6×-fold gain of ninth harmonic (H9) with

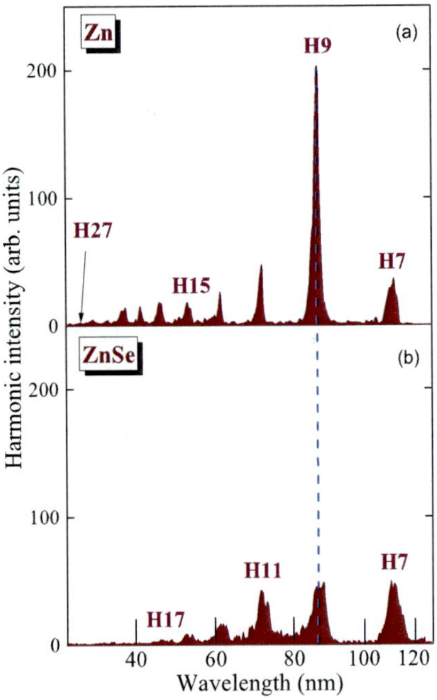

Fig. 8.7 HHG from Zn-containing components of laser-induced plasma (LIP) using 806 nm driving pulses. Targets were ablated by 370 ps pulses. **a** Harmonic spectrum generated in Zn plasma using the fluence of heating F = 2 J cm^{-2}. **b** Harmonic spectrum generated in ZnSe plasma (F = 2.3 J cm^{-2}). Dashed line shows the position of $3d^{10}4s$–$3d^{9}4s4p$ transition. H7–H27 correspond to the harmonic orders. Reproduced from [38] with permission from MDPI

regard to the neighboring (H7 and H11) harmonic orders was obtained. The observation of such an enhancement of the single harmonic cannot be attributed to the predominance of lower orders over higher ones at the beginning of the plateau-like range of the harmonic distribution. In our experiments, the lower harmonic (H7) was approximately six times weaker than H9. In the meantime, the application of different plasmas (Ag, Au, Mo, etc.) showed that H9 was approximately equal to H7.

The wavelength of H9 ($\lambda = 89.6$ nm) lies close to the Zn II transition ($3d^{10}4s$–$3d^{9}4s4p$, 89.3 nm). Increasing the fluence of the heating pulses did not lead to disappearance of the resonance-related amplification effect, while reducing the gain of this harmonic. The maximum achieved harmonic cut-off under these conditions was H37 (not shown in the upper panel of Fig. 8.7).

The application of ZnSe plasma for HHG significantly changed the envelope of the harmonic spectra (bottom panel of Fig. 8.7) compared to that obtained from Zn ablation. Here we show the spectrum of harmonics obtained under optimal plasma formation conditions on the ZnSe bulk surface ($F = 2.3$ J cm^{-2}). The term "optimal plasma formation" refers to the highest harmonic output from this LIP. One can see a featureless, gradually decreasing harmonics starting from H7 up to the cut-off range (H17 for ZnSe).

In Fig. 8.7, the lines marking the position of the transition ($3d^{10}4s$–$3d^{9}4s4p$, 89.3 nm [40]) responsible for the enhancement of single harmonic in Zn LIP are included. To further distinguish a difference between the HHG spectra from two plasmas under consideration, we included the same line in other figures (Figs. 8.8.

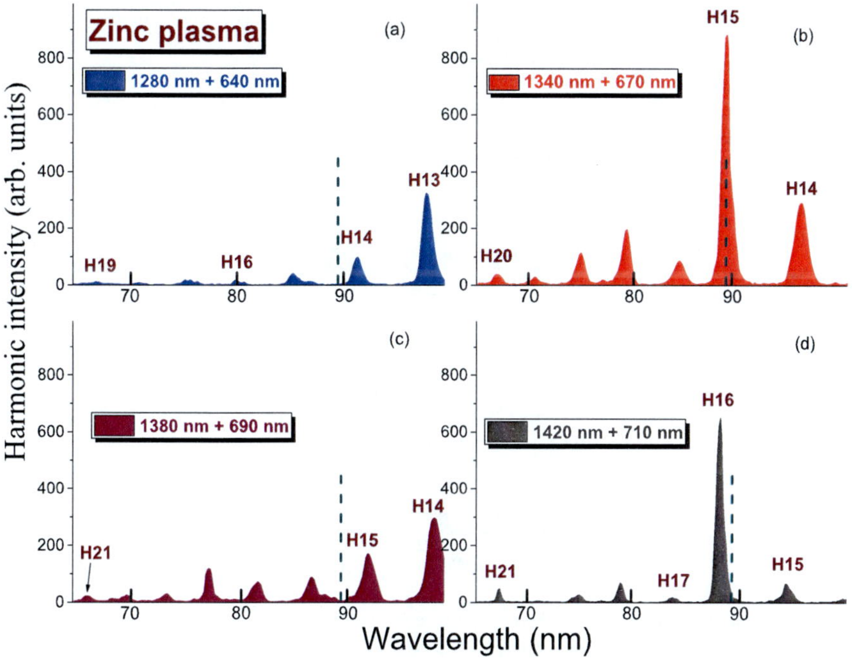

Fig. 8.8 Harmonic spectra obtained during propagation of spectrally tunable pulses through the zinc LIP. A two-color pump (TCP) was used for determination of the resonantly enhanced harmonic orders matching with the wavelength of strong ZnII transition (89.3 nm, dashed line). Four groups of HHG spectra were obtained using **a** 1280 nm + 640 nm, **b** 1340 nm + 670 nm, **c** 1380 nm + 690 nm, and **d** 1420 nm + 710 nm pumps. Dashed lines show the position of the $3d^{10}4s$–$3d^{9}4s4p$ transition. H14–H21 correspond to the harmonic orders. Reproduced from [38] with permission from MDPI

and 8.9) as well. Also note the observed closeness of two $3d^{10}4s$–$3d^{9}4s4p$ transitions of Zn II (88.1 and 89.3 nm) with a wavelength of 9th harmonic (λ = 89.6 nm) of 806 nm driving radiation.

The application of tunable NIR pulses from OPA allowed demonstration of the variation of single harmonic enhancement in the zinc plasma. Figure 8.8 shows four spectra of harmonics generated using different driving pulses (1280, 1320, 1380, and 1420 nm) and their second harmonics. The application of a 1280 nm + 640 nm pump did not allow generation of enhanced harmonics in the vicinity of strong ZnII transition (89.3 nm) marked by green dashed line (left upper panel). The nearest harmonic (H14) was not enhanced compared with the neighboring harmonic orders due to the detuning out from the above transition. The shift of the wavelength of driving pulses towards 1340 nm allowed observing very strong H15 (λ = 89.3 nm) coinciding with the strong 89.3 nm ionic transition (right upper panel). Other odd and even orders were 3–7 times weaker than the resonance-enhanced harmonic.

Further shift of the driving wavelength towards the longer wavelength range (1380 nm, bottom left panel) again resulted in detuning of neighboring harmonics

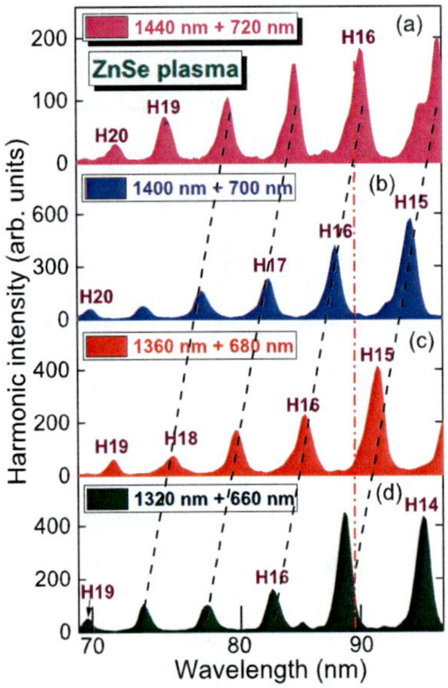

Fig. 8.9 Harmonics generation in ZnSe plasma using the tunable NIR pulses. The wavelength of driving pulses was gradually tuned between 1320 and 1440 nm. No enhancement of odd and even harmonics in the vicinity of 89.3 nm was observed. Inclined dashed lines show the tuning ranges of odd and even harmonics. H14–H20 correspond to the harmonic orders. Dash-dotted line shows the position of the $3d^{10}4s$–$3d^{9}4s4p$ transition. The two-color pump schemes were used during these studies with the following wavelengths of the interacting pulses: **a** 1440 nm and 720 nm, **b** 1400 nm and 700 nm, **c** 1360 nm and 680 nm, and **d** 1320 nm and 660 nm. Reproduced from [38] with permission from MDPI

(H15 and H16) out from the position of resonance (dashed line). Finally, the application of 1420 nm emission and its second harmonic (bottom right panel) again created conditions for enhancement of the single harmonic. In that case the even (H16) harmonic was enhanced due to closeness with the ionic transition. Thus the discussed studies using different lasers showed the enhancement of single harmonic once the conditions of closeness of the transition possessing strong oscillator strength and some harmonic order. In the case of the laser with fixed wavelength of driving pulses ($\lambda = 806$ nm), the enhancement of the single harmonic (Fig. 8.7, upper panel) was similar to the one shown in Fig. 8.8 using properly tuned wavelength of the driving field (see the spectra obtained at 1340 + 670 nm and 1420 + 710 nm).

Another pattern was observed in the case of ZnSe LIP. The tuning of the driving pulses between 1320 and 1440 nm in the case of TCP of this plasma caused a shift of harmonic wavelengths (see inclined dashed curves in Fig. 8.9). However, the envelope of the harmonic distribution was approximately similar in each of these cases. No enhancement of the harmonics approaching the $\lambda = 89.3$ nm was observed. These experiments were carried out using the fluence of heating 370 ps pulses $F = 2.3$ J cm^{-2}. The conditions of experiments were modified by either increasing the fluence of heating pulses up to the value when no strong plasma emission appears ($F = 3$ J cm^{-2}), or increasing the intensity of driving pulses. However, no strong coherent emission in the vicinity of 90 nm was observed. Notice a similarity and gradual decrease of odd and even harmonics, which shows the sufficient overlap of NIR and second harmonic pulses in the plasma area in spite of using relatively thick non-linear crystal for the formation of a second field.

Numerous experiments using pulses in different time scales (between 4 and hundreds of femtoseconds) have shown the efficiency of harmonics generation in LIPs. Spectral broadening in the case of ultrashort pulses allows generation of subfemtosecond pulses during HHG in plasmas. In particular, the application of the few-cycle pulses for HHG in LIPs allowed almost entire suppression of surrounding harmonics during resonance enhancement of the notably broadened single (33rd) harmonic in the manganese plasma [41].

Previously, different methods, like application of the photodiodes calibrated in XUV range, calibration of microchannel plate using the measurements of luminescence in different spectral ranges, etc., were implemented to quantitatively determine the harmonic yield from different plasmas. Those quantitative measurements of harmonic yield still suffer in accuracy, though some comparative estimates and measurements show that harmonics generation in plasma is at least a few times higher than in gases at similar experimental conditions. Although it is difficult to compare harmonic data in the case of different facilities, the conversion efficiency up to 10^{-5} was reported in the case of silver plasma. Even higher conversion efficiencies were reported in the case of a resonantly enhanced single harmonic in indium plasma, as well as harmonics generating in nanoparticle-containing LIPs. In the discussed case, the 10^{-6} conversion efficiency was estimated in the plateau range of harmonic distribution, while the resonant harmonics were up to 10 times stronger (i.e., their conversion efficiency was in the range of 10^{-5}).

8.2.4 Discussion

The following characteristics of two studied materials related to the topics of present studies can cause a difference in the harmonic emission. Zinc is a chemical element belonging to the transition metals with the atomic weight 66. Its electron configuration ($3d^{10}$ $4s^2$) coincides with Ar. Other important parameters are the melting point (~700 K), boiling point (~1200 K), heat of fusion and vaporization (7.3 and 115 kJ/ cm^{-3}), and ionization energies (1st: 906.4 kJ/mol, 2nd: 1733.3 kJ/mol, 3rd: 3833 kJ/mol). Zinc selenide is a light-yellow, solid compound comprising zinc (Zn) and selenium (Se) with the atomic weight 144. ZnSe can be made in both hexagonal and cubic crystal structure. It is a wide-bandgap semiconductor of the II-VI semiconductor group (since zinc and selenium belong to the 12th and 16th groups of the periodic table, respectively). Its melting point and band gap are ~ 1800 K and 2.8 eV, respectively. Our studies showed that zinc ablates at lower fluence compared with zinc sulfide. Nevertheless, in spite of the difference in the physical properties of these two solids, the ablation at optimal fluence of each of these materials allowed generation of harmonics with approximately similar conversion efficiency. The main distinction between two HHG spectra from those plasmas is related to the envelopes of harmonic distribution (compare Figs. 8.7, 8.8 and 8.9) demonstrating the resonance effect in the case of atomic plasma.

There are no data describing the ZnSe transitions in the XUV range possessing large *gf* s, especially in the vicinity of the harmonics of either fixed (806 nm) or tunable NIR pumps. The unavailability of this information does not tell us conclusively about their absence. They may either be lower than that responsible for the enhancement of a single harmonic generating in atomic/ionic Zn plasma in the vicinity of 90 nm, or can be tuned out from the initial position assigned in the case of atomic medium. The most probable reason is the former since the latter assumption was analyzed by tuning the pump wavelength in the case of a tunable NIR source (Fig. 8.9), while no enhancement of harmonic, contrary to that observed in the case of ablated zinc (Fig. 8.7), was achieved.

The cutoffs of the harmonics generating in zinc selenide plasma were in the range of ~50 nm in the case of the 806 nm pump (Fig. 8.7) and 60 nm in the case of the 1400 nm + 700 nm pump (not shown in Fig. 8.9). The discrepancy between the experiment and the expectations in the extension of this parameter for the longer-wavelength pump ($E_{cutoff} \propto \lambda^2$, E_{cutoff} is the harmonic cutoff and λ is the driving field wavelength) can be attributed to the difference in the beam sizes in plasma area and smaller conversion efficiency in the case of tunable NIR pulses. One can recall that the efficiency of the harmonics produced by shorter wavelength source becomes higher due to the strong wavelength-dependent yield ($I_{harm} \propto \lambda^{-5}$ [42]; I_{harm} is the harmonic intensity).

The mechanism of resonant amplification in atomic medium is still debatable. There are two possible approaches: one relies on a better recombination cross-section through the AIS in a single-atom response, and the other relies on the improved phase matching conditions in the vicinity of the resonance. Which of them provides stronger influence on the unusual distribution of harmonics depend on the medium used, laser pump conditions, and coincidence of harmonic and ionic transition wavelength, as well as the oscillator strengths of those transitions. Another additional factor is the elemental consistence of plasma components.

The existing theories of the microprocess of resonance-induced enhancement of a single harmonic in atomic plasmas are mostly based on (i) the four-step model, (ii) the harmonic generation in the presence of a shape resonance using time–frequency analysis of the intensity and phase, which underlined the resonance enhancement irrespective of the pulse length and supported the four-step model, and (iii) the approach generalized for the case when the capture into an AIS is replaced by the field-induced excitation of the ground state into this state. The variation of target excitation broadens analysis of the non-linear response of such plasma by comparing the enhancement of different harmonics provided they coincide or stay close to those resonances. In particular, the application of this approach during numerical study of HHG in Zn ions allowed the resonance effect in harmonic generation to be observed due to transitions to the highly excited bound states, rather than to the AIS. However, these bound states are broader due to photoionization in the intense laser field. Thus in this sense there is no fundamental difference between transitions to AIS (enhancing above-ionization threshold harmonics) and transitions to highly excited bound states (enhancing below-threshold harmonics). Moreover, the generation of the below-threshold harmonics also includes the quasi-free electronic motion as one of the steps of the process, and therefore the resonant enhancement of this generation can be described within the similar four-step model. One has to take into consideration the opportunity in competition of the micro- and macro-processes in the LIP.

The discussed experimental results in the case of atomic Zn plasma well match with the theoretical consideration of this process described in the above refereed theoretical studies independent of the wavelength of the pumping laser, thus confirming the generalized approach of those theoretical studies. In particular, note the observed closeness of two $3d^{10}4s$–$3d^{9}4s4p$ transitions of Zn II (88.1 and 89.3 nm) with a wavelength of the 9th harmonic of 806 nm driving radiation. These transitions also influenced the harmonic yield in the case of our HHG in Zn plasma using 1340 nm (H15) and 1420 nm (H16) pump radiation (Fig. 8.8). Those harmonics (H15 and H16), being positioned near the above ionic transitions of singly-ionized zinc, were enhanced compared with the neighboring ones.

The absence of resonance enhancement in ZnSe points out the diminished role of the spectrally shifted transition in this process. The detuning of this transition in the case of molecules out from the atomic transition can be compensated for by tuning the wavelength of harmonic, as in the case of discussed experiments with the OPA pumping system. However, these experiments showed that the fine tuning of the harmonic along the expected strong transition did not result in the enhancement of some specific harmonics in the vicinity of this resonance (Fig. 8.9). The attempt to decay and disintegrate this molecule at higher fluencies ($F = 3$ J cm^{-2}), did not result in the enhancement of the single harmonic caused by the presence of a sufficient amount of Zn ions in such a plasma. Stronger fluencies ($F = 3.5$ J cm^{-2}) caused the intense emission of incoherent radiation from plasma, as well as the formation of a large amount of free electrons, which fully destroy the phase matching conditions, especially for the higher orders of harmonics.

8.3 Summary

In this chapter, we discussed the HHG in the Sn, Mn, Zn, Se, In, Sb, Cr, Ag, Au, Mg, C, and Cd plasmas, which showed numerous cases of the coincidence of harmonics with some emission lines of corresponding ions. However, this coincidence did not result in the enhancement of harmonics. We compared plasma conditions (spectral distribution of emission, excitation of high-charged ions, etc.) and harmonic generation. We compared the cases of enhancement of single harmonic in the vicinity of some ionic transitions and featureless plateau-like HHG spectra from different plasma species to add more information about the role of plasma emission spectra.

The comparative analysis of those two groups of plasma plumes has revealed the relative role of the strong emission lines and the transitions possessing high *gf* s in the enhancement of single harmonics. The analysis of the harmonic spectra in the case of the two abovementioned groups of LIPs has shown that ionic transitions demonstrating high oscillator strengths are responsible for enhanced emission of a single order or the group of harmonics. Meanwhile, plasma emission lines play rather a destructive role in HHG. We analyzed different models describing the resonance-induced enhancement of single harmonics. The multielectron resonance recombination model allowing a qualitative description of observed experimental results was introduced.

We have reconsidered the mechanism of resonant amplification of single harmonic in zinc-containing atomic and molecular plasmas when generating high-order harmonics in laser plasma plumes. It has been shown that selenides of this metal noticeably reduce the amplification of the single harmonic in comparison with purely atomic plasmas, probably due to the decrease of the oscillator strength of the involved ionic transitions. Stronger ablation of these molecular targets did not allow the resonant amplification of the single harmonic in ZnSe plasma to be restored using different laser sources, probably due to the insufficient decay of molecules leading to the appearance of zinc ions.

The discussed research allowed to predict modification of the mechanism of resonant amplification of single harmonics reported in other atomic plasmas (manganese, tin, chromium, selenium, tellurium, molybdenum, indium, and arsenic) being presented in the molecular form. This process can also be reconsidered by comparing the atomic and molecular plasmas containing the above elements, as well as by tuning the wavelength of driving radiation along those resonances that are responsible for the observed amplification of harmonics.

References

1. P.B. Corkum, F. Krausz, Nature Phys. **3**, 381 (2007)
2. J.F. Reintjes, *Nonlinear Optical Parametric Processes in Liquids and Gases* (Academic Press, Orlando, 1984)
3. R.A. Ganeev, M. Suzuki, T. Ozaki, M. Baba, H. Kuroda, Opt. Lett. **31**, 1699 (2006)
4. J. Rothhardt, S. Hadrich, S. Demmler, M. Krebs, S. Fritzsche, J. Limpert, A. Tunnermann, Phys. Rev. Lett. **112**, 233002 (2014)
5. R.A. Ganeev, J. Appl. Phys. **121**, 133108 (2017)
6. R.A. Ganeev, M. Suzuki, S. Yoneya, H. Kuroda, J. Appl. Phys. **117**, 023114 (2015)
7. A. Tauheed, Hala, Phys. Scripta **85**, 025304 (2012)
8. D.A. Esteves, R.C. Bilodeau, N.C. Sterling, R.A. Phaneuf, A.L.D. Kilcoyne, E.C. Red, A. Aguilar, Phys. Rev. A **84**, 013406 (2011)
9. C. McGuinness, M. Martins, P. Wernet, B.F. Sonntag, P. van Kampen, J.-P. Mosnier, E.T. Kennedy, J.T. Costello, J. Phys. B **32**, L583 (1999)
10. C. McGuinness, M. Martins, P. van Kampen, J. Hirsch, E.T. Kennedy, J.-P. Mosnier, W.W. Whitty, J.T. Costello, J. Phys. B **33**, 5077 (2000)
11. J.B. West, J.E. Hansen, B. Kristensen, F. Folkmann, H. Kjeldsen, J. Phys. B **36**, L327 (2003)
12. R.A. Ganeev, L.B. Elouga Bom, J.-C. Kieffer, T. Ozaki, Phys. Rev. A **75**, 063806 (2007)

13. G. Duffy, P. van Kampen, P. Dunne, J. Phys. B **34**, 3171 (2001)
14. M. Suzuki, M. Baba, R. Ganeev, H. Kuroda, T. Ozaki, Opt. Lett. **31**, 3306 (2006)
15. R.A. Ganeev, H. Singhal, P.A. Naik, V. Arora, U. Chakravarty, J.A. Chakera, R.A. Khan, I.A. Kulagin, P.V. Redkin, M. Raghuramaiah, P.D. Gupta, Phys. Rev. A **74**, 063824 (2006)
16. M. Suzuki, M. Baba, H. Kuroda, R.A. Ganeev, T. Ozaki, Opt. Express **15**, 1161 (2007)
17. M. Suzuki, M. Baba, R.A. Ganeev, H. Kuroda, T. Ozaki, J. Opt. Soc. Am. B **24**, 2686 (2007)
18. R.A. Ganeev, J. Zheng, M. Wöstmann, H. Witte, P.V. Redkin, H. Zacharias, Eur. Phys. J. D **68**, 325 (2014)
19. A. Kramida, Y. Ralchenko, J. Reader, NIST ASD team, in *NIST Atomic Spectra Database (Version. 5.1)* (Gaithersburg, MD: National Institute of Standards and Technology, 2013)
20. R.A. Ganeev, M. Suzuki, S. Yoneya, V.V. Strelkov, H. Kuroda, J. Phys. B **49**, 055402 (2016)
21. V. Strelkov, Phys. Rev. Lett. **104**, 123901 (2010)
22. R.A. Ganeev, Z. Abdelrahman, F. Frank, T. Witting, W.A. Okell, D. Fabris, C. Hutchison, J.P. Marangos, J.W.G. Tisch, Appl. Phys. Lett. **104**, 021122 (2014)
23. K.A. Dick, Can. J. Phys. **46**, 1291 (1968)
24. R. D'Arcy, J.T. Costello, C. McGuinness, G. O'Sullivan, J. Phys. B **32**, 4859 (1999)
25. G. Duffy, P. Dunne, J. Phys. B **34**, L173 (2001)
26. H. Kjeldsen, F. Folkmann, B. Kristensen, J.B. West, J.E. Hansen, J. Phys. B **37**, 1321 (2004)
27. E.S. Toma, P. Antoine, A. de Bohan, H.G. Muller, J. Phys. B **32**, 5843 (1999)
28. M.B. Gaarde, K.J. Schafer, Phys. Rev. A **64**, 013820 (2001)
29. R. Bartels, S. Backus, E. Zeek, L. Misoguti, G. Vdovin, I.P. Christov, M.M. Murnane, H.C. Kapteyn, Nature **406**, 164 (2000)
30. R. Taïeb, V. Véniard, J. Wassaf, A. Maquet, Phys. Rev. A **68**, 033403 (2003)
31. D.B. Milošević, J. Phys. B **40**, 3367 (2007)
32. D.B. Milošević, Phys. Rev. A **81**, 023802 (2010)
33. M.V. Frolov, N.L. Manakov, A.F. Starace, Phys. Rev. A **82**, 023424 (2010)
34. P.V. Redkin, R.A. Ganeev, Phys. Rev. A **81**, 063825 (2010)
35. M. Tudorovskaya, M. Lein, Phys. Rev. A **84**, 013430 (2011)
36. P.V. Redkin, M. Danailov, R.A. Ganeev, Phys. Rev. A **84**, 013407 (2011)
37. Y. Mairesse, J. Levesque, N. Dudovich, P.B. Corkum, D.M. Villeneuve, J. Modern Opt. **55**, 2591 (2008)
38. R.A. Ganeev, H. Kuroda, Photonics **8**, 29 (2021)
39. R.A. Ganeev, P.A. Naik, H. Singhal, J.A. Chakera, P.D. Gupta, Opt. Lett. **32**, 65 (2007)
40. A.M. Crooker, K.A. Dick, Canad. J. Phys. **46**, 1241 (1968)
41. R.A. Ganeev, T. Witting, C. Hutchison, F. Frank, M. Tudorovskaya, M. Lein, W.A. Okell, A. Zaïr, J.P. Marangos, J.W.G. Tisch, Opt. Express **20**, 25239 (2012)
42. P. Lan, E. Takahashi, K. Midorikawa, Phys. Rev. A **81**, 061802 (2010)

Chapter 9
Quasi-Phase-Matching in Laser-Induced Plasmas

9.1 Basics of the Quasi-Phase-Matching in Laser-Induced Plasmas

9.1.1 Introduction

Many methods have been implemented in the high-order harmonic generation in gases and plasmas to improve the conversion efficiency. In the case of harmonics generated in laser-plasma they include the use of nanoparticles and clusters as harmonic sources, the use of various combinations of two- or three-color pumps, the application of extended plasma, the realization of the conditions of resonant amplification of single harmonic, and the formation of quasi-phase-matching (QPM) between interacting waves. The latter of these techniques in combination with two-color pumping of plasma have demonstrated their attractiveness by increasing the conversion efficiency in various ranges of the extreme ultraviolet region [1]. Most previous studies of QPM were performed using conventional Ti:sapphire lasers and their second harmonics as the radiation sources. Meantime, the use of longer wavelengths of the laser sources can lead to a further increase of harmonic yield by applying the multi-jet plasmas compared with the extended homogeneous plasmas. Particularly, the tunable optical parametric amplifiers (OPA) generating in the near-infrared (NIR) region, and their second harmonics will allow generating a wide range of wave sum and difference harmonics.

The phase mismatch between the interacting waves suppresses the HHG in the gas and plasma media. The dephasing between the propagating harmonic wave and laser-induced polarization is mainly caused by the dispersion of the medium and becomes essential in the presence of a significant number of free electrons. Due to the difference in the velocity of the waves, at some distance from the beginning of the medium, the phase shift becomes close to π. Beyond this distance, called the coherence length, the constructive accumulation of harmonic photons is reversed and

R. A. Ganeev, *High-Order Harmonics Generation in Plasmas*, Springer Series on Atomic, Optical, and Plasma Physics 122, https://doi.org/10.1007/978-3-031-09040-0_9

the harmonic energy starts to decrease. QPM of interacting waves can significantly reduce the negative impact of phase mismatch as well as increase the conversion efficiency in different parts of the XUV spectrum.

QPM has been demonstrated in gases [2–7] and multi-jet plasma formations [8]. Though not new for the case of gas harmonics, this approach recently allowed a significant increase of harmonic yield in the case of modulated plasmas. Moreover, the so-called 'self-probing' schemes to extract structural and dynamical information about the generating medium from the intensity measurements of HHG, which have already been reported in the case of gas targets, are extended to plasma targets. The implementation of this concept for the case of plasma harmonics widely broadens the field of study due to the overwhelming prevalence of the ablated solid species used in the latter case over the few gas types routinely exploited in conventional HHG.

The key enabling technology for this research is the capability of laser ablation to put atoms and molecules from the solid-state into the gas phase at appropriate densities [8]. This method of plasma formation also allows the analysis of the relative role of micro- and macro-processes in the enhancement of harmonic yield [9]. Such a concept can be useful for the formation of a group of harmonics when the phase-locking of the harmonics leads to the production of an attosecond pulse train. The comparative analysis of different sources of pump radiation (e.g. fixed-wavelength ones, such as Ti:sapphire lasers and their second harmonics, or those with tunable wavelengths, such as OPA and their second harmonics) allows defining the best schemes for enhancement of HHG in different spectral regions of XUV.

In this section, we discuss the experimental observations of the QPM in the set of plasma jets [10]. The behavior of the phase mismatch as a function of the propagation distance in the modulated plasma plume is analyzed. We also discuss the application of tunable mid-infrared pulses and fixed wavelength 800-nm-class laser for the QPM of harmonics. We demonstrate various peculiarities of QPM.

9.1.2 Experimental Arrangements

Two laser sources were used for these studies. In both cases, the uncompressed radiation of the Ti:sapphire laser was used as a heating pulse (central wavelength $\lambda =$ 806 nm, pulse duration 370 ps, pulse energy up to $E_{\mathrm{hp}} = 4$ mJ) for plasma formation using a 200 mm focal-length cylindrical lens, which focused the pulse inside the vacuum chamber containing an ablating target to create the extended laser-induced plasma above the target surface (Fig. 9.1a). The focusing of the heating pulse on the target surface produced the extended non-perforated plasma. The intensity of the heating pulses on a plain target surface was varied up to 5×10^9 W cm^{-2}.

In the first set of studies, the compressed driving pulse from the same laser with the energy of up to $E_{\mathrm{dp}} = 5$ mJ and 64 fs pulse duration was used, 45 ns from the beginning of ablation, for harmonic generation in the plasma plume. The driving pulse was focused using a 400 mm focal-length spherical lens onto the prepared plasma from a perpendicular direction, at a distance of ~100 μm above the target surface.

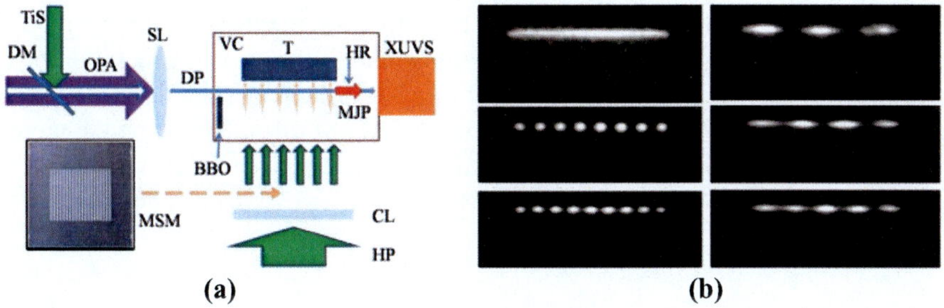

Fig. 9.1 **a** Experimental setup for the HHG in the multi-jet plasmas. HP, heating pulse; DP, driving pulse; CL, cylindrical lens; SL, spherical lens; MSM, multi-slit mask; DM, dichroic mirror; TiS, femtosecond radiation from Ti:sapphire laser; OPA, femtosecond radiation from optical parametric amplifier; VC, vacuum chamber; T, target; BBO, nonlinear crystal for second harmonic generation; HR, harmonic radiation; MJP, multi-jet plasma; XUVS, extreme ultraviolet spectrometer. **b** Shapes of silver plasma plumes produced at different modes of excitation of the target. Left picture shows the extended (5-mm-long) homogeneous plasma (upper panel), eight-jet plasma produced during installation of the MSM between cylindrical lens and target (middle panel), and nine-jet plasma produced during tilting the MSM at $\theta = 25°$ (bottom panel). Right picture shows the images of three-, four- and five-jet plasmas produced during placement of the MSM in different positions inside the telescope installed in front of the cylindrical lens. Reproduced from [10] with permission from Optical Society of America

The confocal parameter of the focused driving beam was 18 mm. The intensity of the driving pulse in the focal area was varied up to 9×10^{14} W cm^{-2}. The harmonic emission was analyzed by an XUV spectrometer.

In the second set of studies, the OPA was pumped by 10 mJ, 806 nm, 64 fs pulses. Signal and idler pulses from the OPA allowed tuning within the 1170–1620 nm and 1580–2650 nm NIR ranges respectively. The experiments were carried out using ~1-mJ, 70-fs, 1310-nm signal pulses. The spectral bandwidth of these pulses was 50 nm. The intensity of the 1310-nm pulses focused by 400-mm focal length lens inside the extended plasma was ~2×10^{14} W cm^{-2}. Most of the experiments in this configuration were performed using a two-color pump of the LIP. The choice of the double beam configuration was due to the small energy of the driving pulse (~0.7–1 mJ depending on the wavelength of signal radiation). The $I_H \propto \lambda^{-5}$ rule (I_H is the harmonic intensity and λ is the driving field wavelength [11, 12]) did not allow the observation of strong harmonics from the ~1310-nm pulses. Because of this, the second-harmonic generation of the signal pulse was used to implement the two-color pump scheme of plasma HHG. 0.5-mm thick BBO crystal was installed inside the vacuum chamber in the path of the focused signal pulse. The conversion efficiency of the 650-nm pulses was ~20%. Two pulses overlapped both temporally and spatially in the extended plasma and allowed a significant enhancement of odd harmonics, as well as generation of even harmonics with intensity similar to that of the odd ones.

Silver, indium, and manganese were used as the targets for ablation. The reasons for choosing these samples are described in the following subsection. The size of the targets where the ablation occurred was 5 mm. To create multi-jet plasmas a multi-slit mask (MSM) was used. The slits of 0.3 mm width were spaced at 0.3 mm (d = 0.3 mm and D = 0.6 mm, see Fig. 9.1). The energy of the heating pulse decreased after propagation through the MSM. However, the fluence of this radiation on the target surface remained unchanged, since the size of the ablated area was also decreased. It means that the electron and plasma densities in the cases of extended homogeneous and multi-jet plasmas were almost equal.

9.1.3 Enhancement of the Groups of Harmonics

In the case of homogeneous plasma target, the silver plasma allows generation of almost ideal plateau-like distribution of harmonic intensities: the decrease of the latter starting from the 13th order (for 806 nm pump wavelength) is very slow up to the cut-off region (~H61). Therefore the multi-jet target setup for silver plasma allows a clear observation of the quasi-phase-matching effect in different regions of XUV by variation of the plasma modulation parameter. This parameter actually represents the size of a single jet in the multi-jet plasma formed by the installation of the MSM between the focusing cylindrical lens and the target (inset in Fig. 9.1a).

Initially, it can be shown that the plasma shapes used for harmonic generation in the silver multi-jet plasmas are produced at different modes of excitation of the target. The left picture in Fig. 9.1b shows the extended (5-mm-long) homogeneous plasma (upper panel), the eight-jet plasma produced with MSM installed between the cylindrical lens and the target (middle panel), and the nine-jet plasma produced as the MSM is tilted at $= \theta 25°$ (bottom panel). The picture on the right shows the images of three-, four- and five-jet plasmas produced as the MSM is placed in different positions inside the telescope installed in front of the cylindrical lens.

Figure 9.2a shows the harmonic spectra generated using the 806 nm pump in the extended (thin curve) and eight-jet (thick curve) silver plasmas produced by the installation of the MSM ($\theta = 0°$). In the latter case, one can see an enhancement of the group of harmonics centered at H33. The enhancement factor for the H33 was measured to be 11.

Stronger enhancement was observed in the case of a two-color pump using the 1310 nm + 655 nm radiation (Fig. 9.2b). The two spectra shown in this figure represent the harmonic distributions captured under similar experimental conditions (energy of heating pulse, collection time of the harmonic spectra). The thin curve demonstrates the harmonic distribution obtained in the case of 5-mm-long silver plasma. The application of the MSM tilted at 35° drastically changed this distribution (thick curve). The group of harmonics centered on H37 was notably stronger compared with the previous case (the enhancement factor of the maximally enhanced H37 was ~18), whereas the intensity of the lower-order harmonics decreased due to worsened conditions of the phase matching.

Indium plasma has an attractive feature allowing the observation of the joint influence of micro- and macro-processes on the harmonic efficiency. From the very

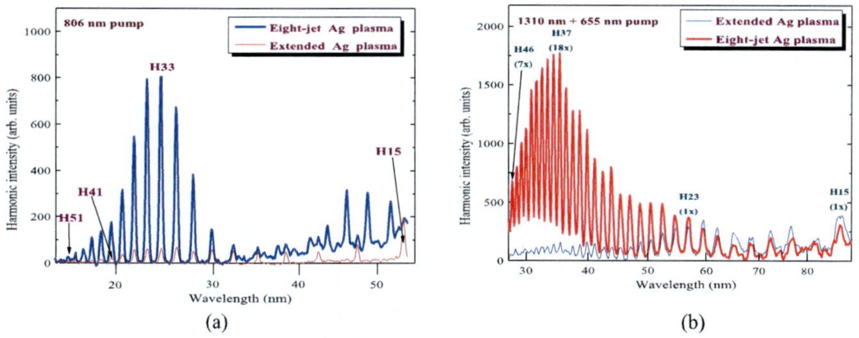

Fig. 9.2 **a** Harmonic spectra generated using the 806 nm pump in the extended silver plasma (thin red curve) and eight-jet plasma produced by installation of the MSM containing 0.3 mm slits (thick blue curve). **b** Two-color pump using the 1310 nm + 655 nm radiation of the extended (thin curve) and eight-jet (thick curve) plasmas. The enhancement factors are shown below the harmonic orders. Reproduced from [10] with permission from Optical Society of America

beginning of plasma HHG studies, this plasma demonstrated the largest enhancement of a single (H13 of the 800-nm-class lasers) harmonic. The reported enhancement factors were close to 80 [13, 14]. In the discussed studies [10] this plasma was used to demonstrate the QPM effect for the harmonics far from the spectral region (61 nm) where the resonance-induced harmonic occurred.

In the discussed studies, the variation of this spectrum was analyzed while introducing the modulation of the extended (5 mm) indium plasma, similarly to the method described in the case of silver plasma. Particularly, a proper choice of the plasma conditions and pump wavelength can further optimize the QPM process using the described configuration. It is possible to increase the enhancement factor of QPM harmonics by applying the wavelength of OPA allowing stro nger emission (1310 nm) while exciting the indium target in such a manner that allowed to maintain the QPM conditions for the group of lower-order harmonics (Fig. 9.3). The two-color pump (1310 nm + 655 nm) of this modulated plasma containing eight jets resulted in the enhancement factors of 40–50 for the group of harmonics centered at H35 (λ = 37.4 nm). Some other features of the influence of micro- and macro-processes on the high-order harmonic generation in the laser-produced plasmas were analyzed in Ref. [9].

Finally, we discuss the observation of the QPM effect in the manganese plasma. The peculiarity of this material is the closeness of the QPM- and resonance-enhanced harmonics, which is in contrast to the case of indium plasma. The advantage of Mn plasma from the point of view of the highest harmonic cut-off (i.e. above 100th orders) was reported in Refs. [15, 16], though the reasons for this peculiarity were not yet clearly explained. Note that the harmonic yield from this plasma was not as high as from some other plasma species. The use of few-cycle pulses (3.5 fs) [17] has allowed the observation of a single (33rd) broadband harmonic.

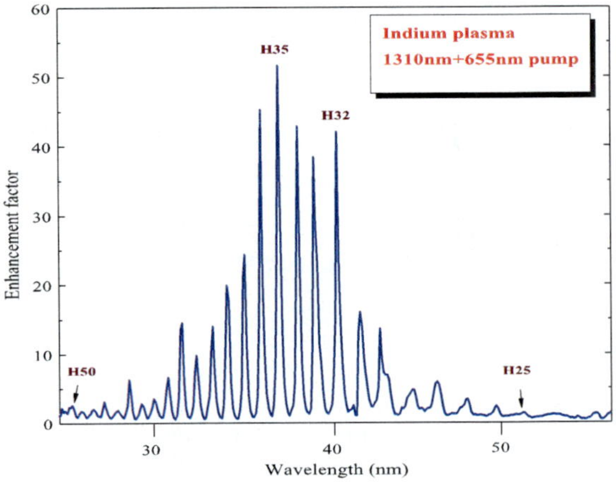

Fig. 9.3 The enhancement factor of the harmonics generated during two-color pump (1310 nm + 655 nm) of the modulated indium plasma containing eight jets. Reproduced from [10] with permission from Optical Society of America

In the discussed studies, the Mn plasma produced by 370 ps pulses allowed the generation of harmonics of 806 nm radiation with orders up to the nineties. A weak excitation of target by 370 ps heating pulses ($F_{\mathrm{hp}} \approx 0.2$ J cm^{-2}) led to plateau-like harmonic generation up to the cut-off (27th harmonic) similar to the one defined from the three-step model of HHG [18] for the singly charged Mn ions. Earlier, a growth of the harmonic intensity for the harmonics above H31 was attributed to the influence of the giant 3p–3d resonances of manganese ions [15]. H27–H31 were significantly suppressed compared with the higher-order harmonics starting from H33.

The increase of the heating pulse fluence up to 1.0 J cm^{-2} caused a significant change of harmonic distribution in the case of imperforated 5-mm-long manganese plasma. The thin curve in Fig. 9.4 shows the generated spectrum, which points out the presence of only low-order harmonics (up to H29) in the case of imperforated Mn plasma, with some weak traces of higher orders. Replacement of the single extended nonlinear medium with the group of separated components of this medium (i.e.

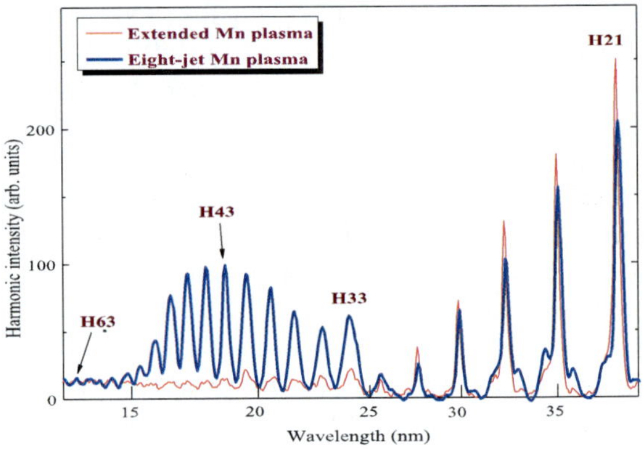

Fig. 9.4 Harmonic spectra generated during propagation of 806 nm pulses through the extended Mn plasma (thin curve) and eight-jet Mn plasma (thick curve). Reproduced from [10] with permission from Optical Society of America

eight 0.3-mm-long plasma jets) drastically modified the generated harmonic spectrum (thick curve in Fig. 9.4). A group of strongly enhanced harmonics (H33–H51) prevailed in this spectral distribution. Notice that the resonance-enhanced harmonic (H33) was also enhanced, to some extent, due to QPM. The total enhancement factor of this harmonic was only ~4 (a more accurate measurement of this factor was problematic because of the low signal from the extended plasma target). Nevertheless, one can see that H33 is more intense than the neighbor harmonics (H31 and H35). The enhancement of the single harmonic is typical for resonance-induced enhancement. Thus the enhancement of H33 is due to both micro- and macro-processes.

9.2 Application of Quasi-Phase-Matching Concept for Enhancement of High-Order Harmonics of Ultrashort Laser Pulses in Plasmas

To diminish the role of phase mismatch in laser-induced plasmas on HHG efficiency the quasi-phase-matching concept has been introduced when the growing difference between the phases of two waves is nullified by using the modulated shape of LIP. As it was described in previous sections, the method is based on the division of extended plasma onto the set of equidistantly separated plasma jets by inserting the multi-slit mask (MSM) in front of ablating surface [8], or by using the perforated targets [19]. Both methods have demonstrated the QPM concept in plasma, while the former approach has shown flexibility compared to the latter one in modifying the pattern of spatially modulated plasma using simple manipulation with the positioning and tilting the MSM.

The conditions of QPM in multi-jet plasmas are similar to those reported in the case of gaseous multi-jet targets. The principles of QPM developed in those studies, as well as the theoretical approaches proposed in the case of QPM in plasmas, allow describing the main peculiarities of this approach and comparing the theoretical predictions and experimental observations of QPM. In the study of QPM conditions, it is usually assumed that the target is in some fixed state with some given degree and level of ionization. This approximation is more or less suitable to describe the propagation of the driving pulses; however, it is not suitable to describe the interaction of harmonics with plasmas, since harmonics are emitted in different spatial points of the nonlinear optical medium during the whole action of driving laser field. The model that takes into account the propagation of the driving laser field and at the same time includes the quantum mechanical description of atomic electron evolution at QPM conditions has been proposed in [20].

In this section, we analyze different schemes allowing modifications of harmonic spectra at the conditions of QPM in silver, indium, and gold multi-jet plasmas using single- and two-color femtosecond pumps at the peak intensities of 800 and 1320 nm pumps in plasma area of 5×10^{14} and 1.4×10^{14} W cm^{-2}, respectively [21]. We analyze the role of the number of coherent zones, sizes of plasma jets and distance between them, plasma formation conditions, and characteristics of fundamental radiation on the HHG efficiency at QPM conditions in different ranges of XUV. We discuss the two-color pump-induced QPM at different regimes of laser-plasma interaction.

9.2.1 Experimental Arrangements

In these HHG studies, the 810 nm, 50 fs, 10 Hz and 810 nm, 340 ps, 10 Hz pulses from Ti:sapphire laser were used as the driving and heating radiation. The 70 fs pulses tunable in the near-infrared were also used for pumping the modulated plasmas and harmonic generation. The 340 ps pulses were focused using 300-mm focal length cylindrical lens on the surface of ablating target placed in the vacuum chamber to form the homogeneous extended plasma ($l = 5$ mm). The ablation of the target was accomplished 35 ns prior to propagation of the driving femtosecond pulses through the plasma. MSM was installed between the cylindrical lens and target to produce MJP (Fig. 9.5). Notice that no difference in the position of MSM between focusing lens and target to form suitable jets was observed. The width of each slit was varied between 0.3 and 0.8 mm with a similar distance between slits.

Fundamental radiation (810 nm or tunable NIR radiation) and its second harmonic produced during propagation of fundamental beam through the 0.3-mm thick barium borate (BBO) crystal installed inside the vacuum chamber were focused onto the MJP. The same BBO was used to double both 810 and NIR radiation at the appropriate phase-matching angle. The duration of driving pulses was sufficiently long to provide the phase-matching bandwidth in this crystal for both wavelengths. The position of the focused femtosecond beam was approximately 0.2 mm away from the surface of the target. The Rayleigh range of femtosecond beam was 4 mm, i.e. close to the whole length of plasma, thus assuming that the driving beam propagated through the 5 mm

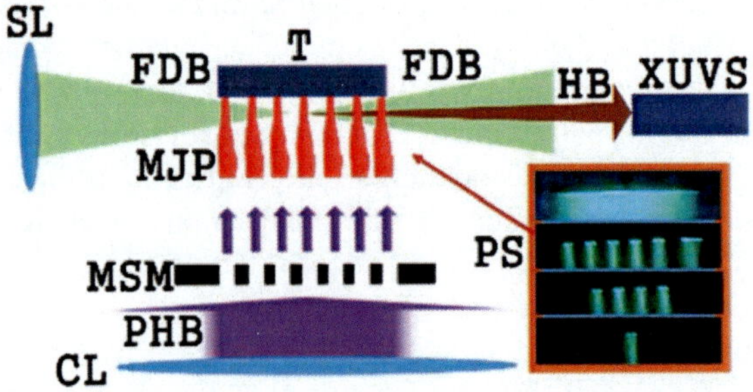

Fig. 9.5 Experimental setup for HHG in MJP. CL, cylindrical lens; PHB, picosecond heating beam; MSM, multi-slit mask; MJP, multi-jet plasma; T, target; SL, spherical lens; FDB, femtosecond driving beam; HB, harmonic beam; XUVS, extreme ultraviolet spectrometer. Inset: The extended, six-, four-, and single-jet plasma images are shown from the top to bottom. PS, plasma shapes. Reproduced from [21] with permission from MDPI

long plasma at the condition of almost plane wave—plasma interaction, which was exactly simulated during this study. Thus the role of the Gouy phase was assumed as insignificant in these experiments. The peak intensities of 800 and 1320 nm pumps in the plasma area were 5×10^{14} and 1.4×10^{14} W cm^{-2}, respectively. The generated harmonics were analyzed using a homemade XUV spectrometer consisting of a gold-coated cylindrical mirror, flat-field grating (1200 groves/mm), and a microchannel plate.

The group velocity dispersion in the BBO crystal leads to some temporal separation of two pulses [22]. However, this temporal mismatch in the area of laser-plasma did not prevent converting them to the odd and even harmonics. The appearance of all even harmonics pointed out the sufficient overlapping of the 1320 and 660 nm pulses in the plasma area. In the case of the two-color pump (TCP) using second harmonic generation the ratio of the pulse energies of second harmonic and fundamental waves of Ti:sapphire laser was in the range between 0.05 and 0.2 for 800 nm and 1320 nm fundamental waves. TCP of plasmas using conventional scheme (800 nm + 400 nm) has been analyzed for the first time in [23] where the notable enhancement of harmonic yield compared with the single-color pump (SCP) was achieved. With suitable control of the relative phase between the fundamental and the second harmonic radiation, the latter field enhances the short path contribution, resulting in a clean spectrum of harmonics [24].

Due to the group velocity dispersion in the type-I BBO crystal, the 800-nm pulse (ω) becomes delayed in the 0.3-mm-long BBO by $\Delta_{\text{cryst}} \approx 57$ fs with respect to the 400-nm pulse (2ω). This delay would decrease the ratio between the overlapped second harmonic and driving radiation in plasma and diminish the influence of the 400-nm wave on the output spectrum of generating harmonics. Nevertheless, as will be shown below, even at these unfavorable conditions the efficiencies of even and odd harmonics were comparable in the cases of the 0.3-mm long crystal. Notice that the application of longer crystals (type-I BBO, $d = 0.7$ mm and beyond) led to a considerable change in the role of the second-harmonic wave. While conversion efficiency toward the second-harmonic wave in these cases (11% for 0.7-mm long BBO) was higher than in the cases of shorter (0.3-mm long) BBO crystal, the much faster-growing delay between pump pulses allowed only insignificant partial overlap in the plasma (0.01 in the case of 0.7-mm long crystal). The calculations showed that, for 0.3 mm long BBO crystal at 5% conversion efficiency and 72 fs duration of broadened second harmonic pulse and delay between the maximums of two pulses on 57 fs, the ratio between overlapping 2ω and ω waves becomes equal to 0.03 [25]. The same features were observed in the case of NIR (1320 nm) pulses, though at rather favorable conditions with regard to the 800 nm pump due to smaller group velocity dispersion in BBO in the longer-wavelength range.

9.2.2 QPM in LIPs: Experiment

Visual presentation of spectra and clear definition of the difference between the so-called plateau-like harmonics and the group of enhanced harmonics in the plateau range motivated to analyze the images of generating spectra appearing on the screen of the monitor. Below, the raw images of harmonic spectra from different LIPs are shown. The saturated images of harmonics are intentionally chosen to present the spectra for better viewing. Notice that for the line-outs of the HHG spectra the unsaturated images were used.

Firstly, we analyze how the harmonic spectra look like in the case of different extended (5-mm-long) homogeneous plasmas. Figure 9.6a shows the plateau-like shapes of the odd harmonics generated in indium, silver, and gold plasmas in the case of 810 nm, 50 fs driving pulses. The harmonics between the seventeenth (H17) and thirty-third (H33) orders are shown. One can see the approximately similar distribution with almost equal intensities of these harmonics.

This plateau-like shape of harmonic spectra drastically changed once one inserts the MSM on the path of the heating picosecond radiation. The examples of three strongly modulated and enhanced spectra of the groups of odd and even harmonics of two-color NIR pump (1320 nm + 660 nm, 70 fs pulses) compared to the case of extended homogeneous indium plasma are shown in Fig. 9.6b. Three upper spectra containing significantly enhanced groups of harmonics centered at H34, H31, and H28 were observed while tilting the MSM at angles of 35°, 20°, and 0° and creating the 11-, 9-, and 8-jet indium plasmas, with the sizes of a single jet of 0.23, 0.28, and 0.31 mm respectively. Meanwhile, the bottom spectrum shows the HHG from the homogeneous extended indium plasma produced at similar conditions.

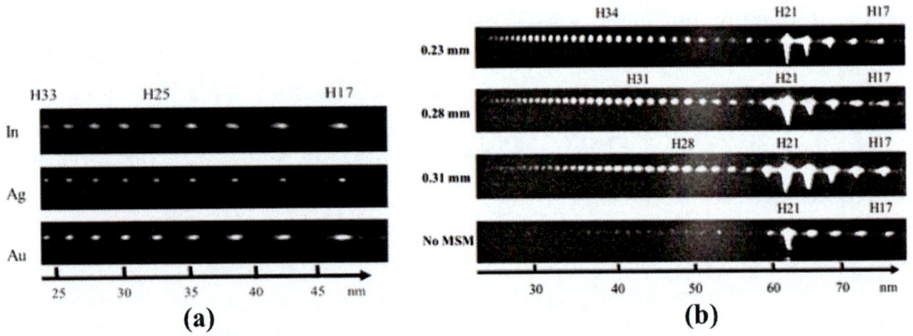

Fig. 9.6 Raw images of harmonic spectra from **a** extended homogeneous (top to bottom) indium, silver and gold LIPs pumped by 810 nm, 50 fs pulses and **b** multi-jet indium LIP produced using tilted MSM with the single plasma jet sizes of (top to bottom) 0.23, 0.28, and 0.31 mm pumped by two-color NIR + visible (1320 nm + 660 nm), 70 fs pulses. The peak intensities of 800 and 1320 nm pumps in plasma area were 5×10^{14} and 1.4×10^{14} W cm^{-2}, respectively, while the intensity of the second harmonic of 1320 nm pump was ~ 1×10^{13} W cm^{-2}. Maximally enhanced harmonics (H34, H31, and H28) are clearly distinguished for each of these plasma configurations. Bottom spectrum corresponds to the HHG from the homogeneous extended indium plasma produced at similar conditions. Reproduced from [21] with permission from MDPI

Notice that collection times of these four spectra and energy fluencies of target ablation were similar to each other. The enhancement factors of these harmonics (three upper spectra) with regard to those generated in the case of imperforated plasma (bottom spectrum) were measured to be 38× , 36× , and 27× respectively. A two-fold decrease of the whole length of multi-jet plasma compared with extended imperforated plasma (2.5 and 5 mm respectively) did not play important role in harmonic yield, which underlines the importance of collective processes over single-atom ones.

The bottom spectrum of Fig. 9.6b, as well as three upper spectra, shows the strong resonance-induced enhancement of H21 and H20. In the meantime, the formation of QPM conditions (three upper spectra) demonstrates the significant enhancement of the groups of harmonics surrounding the one (H_{qpm}) for which conditions of QPM became most favorable.

A similar pattern, though without resonance-induced enhancement of harmonics, was observed in the case of the silver plasma pumped by 1310 nm + 655 nm radiation (Fig. 9.7). Again, in the case of imperforated 5-mm-long Ag plasma, barely seen plateau-like harmonics were obtained (bottom image). The groups of strong harmonics appeared once the MSM was inserted between the target and heating pulse. They were centered at H43, H36, and H31 (see Fig. 9.7 from top to bottom) and corresponded to the MJP structures with the sizes of a single jet of 0.25, 0.31, and 0.35 mm. Notice that in the case of application of MSM, the whole energy of heating pulses interacting with the target was twice smaller compared with the case of extended homogeneous plasma, while the energy fluencies of heating radiation on the target surface were similar to each other.

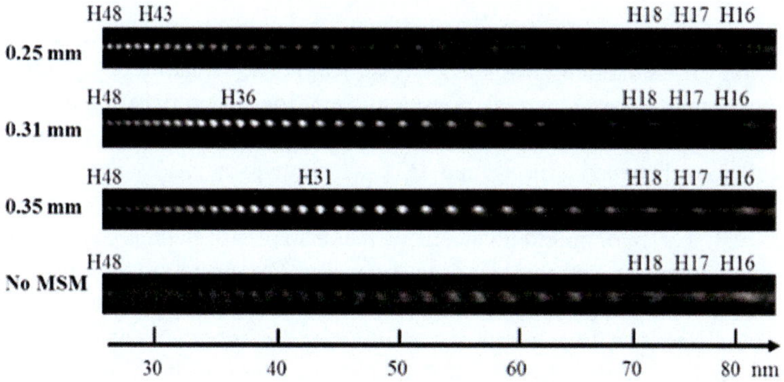

Fig. 9.7 Raw images of harmonic spectra from the multi-jet silver LIP produced using tilted MSM with single plasma jet sizes of (top to bottom) 0.25, 0.31, and 0.35 mm pumped by two-color (1310 nm + 655 nm), 70 fs pulses. The intensities of 1310 nm and 655 nm pumps in plasma area were 1.3×10^{14} and 1×10^{13} W cm^{-2}, respectively. Maximally enhanced harmonics (H43, H36, and H31) are determined for each of these plasma configurations. Bottom spectrum corresponds to homogeneous extended silver plasma. Similar conditions of plasma ablation using the same energy fluence of heating pulses and data collection times are applied for these four cases. Reproduced from [21] with permission from MDPI

The analysis of the intensity of harmonics as the function of the number of plasma jets (n) can provide proof of the decisive role of the QPM in the variations of the harmonic spectrum. The coherent accumulation of harmonic yield along the whole length of the divided nonlinear optical medium is demonstrated in the discussed studies. The indium target was used for ablation and the two-color pump (1300 nm + 650 nm) for HHG in indium plasma. The heating areas on the target (shown in the inset in Fig. 9.5) were shielded one by one to allow the generation of harmonics in different numbers of plasma jets. The number of jets was increased one by one from a single jet to 8 jets. The width of each jet was 0.3 mm and the empty area on the target surface between jets was also equal to 0.3 mm. The whole length of the 8-jet structure was ~5 mm.

Harmonic spectrum in the case of a single jet showed a plateau-like shape of weak harmonics. Each addition of the next jet caused the growth of the group of harmonics centered at around H34. This harmonic had the coherent length of ~0.3 mm at the used fluence of heating pulses on the surface of In target (1.2 J cm^{-2}) and the intensity of driving NIR pulses (1.4 × 10^{14} W cm^{-2}).

The coherence length (in mm) at the conditions of using the 800-nm driving laser could be presented as $L_{\mathrm{coh}} \approx 1.4 \times 10^{18}/(N_e \times H_{\mathrm{qpm}})$ [4]. Here N_e is the electron density in cm^{-3}. H34 has the coherence length of ~0.3 mm at the used fluence of heating pulses on the surface of In target (1.2 J cm^{-2}) and the intensity of driving NIR pulses (1.4 × 10^{14} W cm^{-2}). In the experiment, the MSM containing slits with the sizes of 0.3 mm was used, which allowed exciting the target in such a manner that electron density satisfies the above relation for the H34 at the length of a single jet of 0.3 mm. Correspondingly, at another excitation of target, the electron concentration of plasma may satisfy another harmonic order at the fixed sizes of the single jet in the MJP.

The yield of H34 was grown as $I_{\mathrm{H34}} \sim n^2$ up to $n = 4$. Further addition of jets led to deviation from quadratic dependence (Fig. 9.8a). The violation of equality of the conditions for each next added jet, non-ideal conditions of propagation through jets, and/or low-order nonlinear optical properties of plasma may change the optimal phase relations between driving and H34 waves, which causes the deviation of $I_{\mathrm{H34}}(n)$ dependence from the quadratic one with the growth of the number of jets above $n = 4$. The maximum enhancement factor may also decrease from the ideal value at the conditions when absorption processes are turned on, or in the case of unequal properties of the jets, which can arise from the heterogeneous excitation of the extended target. Even at these conditions of the imperfect coherent summing of nonlinear optical response, it was possible to observe the 40-fold growth of H34 in the case of 8-jet plasma compared with the single-jet one. The surrounding harmonics were also notably increased, though at a lower enhancement factor as compared to H34. Small dispersion of low-dense plasma allowed achieving the enhancement for ~ten harmonics at around H34.

H34 was perfectly suited for comparative analysis of the role of the number of separated jets on the variation of the intensity of generated harmonics. The coherence length in these experiments has a direct relation with the coherent addition of harmonic emission after propagation of H34 through each next jet. As it was shown,

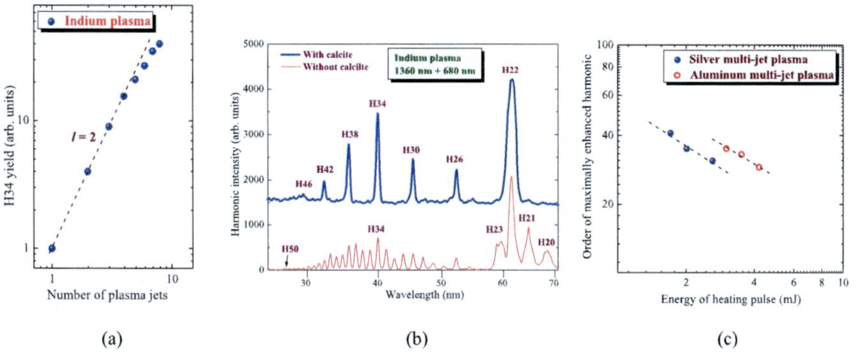

Fig. 9.8 **a** Dependence of H34 yield on the number (n) of jets of the indium MJP. The multi-jet indium LIP was produced using the tilted MSM with the single plasma jet sizes of 0.23 mm and pumped by two-color NIR + visible (1320 nm + 660 nm), 70 fs pulses. The peak intensity of 1320 nm pump was 1.4×10^{14} W cm^{-2}. Dashed line shows the slope $l = 2$ of the experimental dependence up to $n = 4$. **b** Harmonic distribution obtained from indium MJP in the case of propagation of driving pulses (1360 nm + 680 nm) through the plasma [bottom spectrum, similarly to the case shown in the upper panel of Fig. 9.2b] and calcium quarter-wavelength plate and plasma (upper panel). **c** Influence of heating pulse energy on the order of maximally enhanced harmonic in the case of silver (filled spheres) and aluminum (empty circles) MJPs. Reproduced from [21] with permission from MDPI

even at these favorable conditions for coherent summing of harmonic waves in each next jet, the additional impeding processes did not allow following the theoretically predicted quadratic rule once the number of jets exceeded four coherence zones. As for non-phase matched harmonics, for example, H18 of 1320 + 660 nm pump, this consideration (i.e. quadratic growth of harmonic efficiency with the growth of the number of jets) is not applicable, since phase mismatch between harmonic and fundamental waves does not allow the growth of harmonic yield.

Below we analyze the QPM-enhanced spectrum from indium MJP using a 1360 nm + 680 nmtwo-color pump at the conditions when the calcite quarter-wave plate was inserted before the BBO crystal. In that case, the polarization of the driving 1360 nm pulses was changed from linear to circular. The pump radiation after propagation of the BBO crystal consisted of a circularly polarized 1360 nm pump and a linearly polarized 680 nm pump. The variation of polarization conditions caused the generation of high-order harmonics only from the 680 nm pump. The generated spectrum of harmonics is seen in the upper panel of Fig. 9.8b showing the strong odd harmonics of 680 nm radiation. In the case of the absence of calcite, the spectral pattern (bottom graph) contained a whole set of two groups of enhanced harmonics, as has already been shown in the above-described studies (Fig. 9.8b).

The optimization of QPM for shorter wavelengths at lesser excitation of the target during the formation of multi-jet plasmas can be observed in different materials. Particularly, the dependence of the highest enhanced harmonic order on the energy of the heating pulse producing MJPs on the surfaces of silver and aluminum was analyzed (Fig. 9.8c). This dependence underlines the influence of over-dense plasmas and electrons on the formation of QPM conditions in different plasmas. Tuning of the maximum of the spectral envelope towards the higher-order harmonics with a decrease of the fluence of heating radiation was observed. A decrease of electron concentration at weaker excitation of target led to optimization of the QPM for higher orders of harmonics to keep the product $H_{qpm} \times N_e$ unchanged at the fixed spatial characteristics of plasma jets. Figure 9.8c showing the log–log dependence of H_{qpm} on the energy of heating pulses for two different targets (Ag and Al) confirms the relation $H_{qpm} \times N_e \propto$ const.

9.2.3 Theory

To support experimental investigations the 1D model described in Ref. [20] was used. Within this model, the nonlinear medium is simulated by a 1D chain of atoms oriented along the propagation direction of laser pulses. The response of each atom in the chain is calculated using the non-perturbative theory [26, 27]. The two-color laser field parameters that were chosen for the single-atom response calculations on the one hand coincide with the ones used in the experiments, and on the other hand depend on the position of the atoms in the chain. Generated harmonic radiation from each atom is calculated in the far far-field zone by a numerical summation over the chain of atoms [28, 29]. As a result, the spatial distribution of generated radiation is calculated, which is integrated for analysis of the overall intensity of the harmonic radiation.

The numerical calculations were dealt with silver plasma having the following parameters related to the above described experimental conditions: density of plasma is equal to ~2×10^{17} cm^{-3}, the density of free electron is equal to ~2×10^{16} cm^{-3}. Other parameters, such as the intensities of two-color pulses, their durations, and the length of the nonlinear medium were chosen to be the same as in the experiment.

Here we analyze the dependence of harmonic spectra on the size of the individual jet under conditions of the constant length (~5 mm) of nonlinear medium. The number of jets changed with variations of their length. The harmonic spectra are presented in Fig. 9.9a. It is seen that decreasing of d leads to the shift of H_{qpm} towards the shorter wavelength region. The position of H_{qpm} as a function of d is shown in Fig. 9.9b. The results of numerical calculations are shown by open circles. One can see that the positions of the center of enhanced groups of harmonics are inversely proportional to the d. Empty and filled squares correspond to experimental data for indium (Fig. 9.6b) and silver (Fig. 9.7) plasmas, respectively. Here some other reported data [30–32]in the case of silver plasma are also shown. The slopes of dotted lines ($l = -1$) correspond to the H_{qpm}(Ag) $\propto 1/d$ dependence. The data in the case of indium plasma show a similar tendency [(H_{qpm}(In) $\propto 1/d$], while the graph is slightly shifted with regard to the data due to the different role of two groups of free electrons in phase mismatch at specific ablation conditions for these two metals.

We also discuss here the influence of the number of jets on the characteristics of the generated radiation. In the numerical investigations $d = 0.23$ mm was used.

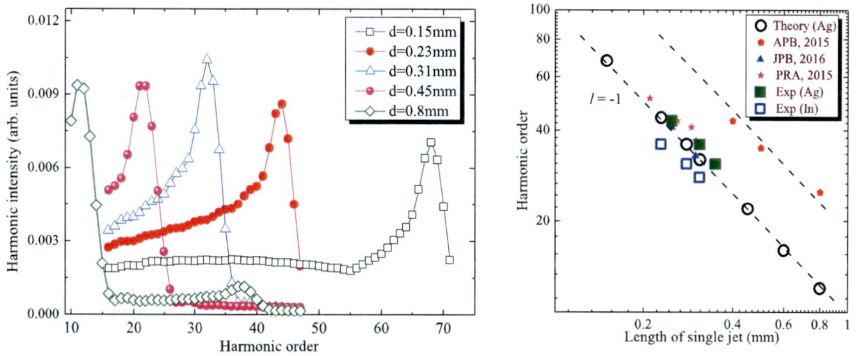

(a) (b)

Fig. 9.9 **a** Harmonic distribution from silver MJP in the case of propagation of two-color driving pulses (1310 nm + 655 nm) through the plasma of ~ 5 mm length containing the jets with the sizes (*d*) equal to 0.15 mm (black empty squares), 0.23 mm (red filled circles), 0.31 mm (blue empty triangles), 0.45 mm (pink filled spheres), and 0.8 mm (green empty rhombs). **b** Influence of the jet sizes on the order of maximally enhanced harmonic H_{qpm} in the case of silver MJP. Black empty circles correspond to theoretical calculations and green filled squares correspond to experiment with Ag plasma. Triangles, stars and pentagons correspond to earlier reported data for silver plasma [65–67]. Here the experimental data for indium plasma (blue empty squares) are also shown. The slopes of all dependences correspond to $l = -1$. Reproduced from [21] with permission from MDPI

Calculated spectra are presented in Fig. 9.10a. One can see that the harmonic's yield increases along with the growth of *n*. At the same time, the maximally enhanced harmonic depended on the *n*. To show it more precisely the position of H_{qpm} and its intensity as a function of *n* was analyzed (see blue filled circles and red empty

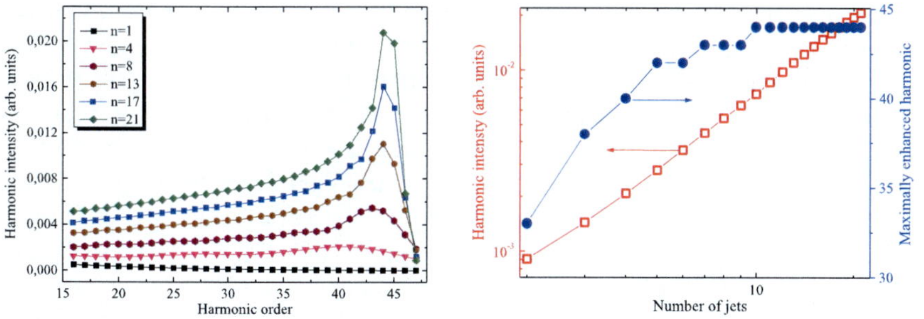

(a) (b)

Fig. 9.10 **a** Harmonic distribution in the case of silver MJP during propagation of two-color pulses (1310 nm + 655 nm) through the plasmas consisting of different number of jets. Single jet sizes were 0.23 mm. **b** H_{qpm} and its intensity as the functions of *n* (blue filled circles and black empty squares). Reproduced from [21] with permission from MDPI

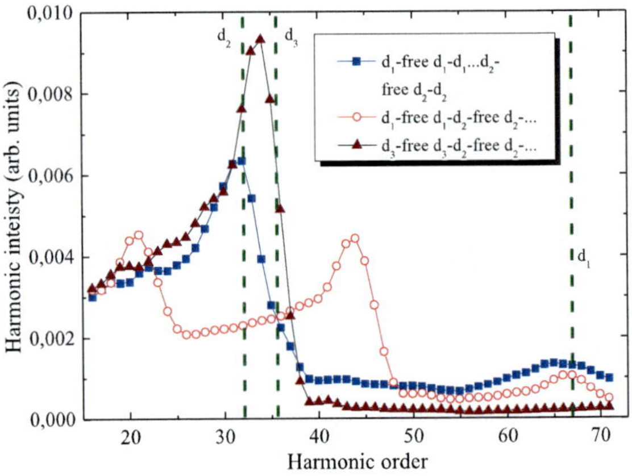

Fig. 9.11 Harmonic distribution in the case of propagation of two-color pulses (1310 nm + 655 nm) through silver MJP of 5 mm length consisting of different combinations of two groups of jets (see text). Reproduced from [21] with permission from MDPI

squares in Fig. 9.10b). It is seen that the position of H_{qpm} shifts towards the higher-order harmonic region, and the value of the harmonic intensity quadratically grows with the increase of the number of jets. Notice that in the experiment the shift of H_{qpm} with the growth of n was also observed, though this shift was less pronounced in comparison with calculations.

Finally, new possibilities for the control of H_{qpm} and peak intensity in the case of the two-color pump of MJP were simulated. It assumes the application of the jets of two different lengths (d_1 and d_2). In all calculations, the whole length of the nonlinear medium was kept the same (5 mm). The above parameters were taken to be $d_1 = 0.15$ mm and $d_2 = 0.31$ mm and the HHG was simulated in the nonlinear medium consisting of 8 jets of d_1 and 4 jets of d_2.

The resulting spectrum is shown in Fig. 9.11 by blue-filled squares. One can see that the HHG spectrum has two maxima (green dotted lines) at the same places as for similar jets (H32 and H67; compare with Fig. 9.9a). The decrease in the amplitude of the peaks in comparison with the cases shown in Fig. 9.9a is explained by the fact that in the former case the number of jets of different ds is twice smaller. The blocks of two groups of jets separated by the free space of the same sizes using d_1 and d_2 jets (i.e., d_1 jet—d_1 free space—d_2 jet—d_2 free space—d_1 jet—d_1 free space...) were also constructed. This group of jets forms the nonlinear medium with the same length as in the previous case (i.e., 8 jets of d_1 and 4 jets of d_2). The HHG spectrum, in this case, is presented in Fig. 9.11 by red empty circles. It can be seen that this HHG spectrum has 3 peaks. The peak corresponding to the d_2 became divided into two peaks, at H21 and H44, as well as the third peak remained in the same place and corresponded to d_1 (H67). To investigate the observed changes of the spectrum more precisely the value of one sort of jet was changed from d_1 to $d_3 = 0.28$ mm while forming the nonlinear medium in the same way as in the previous case. The resulting harmonic spectrum is presented by the blue-filled triangles in Fig. 9.11. It is seen that the main peak of the spectrum became centered at H34 and the harmonic intensity increased by a factor of 2 in comparison with previous cases.

9.2.4 Discussion

The practical aspects of this approach require the answer to a few questions to provide a reliable pattern of the QPM in plasmas.

(1) For practical purposes, one would like to know: what is the most optimal number of jets for a given harmonic? The meaning of the most optimal number of jets cannot be clarified without understanding the impeding processes, which stop the quadratic growth of harmonic yield with the increase of the number of jets. The discussed studies revealed that this deviation from anticipated quadratic dependence occurs at $n > 5$. Notice that this quadratic rule fulfills only for maximally quasi-phase-matched harmonic (i.e. the one, which has the highest enhancement). For non-phase matched harmonic, the enhancement is not realized at all.

(2) Another useful question is: what would be the corresponding jet characteristics in the case of "optimal" (i.e. most effective) QPM? The characterization of jets requires knowledge of the ions/neutrals and free-electron densities. The latter parameter defines maximally enhanced harmonic at a given jet length. For a given size of the individual plasma jet, the QPM for maximally enhanced harmonic could be maintained at a fixed product $H_{qpm} \times N_e$ (since $L_{coh} \approx 1.4 \times 10^{18}(H_{qpm} \times N_e)^{-1}$ for 800 nm driving pulses). The electron density depends on the ablation conditions and the ionization of target species by heating pulse, as well as on the tunneling ionization by femtosecond driving pulse. The former process is well adjustable and can be used for variations of N_e and correspondingly H_{qpm}.

(3) What would be the most optimal material for this purpose? Currently, the most optimal material is silver. The reasons for that include the extended plateau-like shape of the harmonic distribution in the non-QPM regime, relatively high conversion efficiency, easiness in stable Ag plasma formation, etc. Other materials, which also allow observation of this effect, are indium, manganese, gold, and arsenic plasmas.

(4) How much is the most attainable gain? Until now, the $20\times$ to $40\times$ enhancement factors for QPM harmonics were reported. This is the ratio of the intensities of QPM-enhanced harmonic propagated 8 to 11 plasma jets and of the same harmonic propagated through the extended imperforated (i.e. homogeneous) plasma of the same length as the whole length of jets.

(5) Can we use the measurements here in combination with the theory as a way to learn/derive the density-dependent complex refractive index of plasmas? The refraction index of plasma is related to the influence of electrons and all ion species. In some instances, one can neglect the role of ions because their contribution is much smaller than the contribution from electrons. At the same time, the dispersion relation for electromagnetic ion cyclotron waves is almost entirely dominated by ion contributions. There are numerous plasma modes that rely heavily upon ion contributions. In the meantime, one can see from some relations of plasma frequency that the refractive index becomes

imaginary and the electromagnetic wave cannot travel into plasma. Hence for low-frequency wave ions could be completely blocked by electrons. So as one can see, ion contributions to the index of refraction in plasmas can be very important or negligible and just depend upon the specific circumstance.

Returning back to the discussed studies, since the developed plasma QPM technique allows achieving some groups of notably enhanced harmonics at a fixed N_e one can assume that this technique could be a method to determine the value of electron density of the plasma. This is a well-recognized problem in plasma physics, especially the case of dynamically changed N_e, which has been studied for a long time. Numerous techniques and theories were proposed and realized to resolve this issue. Particularly, in the case of laser-produced plasmas, the HYADES code was used to estimate the electron density of plasma at different regimes of plasma formation. The plasma QPM method allows determining the electron density from direct measurements of two other parameters of above—mentioned relation, coherent length, and maximally enhanced harmonic order.

(6) What is the ultimate limit from the point of view of tuning the maximally enhanced harmonic towards the shorter wavelengths? The aim of discussed study [21] was to demonstrate the effect of different parameters of experiments on the modulation of homogeneous harmonic distribution. It was not aimed at determining the limits of H_{qpm} variations since this task has already been analyzed in previous studies. Particularly, H43 of 800 nm pump was the maximally enhanced harmonic in the case of 11-jet silver plasma structure. A similar pattern is also seen in Fig. 9.15 (upper panel) in the case of indium plasma. However, the enhancement, in that case, was lesser compared with the longer wavelength XUV region. As shown in the upper panel of Fig. 9.11, the maximally available cutoff for harmonics extends towards a shorter wavelength region once optimal conditions for QPM realize. This spectrum does not show the cutoff but rather demonstrates the attractiveness of the method when almost invisible harmonics (see bottom panel) appear as the enhanced ones.

(7) Theory predicts that HHG from near-infrared wavelength (1320 nm in discussed case) produced higher cutoff photon energy compared to 800 nm laser pulses at similar experimental conditions. Meanwhile, it was not the case during these studies. Why does it happen? In the case of NIR pulses, the cutoff energy should be $(1320/810)^2$ times higher compared with 800 nm pulses. In other words, the highest photon energies from the 1320 nm pump are expected to be 2.7 times larger than those from the 810 nm pump. However, in the experiment one can see the lower cutoff energies generated at the 1320 nm pump. One can attribute this phenomenon to notably smaller energy of 1320 nm pulses (~1.1 mJ) and worse spatial distribution in the focal area compared with 810 nm pulses (~3 mJ). Moreover, the $I \propto \lambda^{-5}$ rule of wavelength-dependent harmonic intensity predicts the ~12 times smaller harmonic yield even at similar intensities of these two pumps. Overall, the ~30 to 50 times weaker harmonics from

1320 nm pump compared with 810 nm pump may restrict the observation of higher cutoff in the former case due to technical restrictions in extremely weak XUV radiation registration.

(8) Is it would be useful for practical purposes to quantify the total enhancement/improvement compared to the usual, single jet HHG? The difference between "QPM plasma" and "usual plasma" HHG comprising a single jet is the focusing of the heating beam by cylindrical lens instead of spherical one in the latter case and installation of the multi-slit mask. That is why one can directly see the advantages in practical applications of the proposed approach. Those advantages include: (a) notable (i.e. more than one order) enhancement of harmonics centered around the so-called QPM harmonic, (b) availability to tune the area of enhanced harmonics along the XUV region, and (c) advantages in the calculation of the concentration of free electrons.

9.3 High-Order Harmonics Generation Under Quasi-Phase-Matched Conditions in Silver, Boron, and Silver Sulfide Plasmas of Different Configurations

Higher conversion efficiency and, thus, an increase in the photon flux of the harmonics radiation is a requirement for practical applications of coherent XUV sources. A largely unexplored approach for this purpose is the harmonic generation in plasmas, which provides various options in the nonlinear spectroscopy of materials compared with the well-developed technique of harmonics generation in gases. Two- and three-color pumps of plasma, nanoparticle-induced growth of a whole ensemble of harmonics, resonance enhancement of single harmonic, and nonlinear medium's length extension have been applied to amend this process in laser-induced plasmas.

A dependence of QPM-induced enhancement on the distance between plasma jets was analyzed using the ablation of rotating disks [19]. The importance of fine-tuning and control of the quasi-phase-matching for high-order harmonics generation (QPM-HHG) has motivated to investigate the influence of the spacing between plasma jets on this process at different conditions of multi-jet plasma formation. Additionally, a search for new plasma media suitable for QPM-HHG, such as nano- and microparticles and molecules, broadens the applications of this concept and allows determining better conditions for HHG. Notice that most previous QPM studies were performed in the plasmas produced during the ablation of bulk silver, though other metal plasmas were also probed for these purposes.

In this section, we analyze QPM-HHG in the plasmas containing silver microparticles and atoms, and silver sulfide molecules [33]. We discuss different combinations of MJP formations to determine the best choice for the generation of the groups of enhanced harmonics in the 20–60 nm spectral range.

In those QPM-HHG studies, 30 fs, 800 nm, 1 kHz, and 200 ps, 800 nm, 1 kHz pulses from Ti:sapphire laser were used as the driving and heating radiation (Fig. 9.12). The heating 200 ps pulses (HP) were focused using a 300-mm focal

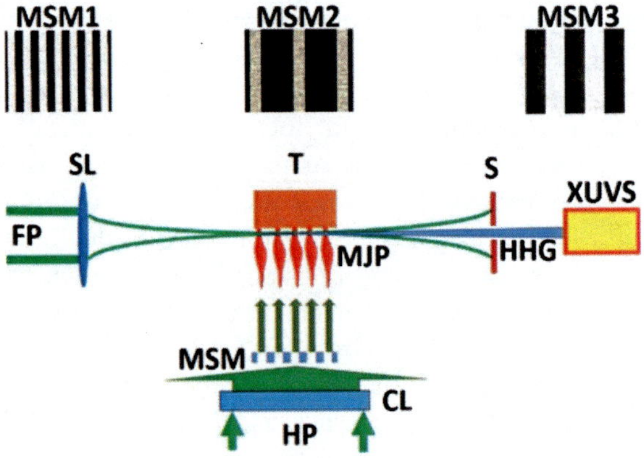

Fig. 9.12 QPM-HHG in MJP. The dark areas shown in the top panels correspond to the openings in the masks. Reproduced from [33] with permission from American Institute of Physics

length cylindrical lens (CL) on the target (T) placed in the vacuum chamber to form the homogeneous extended plasma (l = 5 mm). MSM was installed between the cylindrical lens and the target to produce MJP. Three masks with different slit sizes, two with 0.2 and 0.5 mm slits separated by 0.2 and 0.5 mm from each other (MSM1 and MSM3), and one with slit size of 0.8 mm separated by a 0.3 mm shield between the slits (MSM2, see insets in Fig. 9.12) were used in those studies. The plasma formation was accomplished 40 ns prior to propagation of the 30 fs pulses (FP) focused using the spherical lens (SL) along MJP. The generated harmonics were propagated through the slit (S) in the XUV spectrometer.

Silver microparticles (MPs, 70 μm), boron nanoparticles (NPs, 100 nm), bulk silver sulfide, and bulk silver were used as the targets. The powdered microparticles, nanoparticles, and mixtures of micro- and nanoparticles were pressed to form tablets of 10-mm diameter. Part of the heating pulse was blocked by MSM, while the fluence of this radiation in the ablated area on the target surface (~1 J cm^{-2}) remained almost unchanged compared with the extended plasma.

9.3.1 QPM in Nanoparticle Plasmas

Figure 9.13 shows the plateau-like and QPM-induced shapes of the odd harmonics generated in the plasmas produced on the (a) bulk Ag_2S, (b) bulk silver, and (c) mixture of boron NPs and silver MPs (at the weight ratio of 100:15) in the case of imperforated 5-mm long plasmas and different MJPs using the 800 nm, 30 fs driving pulses. The harmonics between the seventh (H7) and forty-seventh (H47) orders are shown. One can see approximately the same plateau-like distribution of the harmonics generated in the extended 5-mm long plasmas (red dashed curves in all three panels), in accordance with the three-step model of HHG.

The envelopes of the generated spectra of harmonics drastically changed once one inserts different MSMs on the path of the heating picosecond radiation. The examples of three strongly modulated and enhanced spectra of the groups of odd

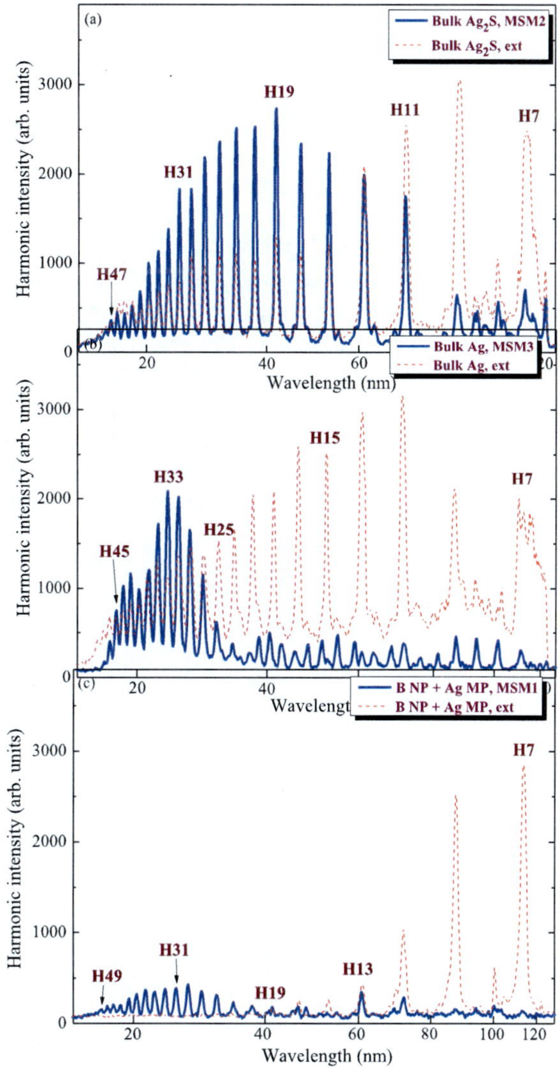

Fig. 9.13 Generation of high-order harmonics in different configurations of LIP. **a** Extended imperforated plasma (red dashed curve) and MJP obtained using MSM2 (blue solid curve); bulk silver sulfide target. **b** Extended imperforated plasma (red dashed curve) and MJP obtained using MSM3 (blue solid curve); bulk silver target. **c** Extended imperforated plasma (red dashed curve) and MJP obtained using MSM1 (blue solid curve); pressed tablet containing boron nanoparticles and silver microparticles. Reproduced from [33] with permission from American Institute of Physics

harmonics compared to the case of homogeneous plasmas are shown as blue solid curves in Fig. 9.13. Particularly, the installation of MSM2 in front of the Ag_2S target significantly suppressed the lower-order harmonics, while the higher orders became enhanced in the range of H19-H31 (Fig. 9.13a). An even stronger difference of these spectra is seen in the case of bulk silver plasma (Fig. 9.13b). The lower-order harmonics from the MJP produced using MSM3 almost disappeared up to H25, while the higher-order harmonics centered at H33 prevailed over those generated in the imperforated plasma. The most prominent difference in the harmonics spectra from the imperforated and perforated plasmas was observed in the case of ablation of the mixture of boron nanoparticles and silver microparticles (B NPs +Ag MPs, weight ratio 100:15, Fig. 9.13c). There were almost no harmonics generated above H19 in the extended plasma (red dashed curve), while the modification of homogeneous plasma shape using the MSM1 allowed the generation of the group of enhanced harmonics far from the above-mentioned cut-off (blue solid curve). Harmonics up to H47 are seen in this spectrum, while the strongest harmonics were centered on H29-H33.

These three groups of QPM-induced spectra contain significantly enhanced high-order harmonics. The twofold decrease of the whole length of MJP compared to the extended imperforated plasma (2.5 and 5 mm, respectively) did not play important role in the yield of harmonics. This reveals the significant advantage of QPM HHG compared to ordinary HHG in extended plasmas.

The energy of the heating pulse producing MJPs on the surface of bulk silver plays an important role once one considers the variations of QPM conditions (Fig. 9.14). Previously, this dependence has demonstrated the influence of a larger concentration of plasmas and electrons on the formation of QPM conditions for different groups of

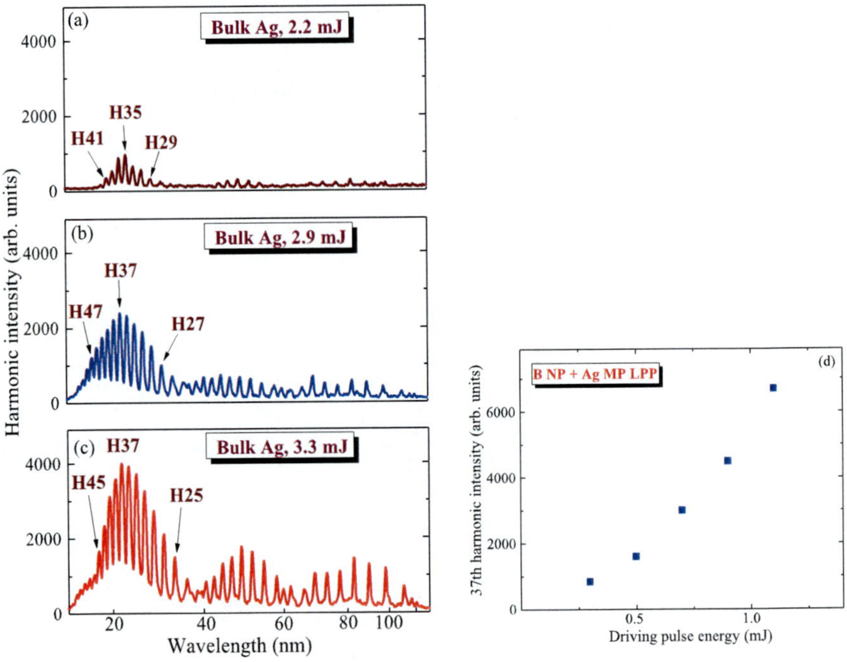

Fig. 9.14 **a–c** Harmonic distributions obtained during propagation of femtosecond pulses through MJP produced on the surface of bulk silver target using different energies of heating pulses. **a** 2.2 mJ, **b** 2.9 mJ, **c** 3.3 mJ. **d** Maximally enhanced harmonic (H37) yield as the function of the driving pulse energy in the case of HHG in the perforated B NP +Ag MP LIP. Maximally enhanced QPM harmonic (H37 or H39) was maintained at the conditions of variable energy of driving 30 fs pulses. Reproduced from [33] with permission from American Institute of Physics

harmonics. Particularly, the tuning of the maximum spectral envelope (H_{qpm}) towards the lower-order harmonics with an increase of the flux of heating radiation has been reported in Ref. [8]. An increase of electron concentration (N_e) in LIP at stronger excitation of target led to optimization of QPM for lower-order harmonics to keep the product $H_{qpm} \times N_e$ unchanged at the fixed spatial characteristics of plasma jets.

Meanwhile, at the conditions when the growth of the heating pulse's energy does not significantly change the concentration of electrons, the maximal order of enhanced harmonics remains almost the same. In the discussed studies, H_{qpm} corresponded to H35 or H37 in the case of Ag bulk plasma produced by pulses of different energy and perforated using MSM1 (Fig. 9.14a-c). The difference in these three spectra is attributed exclusively to (a) a larger amount of harmonic emitters and (b) the formation of better QPM conditions for stronger excitation of the bulk target. The growth of plasma concentration led to enhancement of conversion efficiency once the energy of heating pulses was changed from 2.2 mJ (Fig. 9.14a) to 2.9 mJ (Fig. 9.14b), and to 3.3 mJ (Fig. 9.14c).

Notice almost complete disappearance of the lower-order harmonics in these three spectra. The lines in the longer wavelength part of spectra correspond to the second and third orders of diffraction of the enhanced groups of harmonics. The presence of second- and third-order diffractions of enhanced harmonics is also seen in Fig. 9.13, though not as clear as in the case of Fig. 9.14 when the entire suppression of lower-order harmonics was achieved. This peculiarity allowing the disappearance of some harmonics and amendment of other harmonics is an important finding of the discussed study, which has never been reported. It means that the conditions were determined when the macroscopic nonlinear optical process leading to the selection of the group of phase-matched harmonics drastically overpasses the microscopic process of HHG leading to an almost homogeneous distribution of harmonics.

QPM conditions were also analyzed at different energies of driving femtosecond pulses (0.3–1.1 mJ, Fig. 9.14d). The envelope of enhanced harmonics was centered at around H37 to H39 when the 2.6 mJ, 200 ps heating pulses, and MSM1 with 0.2 mm slits were used. H_{qpm} were almost the same in the case of five different energies of driving pulses since the concentration of the free carriers responsible for QPM for some specific harmonic order maintained unchanged during these experiments due to the same energy of the heating picosecond pulses. With these experiments, it was defined that the free carriers appearing during propagation of the driving femtosecond pulses through the plasma did not strongly affect the QPM conditions. The only difference in these five cases of the growing femtosecond pulse energies was the gradual enhancement of the intensity of H_{qpm} alongside a few surrounding harmonics. The broadening of the envelope of harmonic distribution centered at around H_{qpm} with the growth of the energy of the driving pulse was also observed.

The plasma produced from the tablet containing only boron nanoparticles did not show the creation of conditions when the higher-order harmonics become enhanced. The formation of perforated plasmas on the Ag MPs-containing tablet (i.e. without B NPs) allowed the generation of enhanced harmonics compared with the imperforated plasma in the spectral range where no harmonics from extended plasma were seen (Fig. 9.15).

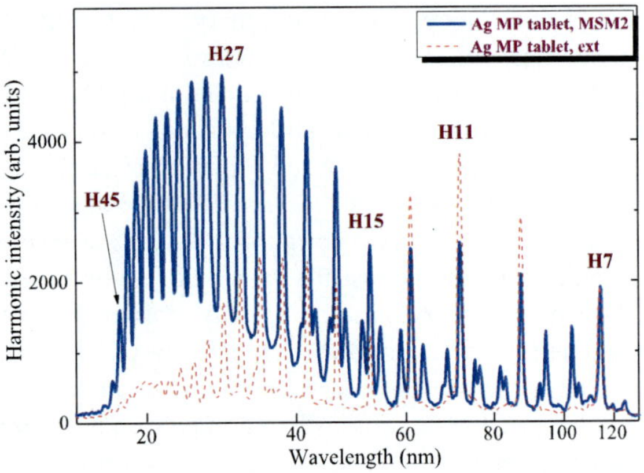

Fig. 9.15 Harmonic intensity distributions obtained in the extended imperforated plasma (red dashed curve) and MJP produced using MSM2 (blue solid curve) on the surface of Ag MP-containing tablet. Reproduced from [33] with permission from American Institute of Physics

The comparison of harmonic yields was carried out under similar conditions of experiments and data collection. Two spectra of harmonics generated with and without the insertion of the MSM on the path of the heating beam were compared. Particularly, Fig. 9.13 shows how the harmonic yield modifies in different spectral ranges during insertion of three MSMs distinguished by different sizes of slits, independently of the material used for ablation. Thus the same intensity of harmonics scales in this figure for the extended imperforated and perforated plasmas. The same can be said about the comparative curves presented in Fig. 9.15.

QPM-HHG was also carried out at the conditions of the two-color pump (TCP) of plasma using 800 nm (97%) and 400 nm (3%) radiation. Figure 9.16 demonstrates the raw images of harmonic spectra obtained using SCP and TCP of extended Ag

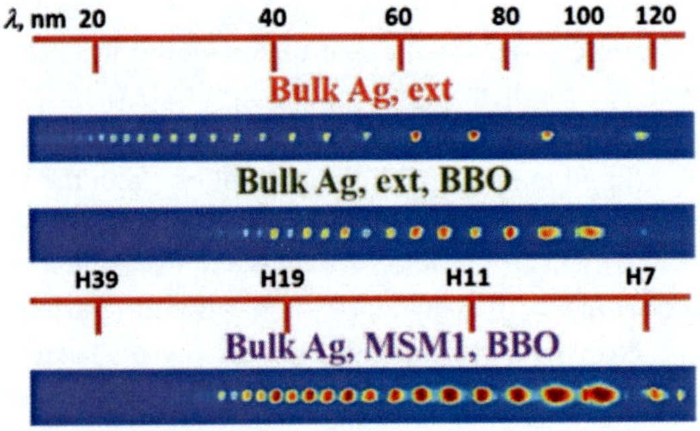

Fig. 9.16 Harmonic spectra obtained using SCP (upper panel) and TCP (middle panel) of extended Ag plasma. Bottom panel shows the harmonic spectrum generated in the perforated plasma produced using MSM1 in the case of TCP. Reproduced from [33] with permission from American Institute of Physics

plasma. The harmonic spectrum generated in the perforated Ag plasma produced using MSM1 in the case of TCP is also shown. The conversion efficiency in the case of TCP (Fig. 9.16, middle panel) was higher than in the case of the single-color pump (SCP, Fig. 9.16, upper panel) of extended plasma. The decrease in the generation of higher-order harmonics did not allow the observation of their enhancement in the case of MJP relative to the case of SCP of perforated plasma (Fig. 9.16, bottom panel).

9.3.2 Discussion

The analysis of the intensity of harmonics as a function of the number of plasma jets has earlier provided proof of the decisive role of QPM in the variations of harmonics spectrum [8, 31]. The analysis of the phase shift of harmonics along plasma using the analytical solution of nonlinear wave equation allowed determining two regimes of phase matching at the conditions of SCP—compensation of dispersion-induced phase mismatch by Gouy phase in extended plasmas and QPM-HHG using the phase change due to free electrons in the space between plasma jets.

The discussed studies showed that the influence of the second harmonic was sufficient to strongly modify the whole harmonic spectrum compared to SCP in the case of extended imperforated plasmas. Meanwhile, the formation of MJP and application of TCP did not lead to strong QPM-induced enhancement of some groups of harmonics and depression of the lower parts of harmonic spectra. The coherence length of the qth harmonic is determined as

$$L_{coh} \approx 4\pi^2 m_e \varepsilon_0 c^2 / q N_e e^2 \lambda \qquad (9.1)$$

where λ is the wavelength of the driving radiation; N_e, m_e, and e are the density, the mass, and the charge of an electron; c is the velocity of light, and ε_0 is the permittivity of vacuum [4]. One can deduce from this relation the dependence between maximally enhanced harmonic and laser wavelength ($H_{\text{qpm}} \propto \lambda^{-1}$). Correspondingly, the use of twice shorter wavelength leads to the two-fold increase of the order of maximally QPM-enhanced harmonic. Thus the observation of QPM at these conditions is possible only in the case when the plasma medium allows generation of twice longer cut-off in the silver plasma (i.e. up to ~100th order). Since this medium cannot provide those harmonics, QPM also cannot be observed.

In the case of homogeneous irradiation of targets, the Ag, B NPs +Ag MPs, and Ag_2S plasmas allowed the generation of the plateau-like distribution of harmonic intensities. Therefore MJP setups led to a clear observation of QPM at SCP. The maximally achieved enhancement factor for H33 in the case of MJP compared to the extended (5-mm-long) plasma was measured to be 22× in the case of B NPs +Ag MPs plasma.

No absolute measurements of harmonic conversion efficiency were carried out at QPM conditions in laser-induced plasmas. Meanwhile, the estimates of conversion efficiency were carried out using the comparison with the earlier reported measurements of this parameter in Ag plasmas at similar experimental conditions. By knowing the conversion efficiency from previous measurements of harmonic generation in the plasmas produced on the surface of bulk Ag (8×10^{-6} [34]), one can determine the conversion efficiency for maximal QPM-enhanced harmonic (3×10^{-5}). The conversion efficiencies in extended Ag, B NPs +Ag MPs, and Ag_2S plasmas in the discussed experiments were estimated to be 4×10^{-6}, 1×10^{-6}, and 3×10^{-6}, respectively. The corresponding conversion efficiencies for maximal QPM-enhanced harmonics were calculated to be 3×10^{-5}, 2×10^{-5}, and 3×10^{-5} for the MJPs produced on bulk Ag, B NPs +Ag MPs, and bulk Ag_2S targets.

The pioneering studies of QPM in plasmas [8] have revealed the potential materials for efficient enhancement of the group of high-order harmonics in different ranges of extreme ultraviolet. Apart from silver, such materials as manganese and indium allowed achieving QPM in laser-induced plasmas. All of these materials have demonstrated the generation of high-order harmonics. Even the weak higher-order harmonics generation is a sufficient requirement for observation of their enhancement at matching the group velocities of the driving and converted waves. In discussed studies, other materials were chosen, which showed potential in the generation of strong coherent emission through HHG. The advantages in the application of multi-atomic species (clusters and nanoparticles) for plasma HHG demonstrated in earlier studies allowed expecting their efficient application during the implementation of the QPM concept.

The choice of multi-slit masks was also based on previous applications of different plasma jets. Earlier, perforated plasma generated using a mask containing 0.3-mm slits allowed achieving the enhancement of the group of harmonics centered at around H30. MSM containing 0.2-mm slits was chosen for demonstration of the shift of maximally enhanced harmonic towards the shorter-wavelength range of XUV assuming the equality of plasma length of each of ten generated jets (0.2 mm) and coherence length of the higher order of harmonic (~H35) at the "optimal" fluence of the heating beam allowing the appearance of the sufficient amount of free electrons. Correspondingly, the application of larger-sized slits, like 0.8 mm, led to the enhancement of the lower orders of harmonics. The implementation of five 0.5-mm slits caused the intermediate harmonics enhancement.

The yet studied issue was the application of the smaller distance between the slits compared with the slit sizes, i.e. the mask having 0.3-mm distance between 0.8-mm slits. One can analyze this issue since some of the assumptions involved in consideration of the role of free electrons in the no-plasma area predicted the independence of this parameter on the harmonic enhancement. The discussed studies have revealed that the distance between slits does not influence the QPM conditions provided the whole plasma area is contained within two Rayleigh lengths of the driving radiation.

Those studies showed that TCP cuts out the high harmonics between 40 nm and less than 20 nmcompared to SCP (Fig. 9.16). Some explanation of this effect can be related to the use of twice shorter wavelength source as one of two pumps in TCP scheme. The relation of the three-step model provides the dependence of the cut-off energy, the ionization potential (E), and the ponderomotive potential (U_p) when the former parameter scales as $E + 3.2U_p$. Because of the λ^2 dependence of the ponderomotive energy ($U_p \propto I_0\lambda^2$), one can assume a decrease of harmonic cut-off with the decrease of the driver wavelength. In the meantime, the integrated harmonic yield of the isolated atom scales as λ^{-5}. With a decrease in driving wavelength, the recombination of the electron wave packet with the parent ion becomes more and more efficient, thus increasing the HHG yield. It has been a common feature for both gas and plasma HHG to achieve a stronger yield of harmonics from 400 nm driving wave with regard to 800 nm radiation, while the cut-off energy decreased in the former case.

The strong first field (800 nm) and the weak second field (400 nm) were used in the discussed studies to pump the plasma plume. It was shown that even the 30:1 ratio between these two fields dramatically changes the spectrum of harmonics, which points out the decisive influence of a weak second field on the whole process of HHG. A similar influence could be expected once one considers the variation of cut-off for TCP versus SCP schemes. The decrease of cutoff did not follow the quadratic dependence on the wavelength, though the almost two-fold decrease of maximally generated harmonic order and stronger odd harmonics in the case of TCP was observed.

9.4 Summary

In this chapter, we discussed the HHG quasi-phase-matching in multi-jet plasma plume for the case of one- and two-color generating fields. Quasi-phase-matching in the vicinity of the resonant harmonic in manganese shows that harmonic yield can be enhanced through both these processes. We have discussed some peculiarities of QPM in LIPs, which allow to clarify both pros and contras of this approach, as well as to further amend the developed technique. Since the first demonstration of this effect in plasma [8], new opportunities and schemes were developed and analyzed. Inevitably, they overlap, to some extent, with each other to explain, in some cases from the very basic rules, the advantages (and in some cases disadvantages) of the QPM in plasmas. This overlap may cause some confusion; however, it would be difficult to clarify some new findings without a description of the basics and practical issues of this method.

Apart from the novelties of this approach studied in discussed work (comparison of single and two-color pumps of QPM plasmas, the role of a number of coherent zones, the competitiveness of macro-processes (QPM) and micro-processes (resonance enhancement of single harmonic), analysis of dimensional parameters of plasma jets, etc.) we have analyzed a new approach in the theoretical description of the process, which shows the peculiarities not mentioned in earlier theoretical studies of the QPM in plasmas.

We have also analyzed different patterns of experimentally observed modifications of harmonic spectra at the conditions of QPM in silver, indium, and aluminum multi-jet plasmas. In particular, we have studied the role of the number of coherent zones, sizes of plasma jets and distance between them, plasma formation conditions, and characteristics of fundamental and heating radiation on the HHG efficiency at QPM conditions. The TCP-induced QPM at different regimes of laser-plasma interaction was demonstrated.

The results of theoretical simulations have shown good agreement with the experimental data. The calculated spectra qualitatively coincide with the experimentally measured ones. A deeper insight into the mechanisms of the investigated phenomena can be obtained from the analysis of relationships between the characteristics of the processes like those shown in Fig. 9.8. Figure 9.8a shows that the yield of H34 grows as $I_{H34} \sim n^2$ up to a number of jets (n) equal to 4. Further deviation of $I_{H34}(n)$ dependence from the quadratic one can be explained by the violation of equality of the conditions for each next added jet (i.e., non-ideal conditions of propagation through jets, and/or low-order nonlinear optical properties of plasma, etc.). One can see that the quadratic dependence is correct in a wide range of harmonic spectra. New possibilities for the control of QPM conditions by using combined multi-jet plasmas are proposed. The results of discussed experimental and theoretical studies of the QPM conditions provide a reliable interpretation of this phenomenon.

Finally, we have analyzed the high-order harmonics generation at quasi-phase-matched conditions in silver microparticle and silver sulfide plasmas of different configurations. We have discussed QPM of the high-order harmonics generated in the plasmas containing silver microparticles and atoms, boron nanoparticles, and silver sulfide molecules. The influence of the flux of heating beam, as well as variation of the geometry of multi-jet plasmas, on the maximally enhanced harmonic, has been demonstrated. The groups of harmonics in the range of thirtieth to fiftieth orders at QPM conditions prevailed over the lower orders of harmonics. In most cases, the lower-order harmonics fully disappeared from the generated spectra at QPM conditions. The enhancement factors of QPM harmonics up to 22× were demonstrated in the case of mixed boron nanoparticles and silver microparticles plasmas for the harmonics in the energy range of 47–67 eV. The conversion efficiencies for QPM-enhanced harmonics were estimated to be 3×10^{-5}, 2×10^{-5}, and 3×10^{-5} for the MJPs produced on the bulk Ag, B NPs +Ag MPs, and Ag_2S targets respectively.

References

1. R.A. Ganeev, M. Suzuki, H. Kuroda, J. Phys. B **47**, 105401 (2014)
2. A. Paul, R.A. Bartels, R. Tobey, H. Green, S. Weiman, I.P. Christov, M.M. Murnane, H.C. Kapteyn, S. Backus, Nature **421**(6918), 51 (2003)
3. J. Seres, V.S. Yakovlev, E. Seres, C.H. Streli, P. Wobrauschek, C.H. Spielmann, F. Krausz, Nature Phys. **3**, 878 (2007)
4. L. Zheng, X. Chen, S. Tang, R. Li, Opt. Express **15**, 17985 (2007)
5. A. Pirri, C. Corsi, M. Bellini, Phys. Rev. A **78**, 011801 (2008)

6. V. Tosa, V.S. Yakovlev, F. Krausz, New J. Phys. **10**, 025016 (2008)
7. T. Fok, Ł. Węgrzyński, M. Kozlova, J. Nejdl, P.W. Wachulak, R. Jarocki, A. Bartnik, H. Fiedorovicz, Photon. Lett. Pol. **6**, 14 (2014)
8. R.A. Ganeev, M. Suzuki, H. Kuroda, Phys. Rev. A **89**, 033821 (2014)
9. R.A. Ganeev, J. Appl. Phys. **119**, 113104 (2016)
10. V.V. Strelkov, R.A. Ganeev, Opt. Express **25**, 21068 (2017)
11. J. Tate, T. Auguste, H.G. Muller, P. Salières, P. Agostini, L.F. DiMauro, Phys. Rev. Lett. **98**, 013901 (2007)
12. P. Lan, E.J. Takahashi, K. Midorikawa, Phys. Rev. A **81**, 061802 (2010)
13. R.A. Ganeev, M. Suzuki, M. Baba, H. Kuroda, T. Ozaki, Opt. Lett. **31**, 1699 (2006)
14. R.A. Ganeev, H. Singhal, P.A. Naik, V. Arora, U. Chakravarty, J.A. Chakera, R.A. Khan, I.A. Kulagin, P.V. Redkin, M. Raghuramaiah, P.D. Gupta, Phys. Rev. A **74**, 063824 (2006)
15. R.A. Ganeev, L.B. Elouga Bom, J.-C. Kieffer, T. Ozaki, Phys. Rev. A **76**, 023831 (2007)
16. R.A. Ganeev, M. Baba, M. Suzuki, S. Yoneya, H. Kuroda, J. Appl. Phys. **116**, 243102 (2014)
17. R.A. Ganeev, T. Witting, C. Hutchison, F. Frank, M. Tudorovskaya, M. Lein, W.A. Okell, A. Zaïr, J.P. Marangos, J.W.G. Tisch, Opt. Express **20**, 25239 (2012)
18. P.B. Corkum, Phys. Rev. Lett. **71**, 1994 (1993)
19. M. Wöstmann, L. Splitthoff, H. Zacharias, Opt. Express **26**, 14524 (2018)
20. S.Y. Stremoukhov, A.V. Andreev, Laser Phys. **28**, 035403 (2018)
21. R.A. Ganeev, S.Y. Stremoukhov, A.V. Andreev, A.S. Alnaser, Appl. Sciences **9**, 1701 (2019)
22. G. Lambert, J. Gautier, C.P. Hauri, P. Zeitoun, C. Valentin, T. Marchenko, F. Tissandier, J.P. Goddet, M. Ribiere, G. Rey, M. Fajardo, S. Sebban, New J. Phys. **11**, 083033 (2009)
23. R.A. Ganeev, H. Singhal, P.A. Naik, I.A. Kulagin, P.V. Redkin, J.A. Chakera, M. Tayyab, R.A. Khan, P.D. Gupta, Phys. Rev. A **80**, 033845 (2009)
24. J. Kim, G.H. Lee, S.B. Park, Y.S. Lee, T.K. Kim, C.H. Nam, T. Mocek, K. Jakubczak, Appl. Phys. Lett. **92**, 021125 (2008)
25. R.A. Ganeev, M. Suzuki, H. Kuroda, J. Opt. Soc. Am. B **31**, 911 (2014)
26. A.V. Andreev, S.Y. Stremoukhov, O.A. Shoutova, Eur. Phys. J. D **66**, 16 (2012)
27. S. Stremoukhov, A. Andreev, B. Vodungbo, P. Salières, B. Mahieu, G. Lambert, Phys. Rev. A **94**, 013855 (2016)
28. S. Stremoukhov, A. Andreev, J. Opt. Soc. Am. B **34**, 232 (2017)
29. S.Y. Stremoukhov, A.V. Andreev, Laser Phys. Lett. **12**, 015402 (2015)
30. R.A. Ganeev, M. Suzuki, S. Yoneya, H. Kuroda, Appl. Phys. B **121**, 307 (2015)
31. R.A. Ganeev, V. Tosa, K. Kovács, M. Suzuki, S. Yoneya, H. Kuroda, Phys. Rev. A **91**, 043823 (2015)
32. R.A. Ganeev, J. Phys. B **49**, 095402 (2016)
33. R.A. Ganeev, G.S. Boltaev, V.V. Kim, P.V. Redkin, C. Guo, J. Appl. Phys. **125**, 153101 (2019)
34. R.A. Ganeev, M. Baba, M. Suzuki, H. Kuroda, Phys. Lett. A **339**, 103 (2005)

Summary

In this book, we discussed various subjects related to the nonlinear optical features of materials and their ability to convert the frequency of laser radiation. We analyzed recent studies of the low- and high-order harmonics generation of ultrashort laser pulses in various media produced during the ablation of solids. To briefly summarize the main achievements described in the present book, below we consider each mechanism involved in discussed studies.

Lowest-order nonlinearities and harmonics generation of laser-induced plasmas. THG in various plasma targets was analyzed using nanosecond and picosecond heating pulses in both vacuum and air. TH from various plasmas was used as a probe to characterize the components of the plasma. TH intensity shows the cubic dependence on the probe pulse energy in both air and vacuum. It was observed in the case of THG in the air that, at the high intensity of probe pulses, TH intensity decreases, which is an indication of free electrons generation in air. TH intensity variations with the distance between the target and the probe pulses reveal the nature of the ejected particles during ablation under vacuum conditions, which was not observed during THG in air. TH intensity measurements with the change of the distance between the target surface and probe pulses position have shown the significant contribution of the smaller-sized particles along with ions and atoms compared to the heavier particles. The modulation of TH intensity with the variation of delay between the heating and probe pulses is attributed to the influence of the masses of ejected particles in accordance with the relation $t_d \propto M^{0.5}$. Among the studied materials under vacuum conditions, the relative TH conversion efficiency is larger for Ag plasma compared to the other species. It was shown that THG in metal plasmas is governed by the atomic or ionic species, contrary to the wide-bandgap semiconductor materials where TH was generated by the small-sized clusters and nanoparticles. TH maxima were observed at the delays in the range of 100–300 ns for metal plasmas as compared to 500–700 ns for semiconducting materials. This difference in optimal delays has revealed that, during LA, the ejected particles from metal targets move faster compared to heavier targets. This study can be generalized

R. A. Ganeev, *High-Order Harmonics Generation in Plasmas*, Springer Series on Atomic, Optical, and Plasma Physics 122, https://doi.org/10.1007/978-3-031-09040-0

to a variety of different materials as the sources of plasma for THG depending on the delay and spatial spreading of the plasma expansion.

The application of a high pulse repetition rate laser allowed increasing the average power of the low-order harmonics generating in air and different plasmas. We analyzed the generation of the third and fifth harmonics of 37 fs pulses using carbon, boron, manganese, and titanium nanoparticle-contained laser-produced plasmas using a 100 kHz class laser. The fluxes of the third harmonic (343 nm) in the plateau region were estimated to be 8×10^{15} and 3×10^{16} photons/s in the case of air and manganese plasma, respectively. We also analyzed third harmonic generation in quantum dot plasmas. We discussed the increase of the average power of coherent ultraviolet harmonics from plasmas by two orders of magnitude compared with previously used 1 kHz lasers. The long-term stability of plasma harmonic generation at this range of pulse repetition rates still remains to be improved. However, the joint application of the rotating target and the up and down movement allows the shift of the heating beam along the target surface thus leading to the improvement of the stability of harmonic yield from 100 kHz class lasers.

Different aspects of high-order harmonics generation in plasmas. We discussed the strong incoherent emission from boron plasma at $= \lambda 6$ nm and coherent emission in the 15–53 nm spectral range. Particularly, we analyzed generation of the 806 nm radiation harmonics up to the 55th order ($\lambda = 14.6$ nm) and of tunable near-infrared radiation (1280–1400 nm) up to the 35th order ($\lambda = 38$ nm) during the propagation of femtosecond pulses through the boron plasma. The incoherent and coherent emissions from this target were compared and analyzed by considering the influence of ionic transitions on the harmonic conversion efficiency. We discussed different schemes of laser-plasma interaction (two-color pump of B plasma, generation of tunable harmonics in the extreme ultraviolet (XUV) range, formation of structured extended plasma, application of boron nanoparticles, laser-induced breakdown spectroscopy, overheating of boron targets, quasi-phase-matching of plasma harmonics, etc.) and we found major characteristics that would prevail during laser ablation and high-order harmonics generation (HHG). In particular cases (i.e. the plasma produced on the three boron-containing targets), we compared three characteristics: (a) stability of harmonic emission, (b) extension of harmonic cutoff, and (c) harmonic yield. The plasma produced on bulk boron can be considered as the best choice among the three ablated targets when it comes to harmonic emission stability and cut-off extension. On the other hand, the B NP plasma showed the best characteristics to produce the highest yield. The fluencies of heating pulses for the HHG experiments were found different from those used in the plasma emission studies. In each case, the "optimal" fluence of heating pulses was chosen. This term refers to the maximal efficiency of harmonic generation or incoherent emission in the studied plasmas plumes. Furthermore, it was found that the conditions of efficient emission of 6 nm radiation from the over-ionized plasma were unsuitable for efficient HHG. We analyzed the plasma during ablation of boron-containing targets and found the optimal conditions allowing the generation of almost a single emission line ($\lambda = 6.03$ nm, B IV) in the 2–20 nm spectral range.

We discussed the calculations of HHG spectra for Li, Li^+, and He using the Lewenstein model. For neutral lithium atom, the cut-off position at 43rd harmonic of 795 nm wave corresponded to simulated intensity 3×10^{14}–3.5×10^{14} W/cm^2. The difference between the theoretical calculations and the experimentally estimated value can be well explained by the measurement errors of laser intensity in the focal plane. Lithium ions will start to contribute only at very high intensities, and in general, it is not always beneficial to look for the material with higher ionization potentials for the extension of the HHG cut-off.

The spectral features of the harmonic generation using the laser pulses subjected to a considerable phase modulation were analyzed. The modulation was induced by transmitting laser pulses to be converted through the argon under the conditions of the formation of filaments. The filamentation was followed by the change in the spectral characteristics of laser pulses. Correspondingly, the harmonic spectra followed those of driving pulses. The blueshift and spectral broadening of fundamental and harmonic radiation also occurred during propagation through the over-dense laser-produced plasmas. The use of the spectrally blueshifted and broadened pulses for the HHG led to an increase in the spectral width of generated harmonics and their blueshift.

HHG in plasmas using high-pulse repetition rate lasers. The application of the two-laser system approach allows analyzing the slowly spreading plasma, which contains molecular species, on a broad time scale by temporal matching with the propagation of the driving femtosecond pulses through the plasma plume. The advantage of HHG in the plasma created by the nanosecond laser system digitally synchronized with the femtosecond laser system is the delay-dependent analysis of the harmonic yield generated from different components of plasma plumes. The discussed experiments demonstrated the dependence of harmonic yield from different materials on the delay between the heating and driving pulses. The optimal delays for B_4C, ZnO, GaP, GaAs, and Ag_2S plasmas were determined to be approximately 200, 300, 350, 500, and 700 ns, respectively. These variations of delays correspond to the $t_d \propto (M)^{0.5}$ dependence for both molecular and atomic species. This diagnostic technique can be applied to the analysis of the presence of nanostructures in laser-induced plasmas (LIPs) through HHG with high spatiotemporal resolution. We discussed the applicability of the proposed method for analysis of the resonance-enhanced harmonics in atomic and molecular plasmas (Mo and MoS_2) and for studies of the large perovskite aggregates as potential emitters of harmonics. This technique can also be useful for the study of the contribution of different evaporated molecular species for HHG using extremely short pulses. Additionally, the joint application of the harmonic generation and mass-spectroscopy techniques can be considered as a highly versatile tool for in situ diagnostics of the laser plasmas containing a large population of clusters, molecules, and nanoparticles.

We analyzed the high-order harmonics generation using 1 and 100 kHz lasers by ablating different rotating targets, as well as discussed the high average flux of coherent short-wavelength radiation using the latter laser while comparing the plasma formation conditions at different pulse repetition rates. The analysis of harmonic stability in the case of 100 kHz experiments has shown the two-fold decay of the 27th harmonic generating in silver plasma after 3.5×10^6 shots, while no decay in harmonic yield was observed in the case of 1 kHz ablation. The two-color pump of plasma, resonance enhancement of single harmonic, and quasi-phase-matching scheme were presented in the case of 1 and 100 kHz pulse repetition rates of target ablation. The formation of modulated multi-jet plasma on the plane and curved surfaces during ablation by 100 kHz pulses has been demonstrated. In the case of the 25th harmonic of 1030 nm radiation ($E = 30$ eV) generated during experiments in carbon plasma, at 100 kHz and 40 W average power of driving pulses, 0.4 mW of average power in the 40 nm spectral range was reported.

Notice that these studies are the first step towards the formation of the stable plasma plume using 100 kHz class lasers, which needs additional efforts for maintaining stable harmonic yield from such laser-produced plasma. We would like to mention the incompleteness in resolving the stability issue for generating harmonics in the case of 100 kHz plasma. The possible scenarios to overpass this issue by dragging the target up and down, alongside target rotation at optimal velocity, are suggested. The preliminary experiments with the manually movable targets (e.g. by pressing and releasing the target holder from the top of the vacuum chamber) showed the perspective in the stability improvement for harmonic generation from high-pulse repetition rate plasma. The following plans include the programmable movement of rotating targets up and down, which can further diminish the instability issue.

We discussed the study of HHG using a 150 kHz laser and the details of the application of the rotating target for improvement of stability of the low and harmonics from plasma in the case of low and high pulse repetition rates of laser-matter interactions. In discussed studies, the harmonics were generated in carbon, boron, manganese, zinc, and titanium plasmas using a 150 kHz laser. The flux of single (H25, $E = 30$ eV) harmonics using carbon LIP was estimated to be 8×10^{13} photons/s. The corresponding fluxes obtained from other LPPs were 4×10^{13} (B NP + Ag MP), 2×10^{13} (Mn, B, and Ti), and 1.5×10^{13} (Zn) photons/s. The discussed studies also demonstrated the resonance enhancement of single harmonic, HHG in nanoparticles, and other harmonic generation effects, using 1030 nm, 150 kHz, 37 fs laser source.

Applications of clusters and QDs for HHG in LIPs. We analyzed different methods of the applications of clusters and QDs for HHG in LIPs. The nonlinear optical studies of Ag_2S QD thin films are discussed. Using the open-aperture and closed-aperture z-scan technique and theoretical fitting, the nonlinear absorption coefficient was calculated to be 4×10^{-5} cm W^{-1} for the 80 nm thick film in the case of 400 nm probe pulses. The nonlinear absorption coefficients and nonlinear refractive indices of 80 and 500 nm thick films were significantly larger than for the Ag_2S QDs suspension. In the case of transient absorption measurements, the two distinct characteristic time constants of 150 fs and 245 fs were obtained for the films of different thicknesses. We also discussed the studies of high-order harmonics emission from the ablation of 80 nm films using two laser sources. The conversion efficiency in the plasmas produced by ablation of the Ag_2S QD thin film (2×10^{-5}) was several times higher compared to the case of bulk Ag_2S ablation. The analyzed approach establishes a novel and efficient way of generating high-order harmonic yield.

We analyzed the possibility and conditions of using QDs as the efficient emitters of harmonics using optimally delayed heating and driving laser pulses. We discussed the method for the aqueous synthesis of Ag_2S, ZnS, and CdS QDs that enable, by their size and dispersion, as well as their structural properties, the effective generation of harmonics in plasma plumes. The study of the morphological properties of the samples demonstrated the formation of ZnS and CdS QDs with mean sizes of 2–3.2 nm in a cubic lattice, and of Ag_2S QDs with mean sizes of 1.7–2.0 nm in a monoclinic lattice, all with a weight fraction of about 20–25% in gelatin.

A comparison of HHG using QDs and ablation of solid-state targets for the same materials (Ag_2S, ZnS, and CdS) was carried out. In the case of QDs, the conversion efficiency into harmonics is higher despite the lower concentration of these species in the plasma compared to the concentration of molecules and ions for bulk ablation. The observed increase of HHG conversion efficiency could be caused by a larger concentration of harmonic emitters and the specific properties of QDs, particularly their plasmonic properties. The maximum HHG conversion efficiency is achieved much earlier than one would expect from the kinetic model of QDs spreading out from the target surface. Therefore similar velocity distributions of molecules and QDs of these metal sulfides are assumed in the ablation process.

HHG using carbon-contained NPs. Various processes during HHG in the plasmas containing carbon clusters and NPs were discussed. The role of small carbon clusters, which inevitably appear during graphite ablation, on the efficiency of HHG was analyzed. The advantage of the proposed method for laser ablation and delay-dependent study of harmonic emission during ablation of graphite target is that plasma formation in the two-lasers configuration of HHG experiments could be efficiently applied for a variation of the delay between heating and driving pulses using electronic methods. It can also be applied for other clustered targets to determine the best conditions for HHG in such plasmas. We analyzed the advanced properties of the carbon plasma produced by nanosecond pulses as a medium for efficient generation of the low- and high-order harmonics. The role of C_5–C_{25} clusters in the enhancement of harmonic generation efficiency was discussed. The discussed studies show that such clusters commonly appearing during time-of-flight mass spectroscopic studies of carbon ablation strongly influence the harmonics yield and thus could be considered as the sources of efficient harmonics in the 40–100 nm spectral range.

We also analyzed the HHG from the plasmas produced on the carbon-containing targets (graphite, fullerenes, and carbon nanotubes) using near-infrared laser radiation. The evolution of different plasma plumes using the extended delays from the beginning of ablation (up to a few tens microseconds) was discussed. We analyzed the role of three carbon-containing nanostructures in the frequency conversion of laser pulses at different conditions of plasma formation. The discussed research shows that all carbon clusters and nanostructures acquire from the very beginning the approximately same kinetic energy and spread out of the surface with the velocity approximately similar to the one that is possessed by a single carbon atom ablated from the bulk material. Additionally, we discussed the single-color pump (SCP, 1030 nm) and TCP (1030 nm + 515 nm) of small carbon clusters and NPs to determine the opportunity of resonance enhancement of even harmonics, as well as analyzed the role of epoxy resin in observed harmonic spectra from fullerenes and nanotubes.

HHG in metal NPs. We discussed the conditions when the plasma plumes created on the surfaces of ZnO thin films in vacuum conditions can be used as the efficient emitters of high-order harmonics of 800 nm, 30 fs laser pulses. We also analyzed the studies of the influence of dopants in nanocrystal-based media on the variation of high-order harmonics spectra produced during propagation of ultrashort laser pulses through the laser-produced plasmas containing ZnO and Gd. The strong 9th harmonic was observed in the case of plasmas produced on the bulk ZnO due to the resonance-induced enhancement of single harmonic, whereas in powered targets (powdered ZnO NCs and ZnO NCs + 12.5% Gd) the insignificant difference between the 9th and the 11th harmonics was observed. We analyzed the influence of dopant on the harmonics spectra at different delays between the heating and driving pulses. Those studies confirmed that doping of ZnO materials with Gd caused variation of transient, morphological, and high-order nonlinear optical properties of ZnO.

The 5- and 40-fold enhancement of harmonic yield was analyzed from ablation of the Au NPs deposited on paper as compared with ablation of the Au NPs deposited on glass and ablation of the Au bulk target. The harmonic cutoffs obtained from ablated Au NPs on paper, Au NPs on glass, and Au bulk target were 29th, 21st, and 21st orders, respectively, in the case of nanosecond heating pulses. In addition, the effects of heating pulse duration, TCP, and laser chirp on harmonic yield were discussed in Au NP-contained LIPs. The enhanced harmonic intensity was obtained from the ablated Au bulk target by decreasing the heating pulse duration. Those studies showed that the application of picosecond and femtosecond heating pulses for plasma formation allowed the generation of 5 and 4 times stronger harmonics with regard to the nanosecond heating pulses. The harmonic cut-offs were 33rd and 39th orders in the case of picosecond and femtosecond heating pulses, respectively. The enhanced harmonic intensity was demonstrated from ablated Au bulk target by applying TCP (800 nm + 400 nm) of LIP. Two-fold enhancement of harmonics was observed using TCP of ablated Au bulk target with regard to SCP. The negatively chirped 130 fs pulses enhanced 9th and 11th harmonics with regard to the chirp-free (35 fs) and positively chirped 130 fs pulses. The MD-based simulations of NPs formation at different pulse durations supported the experimental measurements and revealed the main mechanisms responsible for the size and morphology of the generated particles.

TCP of LIPs. We analyzed different cases of the application of TCP for efficient HHG in LIPs. A search for suitable schemes for TCP of different isotropic media resulting in efficient HHG is of special interest to the laser community. Originally, gases and LIPs were the main subjects of studies for the conversion of ultrashort near-infrared laser pulses towards the XUV region. The comparative studies of carbon plasma and argon gas showed in the past the advantages of the former medium once one became aimed in stronger harmonic emission in the 40–100 nm spectral region of XUV. Carbon plasma did not allow the generation of strong harmonic in the shorter wavelength region. A search of suitable plasma media showing preferable features compared with gases thus became an interesting topic of study. To analyze the potentially efficient plasma media some of them were chosen taking into account earlier reported data on their ability in generating higher-order harmonics compared with carbon plasma,

We discussed stronger HHG in the case of different plasmas (Ag, Ag_2S, stainless steel) with regard to the harmonics generating in argon gas in the cases of SCP and TCP. We analyzed the harmonic yields from gas and gas + plasma media under closely matched experimental conditions. Three- to five-fold growth of harmonic yield in the case of LIP compared with a gas medium at similar experimental conditions was achieved. Those studies showed that, in the spectral range of 15–50 nm, the used plumes are more efficient than argon. Different schemes of HHG, such as SCP and TCP, as well as variable positions of LIP with regard to gas jet, were analyzed, which allowed determining the best configuration of gas + plasma applications for harmonics generation.

We analyzed the high-order harmonic spectra produced during propagation of the femtosecond pulses through the lanthanide-contained laser-produced plasmas. The species under investigation (praseodymium, terbium, lanthanum, and ytter bium) are among the heaviest non-radioactive solid elements of the periodic table. We discussed HHG in the plasmas produced on the surfaces of these lanthanides and their oxides (La, Yb, Pr_6O_{11}, Tb_4O_7) using different techniques (two-color pump, application of chirped pulses, variation of the parameters of heating and driving pulses, etc.). The ablation of samples using femtosecond pulses has proven to be more efficient from the point of view of harmonic yield with regard to the ablation using picosecond pulses in the case of a two-color pump of plasma. It was shown that Yb plasma is the efficient medium for harmonic generation up to the 73rd order, which is one of the largest harmonic cutoffs generated in the laser-produced plasmas.

The novelty of the discussed studies is a wavelength-dependent demonstration of the prevailing role of the partial overlap of two orthogonally polarized waves in laser-produced plasmas and gases over a larger ratio of the second and fundamental photons produced in a thicker nonlinear crystal. We analyzed the influence of two pulses overlapping in a nonlinear medium on the high-order harmonics generation. The overlap of pulses strongly depended on the group velocity dispersion in the nonlinear crystal used for the formation of a second weak field. We discussed the conditions when the application of thicker crystal leading to a higher yield of second harmonic emission causes a different effect in the case of different sources of laser radiation. Even a small difference in the variations of group velocities in BBO for the 806 and 1030 nm laser sources led to the lesser influence of this effect in the case of the latter emission. Correspondingly, the delay between 806 and 403 nm pulses introduced in thick (0.4 mm) BBO resulted in the generation of 2(2n + 1) even harmonics and suppression of 4(n + 1) harmonics in spite of higher ratio of $I_\omega/I_{2\omega}$ compared with the case of thinner (0.2 mm) crystal. Thinner crystal was a preferable choice since it allowed the generation of almost equal odd and even harmonics.

Another pattern was observed in the case of the 1030 nm + 515 nm pump of laser-produced plasmas and gases. The less influential role of group velocity dispersion using a longer-wavelength laser allowed achieving sufficient overlap of fundamental and second harmonic pulses in the nonlinear media used for HHG while applying relatively thick (0.4 mm) BBO crystal. Best conditions of HHG were achieved using this crystal thus showing that the higher conversion efficiency towards the second harmonic ($I_{2\omega(0.4\text{ mm BBO})}/I_{2\omega(0.1\text{ mm BBO})} \approx 3$) at this driving wavelength overpasses the negative influence of lesser overlap of converting pulses in plasmas and gases in the former case.

We discussed different mechanisms of harmonics enhancement in the case of 806 nm driving pulses and second harmonic pulses (quasi-phase-matching (QPM) and resonance enhancement of harmonics using TCP of LIPs). In the discussed studies using 806 nm radiation, the enhancement of single even harmonics was observed in Cr, Mn, and Sb plasmas. QPM of odd and even harmonics in multi-jet Ag plasma allowed shifting the group of maximally enhanced harmonics towards the 24 nm range (H34). Atomic Ar and molecular N_2 and O_2 showed similar features as plasmas once we considered the overlap effects in gaseous media.

Resonance-induced enhancement of single harmonic. We discussed the HHG in the Sn, Mn, Zn, Se, In, Sb, Cr, Ag, Au, Mg, C, and Cd plasmas, which showed numerous cases of the coincidence of the harmonics with some emission lines of corresponding ions. However, this coincidence did not result in the enhancement of harmonics. We discussed the plasma conditions (spectral distribution of emission, excitation of high-charged ions, etc.) and harmonic generation, as well as the cases of the enhancement of single harmonic in the vicinity of some ionic transitions and the featureless plateau-like HHG spectra from different plasma species to add more information about the role of plasma emission spectra. The comparative analysis of those two groups of plasma plumes revealed the relative role of the strong emission lines and the transitions possessing high oscillator strengths (*gf*) in the enhancement of single harmonics. The analysis of the harmonic spectra in the case of the two abovementioned groups of LIPs showed that ionic transitions demonstrating high oscillator strengths are responsible for enhanced emission of a single order or the group of harmonics. Meanwhile, plasma emission lines play rather a destructive role in HHG. We analyzed different models describing the resonance-induced enhancement of single harmonics. The multielectron resonance recombination model allowing a qualitative description of observed experimental results was introduced.

We have reconsidered the mechanism of resonant amplification of single harmonic in zinc-containing atomic and molecular plasmas when generating high-order harmonics in laser plasma plumes. It has been shown that selenides of this metal noticeably reduce the amplification of the single harmonic in comparison with purely atomic plasmas, probably due to the decrease of the oscillator strength of the involved ionic transitions. Stronger ablation of these molecular targets did not allow the resonant amplification of the single harmonic in ZnSe plasma to be restored using different laser sources, probably due to the insufficient decay of molecules leading to the appearance of zinc ions.

The discussed research allowed to predict modification of the mechanism of resonant amplification of single harmonics reported in other atomic plasmas (manganese, tin, chromium, selenium, tellurium, molybdenum, indium, and arsenic) being presented in the molecular form. This process can also be reconsidered by comparing the atomic and molecular plasmas containing the above elements, as well as by tuning the wavelength of driving radiation along those resonances that are responsible for the observed amplification of harmonics.

Quasi-phase-matching in LIPs. We discussed some peculiarities of QPM in LIPs, which allowed clarifying both pros and contras of this approach. Since the first demonstration of this effect in plasma, new opportunities and schemes were developed and analyzed. Inevitably, they overlap, to some extent, with each other to explain, in some cases from the very basic rules, the advantages (and, in some particular cases, disadvantages) of the QPM in plasmas.

We also analyzed different patterns of the experimentally observed modifications of harmonic spectra at the conditions of QPM in silver, indium, and aluminum multi-jet plasmas. In particular, we discussed the role of the number of coherent zones, sizes of plasma jets and distance between them, plasma formation conditions, and characteristics of fundamental and heating radiation on the HHG efficiency at QPM conditions. The TCP-induced QPM at different regimes of laser-plasma interaction was demonstrated.

We analyzed the HHG at QPM conditions in silver microparticle and silver sulfide plasmas of different configurations. We discussed QPM of the high-order harmonics generating in the plasmas contained silver microparticles and atoms, boron NPs, and silver sulfide molecules. The influence of the flux of the heating beam, as well as variation of the geometry of multi-jet plasmas, on the maximally enhanced harmonic, was demonstrated. The groups of harmonics in the range of thirtieth to fiftieth orders at QPM conditions prevailed over the lower orders of harmonics. In most cases, the lower-order harmonics fully disappeared from the generated spectra at QPM conditions. The enhancement factors of QPM harmonics up to 22× were demonstrated in the case of mixed boron NPs and silver microparticles plasmas for the harmonics in the energy range of 47–67 eV. The conversion efficiencies for QPM-enhanced harmonics were estimated to be 3×10^{-5}, 2×10^{-5}, and 3×10^{-5} for the multi-jet plasmas produced on the bulk Ag, B NPs + Ag MPs, and Ag_2S targets, respectively.

Perspectives. Summing up, we would like to underline that a new method for determining the characteristics of materials using HHG can be successfully implemented to retrieve useful information about their nonlinear optical properties in the extreme ultraviolet range.

Among the potential fundamental results of the development of HHG in ablated materials are diagnostics of laser-induced and other plasmas, the search for improvement and understanding of the fine structure of the energy levels of atoms and ions, etc. The practical aspects of this approach include the formation of an intense coherent radiation source in the spectral range, which cannot be achieved by other methods of generating XUV radiation. Other aspects related to the practical use of plasma harmonics may include such competitive fields as biomedicine, biology, microscopy, research on the properties of materials, etc.

The future development of HHG in the LIPs can be focused on the following areas:

(1) Demonstration of effective HHG in LIPs and gases, as well as in mixtures of these media.
(2) Increasing the harmonic yield from the LIPs and searching for new promising plasma media for HHG.
(3) Determination of the ionic transitions possessing large oscillator strengths by analyzing harmonic spectra.
(4) Development of new approaches to determining the atomic and optical characteristics of materials using LAIHOHGS.
(5) Application of advanced methods of growth of the output of higher-order harmonics (HHG in clusters and quantum dots, in a spatially modulated plasmas, by using two—and three-color pumps, by the implementation of strong ion transitions for the modification of the conversion efficiency of HHG, the application of the quasi-phase-matching in modulated plasmas, etc.).
(6) The search for new methods for the determination of the transitions possessing strong oscillator strength using LAIHOHGS.
(7) Nonlinear spectroscopy of solids via HHG in thin solid films containing nanoparticles and quantum dots.

Previous studies allowed us to develop the foundations for a new branch of spectroscopy—high-order nonlinear spectroscopy of materials. The prospects of LAIHOHGS are as follows. The harmonic generation approach is an advanced method of the high-order nonlinear spectroscopy of materials. The propagation of intense laser pulses through gas and liquid media containing atoms, ions, molecules, and electrons makes it possible to analyze their properties by generating coherent XUV harmonics. The information contained in the short-wave spectra of generating harmonics allows us to obtain the characteristics of the spectral and electronic structure, as well as the properties of the generating medium. Characteristics such as the oscillator strength of ions can be determined by analyzing resonantly amplified harmonics, which are difficult to determine using other spectroscopic methods. The combination of processes involving a single atom/ion reaction and a collective reaction using various approaches developed during previous studies of gases and LIP can reveal new methods for increasing the harmonic output in the XUV range. Among them are the use of two-and three-color pumping of generating media, quasi-phase-matching of harmonics in various XUV ranges, the involvement of specially synthesized quantum dots in plasma media when generating high-order harmonics, and others.

All the above-discussed studies discussed in this book allowed the demonstration of new approaches in generation of the short-wavelength radiation in laser-induced plasmas. The advanced approach in applications of HHG in LIPs provides the tools for the optical analysis of recently developed nanostructures. The developments in this field of nonlinear optics are rapidly improving to follow the needs and requirements of this area of research. One can expect further achievements in this direction in the nearest future.

Subject Index

R. A. Ganeev, *High-Order Harmonics Generation in Plasmas*, Springer Series on Atomic, Optical, and Plasma Physics 122, https://doi.org/10.1007/978-3-031-09040-0